Xiaolei Zhang
Dynamical Evolution of Galaxies

Also of Interest

Origin of the Moon. New Concep. Geochemistry and Dynamics
Erik M. Galimov, Anton M. Krivtsov, 2012
ISBN 978-3-11-028628-1, e-ISBN (PDF) 978-3-11-028640-3

General Relativity: The most beautiful of theories. Applications and trends after 100 years
Carlor Rovelli (Ed.), 2015
ISBN 978-3-11-034042-6, e-ISBN (PDF) 978-3-11-034330-4,
e-ISBN (EPUB) 978-3-11-038364-5

Frontiers in Relativistic Celestial Mechanics. Theory
Sergei M. Kopeikin, 2014
ISBN 978-3-11-033747-1, e-ISBN (PDF) 978-3-11-033749-5,
e-ISBN (EPUB) 978-3-11-038938-8

Frontiers in Relativistic Celestial Mechanics. Applications and Experiments
Sergei M. Kopeikin, 2014
ISBN 978-3-11-034545-2, e-ISBN (PDF) 978-3-11-034566-7,
e-ISBN (EPUB) 978-3-11-037953-2

Xiaolei Zhang

Dynamical Evolution of Galaxies

DE GRUYTER

Mathematics Subject Classification 2010 and Physics and Astronomy Classification Scheme 2010
PACS: 98; MSC: 85-08, 85A05, 85A40

Author
Xiaolei Zhang, Ph.D.
George Mason University
Department of Physics and Astronomy
4400 University Drive
Fairfax, VA 22030
USA
xzhang5@gmu.edu

ISBN 978-3-11-052519-9
e-ISBN (PDF) 978-3-11-052742-1
e-ISBN (EPUB) 978-3-11-052544-1

Library of Congress Cataloging-in-Publication Data
A CIP catalog record for this book has been applied for at the Library of Congress.

Bibliographic information published by the Deutsche Nationalbibliothek
The Deutsche Nationalbibliothek lists this publication in the Deutsche Nationalbibliografie; detailed bibliographic data are available on the Internet at http://dnb.dnb.de.

© 2018 Walter de Gruyter GmbH, Berlin/Boston
Typesetting: Integra Software Services Pvt. Ltd.
Printing and binding: CPI books GmbH, Leck
Cover image: Digital Vision/DigitalVision/thinkstock
♾ Printed on acid-free paper
Printed in Germany

www.degruyter.com

Preface

To explore the possibility of morphological transformation of galaxies over cosmic time had been a major motivation of Edwin Hubble, when he devised the galaxy classification scheme bearing his name (Hubble 1936). Galaxy observations in the recent decades have shown that Hubble's original conception of an evolution trend from the early to the late Hubble types, that is, from the more rounded (or bulgy) galaxies to the more flattened (or disky) galaxies, based on Sir James Jeans' theoretical framework (Jeans 1928), was untenable. The possibility of an evolution trend along the reverse direction, that is, from the late to the early Hubble types, on the other hand, has received growing observational support, and is currently an active area of research under the general heading of "secular morphological evolution of galaxies." The phrase "secular evolution" here refers to the slow and long-term transformation of galaxy morphology driven mostly by internal mechanisms.

In the title of the current book, we choose rather to use the phrase "dynamical evolution" because during the past few decades, the phrase "secular evolution" has often been associated with work on gas accretion in galaxies. The dynamical mechanism discussed in this work, on the other hand, is effective for the radial redistribution of both the stellar and the gaseous disk mass. When we use the phrase "secular evolution" in this book as a short-hand for "internal and dynamically driven morphological evolution of galaxies," it also has this broader meaning. We will show that long-term and, for all practical purposes, irreversible dynamical evolution in a nominally Hamiltonian system is in fact possible, when the system under consideration is of many degrees-of-freedom, is open, and is far-from-equilibrium. Such a system allows an *effective* singularity to develop when a self-organized global instability pattern emerges, which further generates an arrow of time.

Viewed in this light, the phrase "dynamical evolution" no longer appears to be an oxymoron[†] even without the assistance of resonant interactions: a complex Hamiltonian system can indeed exhibit dissipation-like, long-term evolution behavior, even though its local dynamics is nominally reversible. Here the paradox is resolved by the overwhelming volume of phase space occupied by the type of initial conditions for an

---

[†] The impression that a Hamiltonian system cannot exhibit true long-term evolution behavior probably results partly from the so-called Poincaré recurrence theorem (Poincaré 1890), which states that many dynamical systems will, after a sufficiently long time, return to a state very close to the initial state. One of the conditions under which the theorem was proved is the preservation of the volume of a phase space element along the particle orbit during the evolution of the dynamical system, which is indeed obeyed by all Hamiltonian systems (since they satisfy Liouville's theorem). However, for open and many degrees-of-freedom systems that are far from equilibrium (including disk galaxies), it can occur that the physical region a system explores expands indefinitely, even though the local phase-space density near the particle trajectory is conserved. This violates a second condition needed in the proof of the Poincaré recurrence theorem, that is, that of a finite upper bound on the total accessible phase space volume.

open, complex, and nonequilibrium system, which, when evolved, would lead to the spontaneous emergence of global instabilities. The coarse-grained entropy is very low for almost any possible choice of initial condition for such a nonequilibrium many-body system, which leaves room for the eventual increase of coarse-grained entropy as a result of global-instability-induced secular evolution, even though the fine-grained entropy in such systems never changes, since all Hamiltonian systems obey Liouville's theorem.

Despite the growing observational support for secular evolution, among published texts on galaxy dynamics there has not been a systematic treatment of the dynamical foundation of the secular evolution process. Past theoretical investigations in galactic dynamics focused on the construction of self-consistent galaxy models in steady states – that is, building up the masses of galaxies employing passive and mostly regular orbits obtained under a rigid, applied potential. These practices, while being successful at obtaining equilibrium models which can then be compared with the observed galaxy mass and light distributions, have been found wanting in predicting the more-detailed characteristics pertaining to the morphology, kinematics, and especially the long-term evolutionary behaviors of galaxies.

The advent of the density wave theory in the early 1960s was a significant step in our understanding of the dynamical processes in disk galaxies containing large-scale spiral and bar patterns. However, a well-known result of the older version of the density wave theory, that of no wave/basic-state interaction except at the wave-particle resonances for quasi-steady patterns, prohibited any serious consideration of a significant change of the morphologies of galaxies over their lifetime.

The original research described in this monograph was among the first to reveal that the suite of classical mechanical approaches traditionally employed in galactic dynamics is inadequate for the treatment of so-called "emergent" phenomena in open, many degrees-of-freedom, nonlinear, and far-from-equilibrium gravitational systems. The essence of these phenomena is the nonequilibrium phase transition process, whose proper treatment requires new dynamical tools properly matched to the emergent new physics.

The idea that new concepts and methods need to be devised to approach the so-called "problems of complexity" in nonlinear and many degrees-of-freedom systems has been around for quite a while. The spirit of this school has been eloquently summarized by the Nobel laureate solid-state physicist P.W. Anderson as "more is different" (Anderson 1972). Specifically, in dealing with complex systems, one can encounter paradoxical situations where "the whole is more than the sum total of its parts," in the sense that one cannot expect to simply pile together the component molecules to make a biological entity "come alive" on its own. An element of history in the development of the organism, which includes the sum total of its past complex interactions with its environment, as well as interactions among its constituent parts, will have to be taken into account. The history of interactions and the history of growth are partly what give an organism its structure and its dynamical vitality. The mystery of the "life-force" therefore originates in the self-organization process itself, the dynamics of which is irreversible and "emergent" from the underlying simple and reversible dynamical interactions.

In this monograph, we will explore the detailed and often subtle aspects of the self-organization process in the context of the dynamics and evolution of galaxies, which are assemblies of many stars plus the interstellar medium, often organized into coherent large-scale patterns now understood to be density waves. These waves, or wave modes, as we will show, not only give galaxies their breathtaking appearance, but also perform important functions in facilitating the long-term morphological evolution of their parent galaxies. The dynamical concepts and tools used for understanding the roles of these patterns, such as symmetry breaking, collective dissipation, entropy production and export, etc., have their analogues in other disciplines of physical and biological sciences as well. Of particular relevance are the studies of "dissipative structures" by Nobel laureate I. Prigogine and associates (Prigogine 1969, 1980; Glansdorff & Prigogine 1971; Nicolis & Prigogine 1977), as well as the studies of "synergetics" (Haken 2004; Fuller 1975). The author was fortunate to have met and discussed her work on applying the theory of dissipative structures to the problem of galaxy evolution with Prof. Prigogine in the year 2000, three years before his sad passing at the age of 86, on an occasion to give a physics department colloquium talk at the University of Texas at Austin, through a special invitation from Prof. Prigogine.

The current monograph serves to fill a void in the published texts on the internal dynamical mechanisms enabling the secular morphological evolution of galaxies. The past works on secular evolution were mostly performed within two kinds of theoretical framework: (1) treating the *gas* component, or the interstellar medium in galaxies, as the major driver of secular evolution (Kormendy 1979; Kormendy & Kennicutt 2004 and the references therein). This approach has its heritage in the linear response theory of nonequilibrium thermodynamics (de Groot & Mazur 1962 and the references therein), which is closely linked to the Navier–Stokes equation set, which in turn is obtained from the first-order expansion of the collisional Boltzmann equation in terms of the ratio of collisional mean free path versus the system size. Employing linear nonequilibrium thermodynamics gives us the various well-known transport processes such as thermal conduction, viscous dissipation, etc.. (2) When secular evolution in the *stellar* component of the galactic disk was considered, it was invariably in the context of the *resonant* interaction of the density waves and the disk stars (Sellwood 2014 and the references therein). This second approach also has a long heritage (i.e., the Lagrange–Laplace investigation of the stability of the solar system), and contemporary research in planetary science offers many good examples of its application.

The self-organized global instability patterns that emerge spontaneously in complex, many degrees-of-freedom, nonlinear, and far-from-equilibrium systems, however, have innate dynamical mechanisms that go far beyond the physical processes addressed by either the linear nonequilibrium thermodynamics of transport phenomena or the passive resonant interactions between an external driving agent and the orbits of the individual particles of the system under concern. The operation of these new dynamical mechanisms is closely linked to the *correlated fluctuations* that emerge and become amplified when a nonequilibrium phase transition takes place in open, many degrees-of-freedom systems. These correlated fluctuations violate one of the key assumptions in the derivation of the collisional Boltzmann equation (through, e. g., the rigorous BBGKY procedure, see Section 7.1.1 for further discussion), that is, the ansatz of "molecular chaos" which mandates that the fluctuations of component particles be totally uncorrelated. It is indeed this very element of correlated fluctuations that introduces the "history" element and gives "life-force" to a self-organized pattern. These correlations are anything but the random processes assumed in the derivation of the Boltzmann equation.

In the past galactic dynamics work, the most often employed fundamental kinetic equation is the collisionless version of the Boltzmann equation, also known as the Vlasov equation or the stellar dynamical equation. This version of the Boltzmann equation has served as the foundation of the early density wave theory explorations. Both versions of the Boltzmann equation, as well as the moment-equation derivatives of the collisional version of the Boltzmann equation (such as the zeroth-order Eulerian equation set, or the first-order Navier–Stokes equation set), are lacking the crucial ingredients that allow correlated fluctuations to spontaneously emerge – even though they can have indications of the necessity of such additional processes through the formation of unstable modes, which are not able to be stabilized within the range of validity of these (approximated) theories themselves. Thus a full understanding of the self-organization process cannot be based entirely on these *approximated* equation sets, even though they have been proven to serve our previous purposes of understanding the equilibrium or close-to-equilibrium phenomena.

The new approach we will adopt in the current work combines the analytical, N-body simulational, as well as observational aspects to bring about a new synthesis. The methodology we develop here may be generalized to the studies of self-organization phenomena in other physical sciences as well: this is so especially because most of the past studies of spontaneous symmetry-breaking were conducted in a model Lagrangian or Hamiltonian context, rather than self-consistently from first principles. Therefore the important process of interaction of the system with the environment, which leads to the irreversible long-term evolution of the environment, was ignored in most of these previous studies. We will show in the main body of this monograph that, as a result of collective processes due to unstable density wave modes, new kinds of closure relations are arrived at when the *global self-consistency* requirement is enforced at the quasi-steady state of the wave modes. The emergence of these new dynamical-equilibrium relations signals the failure of the classical mean-field approaches commonly used for treating the physics of close-to equilibrium (linear and quasi-linear) regimes. Furthermore, the need for replacement of the older governing laws with the new emergent closure relations can be understood by the fact that the fundamental nature of galactic density wave modes is akin to plasma collisionless shocks. It is well-known that traditional differential and mean-field treatments break down at the onset of nonequilibrium phase transitions or at the collisionless-shock discontinuities (Balogh & Treumann 2013), and interparticle correlations become the dominant factor governing the new physical processes that result, which the mean-field theories are known to ignore. We will show that the integral manifestation of secular shock dissipation at the density wave crest is *a characteristic radial distribution of the azimuthal phase shift between the potential and density wave patterns*, which allows the density-wave-assisted galactic secular evolution to proceed in a globally self-consistent and quasi-stationary manner.

The current text is based mainly on the original work by the author and her collaborators in the past two and a half decades, starting from her Ph.D. dissertation work at the University of California, Berkeley in the late 1980s, through her postdoc years at the Harvard-Smithsonian Center for Astrophysics, and up until her most recent collaborative work with Prof. Ron Buta of the University of Alabama on the observational verifications of theoretical predictions. Most of the original contributions described in this book had previously appeared in refereed journals and conference proceedings. When deciding what to include in this book from the author's previously published work, priority is given to including enough details of derivations so that a reader can follow the logical flow of the argument without needing to refer back to the original papers; whereas more extended applications to large samples of galaxies are only briefly mentioned, and interested readers can find these in the referenced papers. Permissions to use material from these previously published works have been obtained from the original publishers. Due to the differing publisher requirements, for some works the permissions are printed explicitly (i.e. for works published in the AAS journals associated with the IOP), some referenced with its DOI (i.e., for work published in the PASP journal), some referenced with the conference proceedings citation link, and still others with journal citation together with a Science Direct link (i.e., for works published in the Elsevier journals). A number of N-body simulations are rerun with the same global parameters as the previously published ones for the purpose of generating higher-quality graphics to meet the publisher's requirement. Even though the random number seeds for initial condition assignment will not be exactly duplicated, this does not affect the global modal characteristics obtained. Like the subject of the collective effects itself, here we hope the whole of the completed manuscript will be more than the sum total of its parts, through the context provided by the organization of the material, and the logical links among the different components. Additional published results of other researchers are discussed as relevant, though by the nature of this monograph a thorough survey has not been attempted. The study of secular evolution of galaxies is still an emerging discipline, so new understandings and new results are expected to occur as the field develops and matures. It is the hope of the author though, that through this monograph a coherent framework is provided for understanding the dynamical principles underlying the secular morphological evolution of galaxies. This in turn should stimulate and facilitate further advances in this area.

Throughout the development of this work numerous colleagues have provided support, help, and feedback in various forms. The author would like to thank especially the following individuals for helpful interactions: Ron Allen, Hong Bae Ann, Lia Athanassoula, Steve Balbus, James Binney, Alberto Bolatto, Ron Buta, Gregory Bothun, Gene Byrd, Francoise Combes, George Contopoulos, Warrick Couch, Victor Debattista, Tim De Zeeuw, Bruce Elmegreen, Bob Ehrlich, Henry Ferguson, Jackie Fischer, Roger Fux, Ken Freeman, Daniel Friedli, Oleg Gnedin, Rosa González-Lópezlira, Preben Grosbol, Jin-Lin Han, Carl Heiles, Richard Hills, Luis Ho, Paul Jaminet, Agris Kalnajs, Sasha Kashlinsky, Rob Kennicutt, Ivan King, John Kormendy, Eija Laurikainen, Youngung Lee, Chia-Chiao Lin, Xiao Liu, Steve Lubow, Shude Mao, Eric Martínez-García, Colin Masson, Stacy McGaugh, Richard Miller, Colin Norman, Panos Patsis, Daniel Pfenniger, Alice Quillen, Alessandro Romeo, Heikki Salo, Paolo Salucci, Shobita Satyapal, Paul Schechter, James Shombert, Jerry Sellwood, Frank Shu, Antony Stark, Mario Tafalla, Peter Teuben, Magnus Thomasson, Scott Tremaine, Alar Toomre, Keiichi Wada, Simon White, Jack Welch, Robert Wilson, Mel Wright, Anatoly Zasov, as well as several referees of the journal articles as well as the current book. She thanks the management at the George Mason University for the Affiliate Faculty position which allowed her the freedom to pursue independent research. She also thanks her husband Brian and her extended family for the support and encouragement throughout the years. Finally, she thanks the staff at the De Gruyter Publishers, especially acquisition manager Chao Yang and editorial managers Astrid Seifert & Nadja Schedensack, as well as production editor Sabina Dabrowski and the production team, for their valuable help in making this monograph possible.

Contents

1 Introduction —— 1
1.1 Observational Background —— 2
1.2 Theoretical Background —— 5
1.3 Organization of the Material —— 12

2 Dynamical Drivers of Galaxy Evolution —— 14
2.1 Motivation and Outline for the Theoretical Approach —— 14
2.2 Density Wave Crest as the Site of Gravitational Instability —— 18
    2.2.1 Local Stability Condition at the Spiral Arm and Interarm Region —— 19
    2.2.2 Length Scale of Spiral Instability at the Solar Neighborhood —— 22
2.3 Potential-Density Phase Shifts for Density Wave Modes —— 24
    2.3.1 Definition of the Potential-Density Phase Shift —— 24
    2.3.2 A Phase Shift Given by the Poisson Integral —— 27
    2.3.3 Phase Shifts in the Eulerian Equations of Motion and in the Linear Periodic Orbit Solution —— 29
    2.3.4 Phase Shift in the Linear and Nonlinear Regimes of the Eulerian Solutions —— 32
2.4 Linear Regime and Quasi-Steady State of the Wave Modes —— 34
    2.4.1 The Process of Reaching the Quasi-Steady State —— 35
    2.4.2 Reconciliation of the Two-Wave Superposition and the Unstable-Mode Points of View —— 40
2.5 Torque Coupling and Angular Momentum Transport —— 46
    2.5.1 Torque Couplings due to Density Waves —— 46
    2.5.2 The Relation Between the Torque Couplings and the Rate of Angular Momentum Change in the Disk —— 48
    2.5.3 The Relation Between the Volume-Torque Integral $\overline{\mathcal{T}}$ and the Torque-Coupling Integrals C_g and C_a —— 51
    2.5.4 A Closure Relation for the Quasi-Steady State —— 53
2.6 Rates of Secular Evolution —— 56
    2.6.1 Secular Change in the Mean Stellar Orbital Radius —— 57
    2.6.2 Secular Heating of the Galactic Disk —— 59
    2.6.3 Secular Mass Flow Rate Determination —— 60
    2.6.4 Secular Change of Disk Surface Density —— 62
    2.6.5 Viscous Accretion Disk Analogy —— 65
2.7 Relation to "Broadening of Resonances" —— 67
2.8 In a Nutshell —— 69

3 N-Body Simulations of Galaxy Evolution —— 70
- 3.1 Overview of N-Body Simulations of Disk Galaxies —— 70
- 3.2 Simulation Codes and Basic State Specifications —— 71
  - 3.2.1 Choice of Basic State Parameters —— 72
  - 3.2.2 Choice of Simulation Parameters (First-Generation Tests) —— 74
- 3.3 Signature of Collisionless Shock in N-Body Spirals —— 77
- 3.4 Modal Nature of a Spontaneously Formed Pattern —— 85
- 3.5 Qualitative Signature of Secular Mass Redistribution —— 90
- 3.6 Longevity of the Spiral Modes —— 93
  - 3.6.1 Faithfulness of the N-Body Simulations —— 93
  - 3.6.2 Three Test Runs Using Different Numbers of Particles —— 95
  - 3.6.3 Lifetime of a Spiral Pattern Inferred from N-Body Simulations —— 98
- 3.7 Role of Gas —— 99
- 3.8 Implication on Orbits as "Building Blocks" —— 103
- 3.9 Second-Generation Tests —— 109
  - 3.9.1 The Critical Roles of the Softening Parameter —— 110
  - 3.9.2 The Impact of Softening on N-Body Simulated Mass Flow Rates —— 111
  - 3.9.3 Four Runs with Differing Softening Parameters —— 114
  - 3.9.4 Grid Noise Associated with the Use of Small Particle Softening Parameter —— 137
  - 3.9.5 Accuracy and Implications of the Secular Radial Mass Flow Rates Obtained in N-Body Simulations —— 143

4 Astrophysical Implications of the Dynamical Theory —— 146
- 4.1 Motivations and General Outline —— 146
- 4.2 PDPS Method for CR Determination —— 149
  - 4.2.1 Dynamical Basis and Practical Considerations for the PDPS Method —— 149
  - 4.2.2 First Application of the Method: NGC 1530 —— 153
  - 4.2.3 Phase Shift in a Pure Spiral Galaxy: NGC 5247 —— 157
  - 4.2.4 Multiple Nested Resonances: NGC 4321 —— 158
  - 4.2.5 Phase Shift in an Interacting Galaxy: M51 —— 161
  - 4.2.6 Regarding the So-Called "Super-Fast Bars" —— 162
  - 4.2.7 Physical Basis Underlying the Validity and Accuracy of the Phase Shift Method —— 164
  - 4.2.8 Implications for the Kinematics and Dynamics of Nearby Galaxies —— 167

4.3	Secular Mass Migration and Bulge Building —— 169	
	4.3.1	Formation and Evolution of Galactic Bulges —— 170
	4.3.2	Secular Mass Flow Rate Determination Using NIR and MIR Images —— 172
	4.3.3	Relative Contributions from Gravitational and Advective Torques —— 179
	4.3.4	Relative Contributions of Stellar and Gaseous Mass Flows —— 181
4.4	Secular Heating and the Age–Velocity–Dispersion Relation —— 187	
	4.4.1	The Age–Velocity Dispersion Relation of the Solar Neighborhood Stars —— 188
	4.4.2	The Cause of Isotropic Velocity Diffusion in Three Dimensions and the Preservation of the Gaussian Velocity Distribution through Time —— 191
	4.4.3	The Origin of Radial Variation of the Stellar Velocity Dispersion with Galactocentric Distance —— 191
4.5	Secular Heating and the Size–Line-Width Relation —— 193	
	4.5.1	History and Motivation —— 193
	4.5.2	Energy Injection into the Star-Gas Two-Fluid through the Spiral Collisionless Shock —— 195
	4.5.3	The Rate of Energy Injection and Rate of Energy Cascade —— 196
	4.5.4	Application: The Carina Molecular Cloud Complex —— 197
4.6	Other Characteristics of the Milky Way Galaxy and External Galaxies —— 202	
	4.6.1	Mass Distribution of the Different Galactic Components —— 203
	4.6.2	Stellar Population and Kinematics in the Thin and Thick Disks —— 204
4.7	Universal Rotation Curve —— 205	
4.8	Formation and Maintenance of Galaxy Scaling Relations —— 209	
	4.8.1	General Considerations —— 209
	4.8.2	Origin and Evolution of the Scaling Relations —— 210
4.9	Butcher–Oemler Effect and Evolution of Cluster Galaxies —— 214	
	4.9.1	Observations and Candidate Mechanisms of the Morphological BO Effect —— 214
	4.9.2	An Infrared Diagnostic Approach for the Star-Formation States of Cluster Galaxies —— 217
	4.9.3	Application of the Infrared Diagnostic Approach —— 224
	4.9.4	Other Cluster Observations in Support of Secular Evolution —— 228

		4.9.5 Further Comments on the Different Proposed Mechanisms for Cluster Galaxy Evolution —— 229
4.10		Secular Evolution and the Origin of Color–Magnitude Relation —— 233
4.11		An Example of Secular Evolution in Interacting Galaxies —— 234
	4.11.1	Background —— 235
	4.11.2	Previous Observations and Simulations of the Leo Triplet —— 237
	4.11.3	CO 1-0 and HI Aperture Synthesis Observations of NGC 3627 —— 237
	4.11.4	Analysis and Discussion —— 239
4.12		Black Hole Mass and Bulge Mass Correlation —— 243

5 Putting It All Together —— 246

5.1 Reexamine the Foundations —— 246
    5.1.1 On the Modal and Quasi-Steady State Hypotheses of Density Waves in Physical and Simulated Disk Galaxies —— 246
    5.1.2 Role of Basic State Specification —— 260

5.2 Broader Implications —— 266
    5.2.1 Self-Organization in Nonequilibrium Systems and the Formation of Singularity Hierarchy —— 266
    5.2.2 Implications on the Cosmological Evolution of Galaxies —— 269

6 Concluding Remarks —— 284

7 Appendix: Relation to Kinetics and Fluid Mechanics —— 287

7.1 Foundation of Kinetic Theory: The Boltzmann Equation —— 287
    7.1.1 Outline of the Derivation of the Boltzmann Equation Through the BBGKY Hierarchy —— 287
    7.1.2 Growth of Instability and the Arrow of Time —— 292

7.2 From Kinetic Theory to Fluid Mechanics —— 294
7.3 Nonequilibrium Phase Transition and Galaxy Evolution —— 296
7.4 The Proper Choice of Analytical Hierarchies —— 297

References —— 299

Index —— 313

1 Introduction

During the past few decades, the dominant view was that the structural properties of galaxies remain largely unchanged over time, unless galaxies were perturbed by violent events such as major or minor mergers (Toomre & Toomre 1972). This view is particularly favored in the currently popular hierarchical clustering/Lambda cold dark matter (LCDM) paradigm of structure formation and evolution (Ostriker & Steinhardt 1995; Liddle 2003 and the references therein; Mo, van den Bosch, & White 2010 and the references therein). Even though the role of mergers during the early phase of galaxy assembly at high redshifts could indeed be important, growing evidence has shown that at least since redshift $z \sim 1$ the rate of merger appears to have been significantly reduced (Cohen 2002; Conselice et al. 2003; López-Sanjuan et al. 2009. See also Conselice et al. 2016 which indicates that the role of merger was already insignificant starting from $z = 2$, when the universe was only a quarter of its current age), and subsequent galaxy morphological transformation is likely to be dominated by the slower internal secular evolution process.

In the late 1970s, photometric and kinematic evidence in the bulges of late-type galaxies, as well as hints from N-body simulations of barred galaxies which incorporated a dissipative gas component, prompted several investigators to speculate that the morphology of late-Hubble-type disk galaxies may be transformed into that of earlier Hubble-types by gas accretion under barred potential (Kormendy 1979, 1982; Combes & Sanders 1981). These initial speculations have since been developed into one version of the secular evolution scenario, which emphasizes the role of dissipative gas accretion in the formation of the so-called pseudo bulges (or disky-bulges) in late-type galaxies (Kormendy & Kennicutt 2004 and the references therein). More careful examination, on the other hand, has since demonstrated that the relevant gas-dynamical process is not powerful enough to transform the morphology of intermediate- to early-Hubble-type disk galaxies, considering the observed fraction of gas and the rate of star formation in these galaxies (Kormendy & Kennicutt 2004).

Beginning in a series of papers published in the late 1990s (Zhang 1996, 1998, 1999), the present author showed that the stellar component, which constitutes the major portion of the galaxy mass in all but the very late-type galaxies, in fact is the main contributor of the secular morphological evolution of galaxies, through a collective interaction process between the spontaneously formed density wave modes and the basic state of the galactic disk[†]. This process had previously been overlooked in galactic dynamical analyses employing the stellar dynamical equation (which is the same as the collisionless Boltzmann equation, or the Vlasov equation), as well as

---

[†] The term *basic state* refers to the axisymmetric galaxy disk from which a density wave mode emerges. It is usually specified as the radial distribution of the galaxy-disk surface density, circular velocity, and velocity dispersion.

the moment equation descendant of the collisional Boltzmann equation, the Eulerian equation set, because the Boltzmann equation itself was derived in a context that ignored collective interactions and interparticle correlations (see the detailed derivation of Boltzmann equation in the Appendix of this monograph), thus is not suitable for applications that focus on the self-organization and collective dissipation processes such as the study of the secular morphological evolution of galaxies mediated by spontaneously formed density wave modes.

In the course of presentation of this monograph, we will make extensive use of the phrases "collective dissipation" or "collective effects." These phrases describe a general tendency for nonequilibrium, complex systems to spontaneously form global patterns, which lead to emergent new dynamics through *coherent* interactions of their component parts. Other phrases used in this context in the literature, ranging in applications from condensed matter physics, fluid dynamics, economics, biological and social sciences, include "cooperative effect," "coupling and interaction among the degrees-of-freedom of a complex system," and "synergetics."

In the case of a galaxy possessing a density wave pattern, the traditional approach employed in constructing galaxy models displays a top-down type of organization, meaning that the stellar orbits in such galaxy models respond *passively* to an *applied* potential. Though still adequate for the study of the density wave modal emergence phase, the top-down approach is entirely powerless to deal with the self-consistent evolution of the galaxy basic state together with the corresponding self-limiting density wave modal set. The density wave modes of galaxies are able to maintain their quasi-steady amplitude *only at the expense* of the dissipative secular evolution of the mass distribution of the basic state of the galactic disk. *So the long-term survival of the mode and the secular evolution of the basic state of the disk are closely linked.*

The secular evolution process, which slowly transforms the morphology of a galaxy over its lifetime, could naturally account for the observed properties of the great majority of physical galaxies, if both the stellar and the gaseous accretion processes are taken into account. As an emerging paradigm for galaxy evolution, its dynamical foundation has been gradually established in the past few decades, and its astrophysical consequences are just beginning to be explored. In the current monograph, we seek to establish that the secular evolution scenario provides a coherent framework for understanding the extraordinary regularity and the systematic variation of galaxy properties along the Hubble sequence.

1.1 Observational Background

Observational data on the characteristics of galaxies have been accumulating for more than a century, but evidence for the importance of internal secular evolution processes has only become apparent in the recent decades, largely as a result of the increased sensitivity, spectral coverage, and angular resolution of the space and ground-based telescopes. The supporting evidence ranges from the characteristics of individual galaxies both in isolated and in group/cluster environments, as well as the statistical properties of galaxy populations as a whole.

The most detailed characteristics of an individual galaxy come from the observation of our own Galaxy, the Milky Way. It has long been known that the observed kinematics of the different age groups of stars in the Milky Way Disk differ systematically, manifesting as the well-known age–velocity dispersion relation of the solar neighborhood stars (Wielen 1977). It is obvious that there exists a dynamical mechanism that heats the Disk stars secularly as they age, which operates smoothly across the entirety of a Hubble time (Gilmore, King, & van der Kruit 1990). Furthermore, the stellar population in the Galactic Bulge region has the well-known stratified distribution, with younger populations closer to the Galactic central region. This stratification trend also extends to the Thick and Thin Disks away from the Bulge region (Gilmore et al. 1990). These distributions also hint at a secular evolution origin for their formation.

Recent deep surveys have found that galaxies in the general field environment similar to that occupied by the Milky Way have undergone significant morphological transformation over the cosmic time. It is found that more field galaxies are of earlier Hubble types in the nearby universe than at higher redshifts (Lilly et al. 1998). There exist also the so-called faint-blue galaxies, which are in fact L_* galaxies having luminosities and sizes similar to the Milky Way, which are found at the intermediate redshifts, but which have all but disappeared in the nearby universe (Ellis 1997). Since the total number density of galaxies of all Hubble types have not evolved significantly between redshifts $z = 1$ and $z = 0$ (Cohen 2002; Conselice et al. 2016), and the merger fraction since $z = 1$ is low (Conselice et al. 2003; López-Sanjuan et al. 2009), the most likely explanation for these observed statistical differences in galaxies between the higher redshifts and the nearby universe is the internal morphological (as well as the accompanying stellar population and color) evolution of galaxies.

Dense cluster was the environment where morphological transformation of galaxies was first hinted at through the so-called Butcher–Oemler effect (Butcher & Oemler 1978a, 1978b). When it was discovered, the Butcher–Oemler effect referred to the bluer colors of galaxies in dense clusters at the intermediate redshifts, compared to similar-density clusters in the local universe which contain mostly red, early-type galaxies. Subsequent Hubble Space Telescope (HST) observations (Couch et al. 1994; Dressler et al. 1994) have been able to resolve the morphology of these intermediate-redshift Butcher–Oemler galaxies, and show that they are mostly late-type disks; therefore the Butcher–Oemler effect is now considered not only a color evolution effect but also a morphological transformation effect. Because of the high-speed nature of encounters of galaxies in dense clusters, mergers are known to be infrequent in the virialized regions of dense clusters, so once again internal dynamical mechanisms are likely to have played a prominent role in the transformation of Butcher–Oemler galaxy morphologies between the intermediate redshifts and the nearby universe.

In the late 1970s, from observational studies of late-type galaxy bulges, J. Kormendy concluded that many of these bulges appear to have disk-like morphological and kinematic characteristics. He subsequently proposed (Kormendy 1979, 1982) that these late-type bulges are evolutionarily linked to disks, and named these bulges "pseudo bulges" to distinguish them from "true" (or "classical") bulges in early-type disk galaxies, which were believed to be formed either primordially, or else through galaxy mergers.

Since it was at the time universally believed that stars "do not dissipate", or that their orbital behavior is adiabatic under a steady potential (see, e.g. Binney & Tremaine 2008), Kormendy proposed that the formation of pseudo-bulges and the associated mild Hubble-type evolution among late-type galaxies were made possible through gas accretion in barred potentials, and these inwardly accreted gas subsequently formed new bulge stars. This proposed pathway for secular evolution, though very influential since then, has a number of problems from comparison with the observed properties of galaxies, as well as from detailed analysis of the physics of the accretion process. First of all, as emphasized by Andredakis, Peletier, & Balcells (1995), the continuity of galaxy properties across the entire Hubble sequence as highlighted for example by the smooth variation of the Sersic index n in fitting the bulge surface density profile, indicates that there is not an apparent break in the formation mechanism between late-type and early-type bulges. However, because of the paucity of gas compared to stars in most galaxies of earlier Hubble types, there simply is not enough of a reservoir of gas to build up the bulges of galaxies such as our own (of type Sb), which has its Bulge mass comparable to the Disk mass, not to say for galaxies of even earlier Hubble types. Secondly, bulges of intermediate- to early-type galaxies, including our own, consist mostly of stars of very old ages (Jablonka, Gorgas, & Goudfrooij 2002), despite also possessing a stratified color distribution (Wirth & Shaw 1983), and could not have been built up by an *extended* process of secular *gas* accretion which subsequently formed stars. A good fraction of these apparently old bulges have stellar kinematics which are also rotation-dominated and are related to the kinematics of their disks, hinting at a secular evolution origin for their formation (Kormendy & Kennicutt 2004 and the references therein). Therefore, a dynamical mechanism which could channel the pre-existing disk stars from the outer to the inner region of a galaxy is needed to explain these rotation-dominated older bulges. Thirdly, recent results of the ATLAS^{3D} team (Cappellari et al. 2013) have shown that the morphologies and other internal properties of spirals, S0s, *as well as disky ellipticals*, form a continuous trend of evolution, which also coincides with the trend of aging of stellar population of their galactic disks. This provides further support for a unified formation history of the majority of the Hubble sequence galaxies through internal processes, which necessarily involves the participation of the stellar component. Fourth, recent results of the COSMOS team (Cisternas et al. 2011a, 2011b) showed that the establishment and evolution of the well-known black-hole-mass/bulge-mass correlation since $z = 1$, on the bulge-building side, was mainly due to the radial mass accretion process on pre-existing *stellar* disks, which were already in place by $z = 1$. The COSMOS team had also excluded merger as a significant contributor to the build-up of the black-hole-mass and bulge-mass correlation since $z = 1$. This further motivates a serious consideration of the internal secular mass redistribution process that *emphasizes the role of stellar mass accretion in bulge building*[†]. Finally, as we will show later in this monograph, even the gas contribution to the secular mass accretion process is through the same collective dissipation process involving large-scale, coherent density wave modes, rather than through the particle-level viscosity commonly attributed to gas dissipation. The mean free path of the collision/scattering of the galactic molecular clouds in a density wave crest region is similar to that of the stars, and both components should be considered together in supporting a star-gas combined two-fluid instability. This common mean free path of the two-fluid instability is many orders of magnitude larger than the microscopic collisional mean free path of the gas. So the roles of stars and gas are now parallel, rather than distinct, in the collective process that leads to the secular morphological transformation of galaxies.

From the above arguments, plus more that will be discussed in the main text, we see that internal secular evolution processes involving the participation of both the stellar and gaseous components appear to be crucial to the transformation of galaxy morphologies along the Hubble sequence at least since $z = 1$, and possibly since $z = 2$. The admittance of the stellar component into the secular mass redistribution process in galaxies, however, cannot be accomplished without a major breakthrough in the dynamical foundation of the theories of galactic structure and evolution.

1.2 Theoretical Background

The ultimate engine for the morphological evolution of galaxies is in a self-gravitating system's tendency to increase its entropy with time. It has long since been known (e. g., Antonov 1962; Lynden-Bell & Wood 1968) that the direction of this entropy evolution for self-gravitating systems is toward configurations with ever more centrally concentrated cores, together with increasingly extended outer envelopes.

[†] Incidentally, the known weaker correlation between the masses of the *late-type* bulges and their corresponding central black holes at any given epoch (Kormendy, Bender, & Cornell 2011) may be on the one hand a result of the quicker onset of accretion events in certain black-hole/active galactic nuclei (AGN) accretion disks compared to the onset of significant mass accretion in their outer galactic disks for bulge-building since the local dynamical time scale is shorter for the smaller AGN accretion disks; on the other hand some galaxies may form shallow pseudo-bulges without a deep central potential well to form a nuclear AGN accretion disk and to build a central black hole, during the early stages of a late-type galaxy's life. So the weaker correlation for late-type bulges and black holes reflect more the inhomogeneity of the galaxy formation process (i. e., the inhomogeneity of the *initial conditions of secular evolution*), rather than the inhomogeneity of the secular mass flow process itself.

Even with the above understanding, it was long held that the intrinsic speed of this evolution process is extremely slow. This slowness is due partly to the well-known pressure and angular momentum barriers in the galactic systems. Through an order of magnitude calculation, one can show that the natural speed of galactic evolution through microscopic transport processes results in a time scale for energy and angular momentum redistribution which is several orders of magnitude longer than the age of the universe (Zhang 1992).

It is clear that in order for secular morphological transformation of galaxies to be a relevant process in the observed history of the universe, so as to explain statistical evolution of galaxy properties with redshifts, dynamical mechanisms other than diffusion or dynamical friction have to be identified. An analogy here is the atmospheric heat flow (or Rayleigh–Bénard convection) problem (Kreuzer 1981). It is well known that when the natural speed of heat conduction is too slow in an atmospheric layer possessing a temperature gradient, an organized macroscopic flow pattern, facilitated by hexagonal convection cells, will spontaneously develop when the temperature gradient exceeds a certain threshold. These convection flow patterns greatly accelerate the speed of reducing the original temperature nonequilibrium in the atmospheric distribution, compared to the speed due to the conduction/diffusion process alone. As we will demonstrate in subsequent text, the new dynamical mechanism in galaxies operates in a similar fashion to the convection process in atmospheric flow. This mechanism will be shown to be closely related to the emergence of density wave patterns (or more specifically, unstable density wave *modes*) in galaxies, and to the maintenance of these modes as a dynamical equilibrium state, balanced by the opposing tendencies of the spontaneous growth of the unstable mode and the irreversible local dissipation process, with the latter process both damps the growth of the wave mode and simultaneously leads to the secular evolution of the basic state of the galactic-disk mass distribution.

Bertil Lindblad was the first astronomer to come up with the idea that spiral patterns in galaxies may be waves of density enhancement. In the 1950s, he made a series of numerical studies using an assembly of point-particles to simulate the appearance of spiral galaxies, but failed to obtain self-sustained patterns. C. C. Lin and F. Shu (1964, 1966) were the first to succeed in obtaining linear perturbative solutions of self-consistent spiral patterns in the tightly wrapped, or the so-called WKBJ (which stands for Wentzel, Kramers, Brillouin, & Jeffries) regime of the density wave parameter space, and the same WKBJ approximation allowed them to simplify the global problem of the large-scale wave pattern on a galaxy disk into a local problem of plane waves. At almost the same time, A. Toomre (1964) obtained the local stability criterion for axisymmetric oscillations in the disk, and A. Kalnajs (1965) independently obtained the dispersion relation for tightly wrapped waves as obtained by Lin & Shu (1964). In Gilmore et al. (1990) Figure 9.3, an illustration originally due to Lin (1967) summarizes the essence of the density wave approach as was practiced at that time: (a) a spiral-formed potential perturbation is introduced onto (i. e., superposed on) the differentially rotating basic state of the disk. (b) This spiral-formed potential perturbation requires a spiral-formed density perturbation in the stellar and gaseous disks to support it, through the Poisson equation (under the local WBKJ approximation, this spiral-density perturbation needed by the Poisson equation is exactly in-phase, or is proportional to, the spiral-potential perturbation). (c) On the other hand, the spiral-formed potential perturbation also leads to the perturbation in stellar orbits and in gas streamlines, which also leads to perturbed spiral-density distributions in the stellar and gaseous disks. (d) By enforcing the equality of the density perturbations from (b) and (c), a so-called *dispersion relation* is obtained, which guarantees (in this case, under the linear and local approximation) the self-consistency of the spiral wave solution.

Subsequent workers began testing the predictions of the newly developed mathematical theories of the density waves against observations, including the observed features of the density wave arms in our own Galaxy (Lin, Yuan, & Shu 1969). These comparisons were limited both by the then poorly known details of the Galactic spiral structure, the poor spatial resolutions obtainable in external galaxies, as well as by the intrinsic limitations of the WKBJ solution itself. When the spiral density wave theory was first proposed, Lin and Shu had sought a *neutral wave solution* (similar to a soliton) that retains a steady amplitude. The lowest-order WKBJ solution is indeed a neutral-wave solution. Toomre (1969), on the other hand, found that the Lin–Shu density waves in fact possess a radial group velocity and (the short-trailing wave branch) will inevitably propagate toward the central region of a galaxy and be absorbed at the inner Lindblad resonance (ILR). There is thus the need to re-generate these transient spiral waves. Lynden-Bell & Kalnajs (1972) discovered that the trailing form of spiral density wave, which is the predominant form among observed wave patterns in galaxies, actually transports angular momentum outward through the combined actions of gravitational and advective torques. Since a density wave rotates slower, or has negative angular momentum density with respect to the basic state of the galactic disk, inside the corotation radius[†], the removal of angular momentum by the wave from the inner disk and the transport of it by the same wave to the outer disk, will promote the spontaneous growth of a density wave train. This was the reason that Lynden-Bell and Kalnajs had named their 1972 paper "On the generating mechanism of spiral structure." They concluded in the same paper, however, that a steady wave exchanges angular momentum with the basic state of the galactic disk only at the wave-particle resonances, therefore the wave maintains a *constant* angular momentum flux *en route* of the outward angular momentum transport between the

---

[†] The corotation radius is a radial location where the density wave pattern rotates at the same speed as the underlying differentially rotating disk matter. For a regular disk galaxy which has higher angular speed in the inner disk and progressively lower angular speed in the outer disk (i. e., a galaxy that possesses *differential rotation*), the corotation radius (designated as r_{co} or CR in the following text) of the dominant mode of the density wave usually occurs somewhere in the mid-disk.

inner and outer Lindblad resonances[‡]. Thus the conclusion from these early density wave studies is that the major portion of the basic state of the galactic disk does not exhibit secular mass redistribution if it contains a density wave of steady amplitude. Lynden-Bell and Kalnajs, like most of their peers at the time, sought the generating mechanism of the wave trains and any associated secular effect only in the resonant interaction between the waves and the particles.

For growing wave disturbances as a result of its continued outward angular momentum transport, there is the need to counter the growth tendency in order to clamp the wave amplitude at a finite value, that is, there is the need of a damping mechanism. Kalnajs (1972), and Roberts & Shu (1972) proposed that the gaseous density waves often found to accompany stellar density waves in galaxies could serve to damp the growing stellar wave, owing to the intrinsic dissipation in the gas component and the resulting azimuthal phase shift between the stellar and the gaseous density wave patterns. However, the asymmetric roles assigned to stellar and gaseous waves meant that gas will be preferentially accreted to the galaxy center compared to the distribution of stars, contrary to what is observed (i. e., early type galaxies, considered to be in the late stages of secular morphological evolution, are known to contain far less gas than late-type galaxies which are the precursors of the secular evolution process). The gas damping mechanism also has difficulty in explaining the so-called anemic galaxies, which have coherent spiral patterns but little or no gas. It seems highly unlikely that the anemic spirals are stabilized through a different dynamic mechanism than those operating in gas-rich disk galaxies.

The problem of damping of the growing wave amplitude to achieve quasi-steady pattern was exacerbated after the discovery of over-reflection mechanisms near the corotation region, that is, the so-called WASER (which stands for Wave Amplification by Stimulated Emission of Radiation) mechanism by Mark (1974, 1976), which is responsible for the amplification of the normal spiral patterns, as well as the SWING mechanism by Toomre (1981) and Zang, which is responsible for the amplification of bar-like patterns, following the earlier work of Goldreich & Lynden-Bell (1965) and Julian & Toomre (1966). These over-reflection mechanisms at corotation, coupled with the wave-reflection/transmission mechanisms at the galaxy center, set up feedback loops for the radially propagating wave trains in the galactic "resonant cavity" between the central region and the corotation region, with a positive amplification factor for each round-trip of wave propagation. Figure 1.1 presents an illustration for the relatively milder version of the over-reflection mechanism WASER

[‡] An inner Lindblad resonance is a galactic radial location where the pattern speed of the density wave Ω_p, the circular speed of the particles according to the galactic rotation curve Ω, and the local epicycle frequency κ satisfy $\Omega_p = \Omega - \kappa/2$ (for a two-armed pattern, which is the predominant form among the observed grand-design density wave patterns). Similarly, an outer Lindblad resonance is where $\Omega_p = \Omega + \kappa/2$. For the dominant pattern of the density wave the inner Lindblad resonance is usually located in the central bulge region if it exists at all, and the outer Lindblad resonance is usually located in the outer disk region external to the corotation radius of the same wave train or mode.

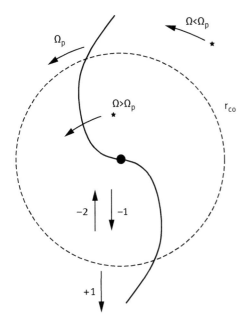

Figure 1.1: Schematic of the wave propagation and amplification in the galactic resonant cavity to form infinitely-growing density wave modes, through the WASER mechanism (Mark 1974, 1976). The symbols Ω and Ω_p denote the galaxy (or disk matter) circular/angular speed and wave pattern speed, respectively. The numbers next to the arrows indicate the normalized amount of wave-train angular momentum relative to the basic state (adapted from Zhang 2016).

(Mark 1974, 1976). For the SWING mechanism (Toomre 1981), the over-reflection factor can be significantly bigger than the factor of 2 illustrated here. The SWING and WASER mechanisms differ by whether the outward-propagating wave train toward corotation is that of the leading or the trailing type, respectively, produced in turn by either the inward-propagating trailing wave tunneling through the central region of the galaxy when no Q-barrier[†] exists, and emerging as a leading wave; or else by the inward-propagating trailing wave being reflected by the Q-barrier when one exists, which produces an outward-propagating trailing wave. The SWING mechanism tends to produce bar-like modes, and the WASER mechanism spiral modes. In this way the superposition of radially oppositely propagating density wave trains produces indefinitely growing *modes* (Lin & Lau 1979 and the references therein).

In addition to the approach for obtaining a modal solution through the superposition of oppositely propagating wave trains, the spontaneously growing spiral/bar modal solutions can also be obtained directly, through solving the density wave perturbation as a normal mode problem under the boundary condition of a given basic state (Lin & Lau 1979). Leaving the general problem of the wave-damping mechanisms undecided, Lin and collaborators went on to solve a complete series of modal morphologies under systematically varying basic state boundary conditions, and the resulting modal series largely reproduce the spiral morphology variation along the Hubble sequence, so the density wave pattern morphologies can now be

---

[†] A Q-barrier is a rapid rise in stellar velocity dispersion, normally occurs in the galactic bulge region, that reflects an inward-propagating density wave train into an outward-propagating wave train.

tied to the corresponding basic state characteristics of the parent galaxies (Bertin et al. 1989a, 1989b), achieving in part the explanation of the observed correlation of these two sets of characteristics which underlies the original Hubble classification scheme.

Throughout the first three decades of density wave studies, the possibility that quasi-steady density wave modes could lead to significant secular morphological evolution of the basic states of galaxy disks was never seriously discussed, even though mild secular changes in certain over-stable basic states possessing *transient* spiral waves were found by, e. g., Sellwood & Carlberg (1984). This impression of the immutable-basic-state for the majority of observed galaxies was further supported by the well-known classical result of the conservation of the Jacobi integral for a single stellar orbit under an enforced, steady, spiral or bar potential (see, e. g. Binney & Tremaine 2008), otherwise known as the adiabatic condition of the stellar orbit. Other relevant studies in this context include that of Goldreich & Tremaine (1979), as well as Goldreich & Nicholson (1989), which found that there is no wave/basic-state interaction in the Eulerian fluid solution up to second order of nonlinear approximation. Bertin (1983) reached a similar conclusion that any potential secular evolution of the basic state induced by the spiral wave happens on a time scale much longer than the age of the universe.

However, these classical work on the density wave theories ignored important collective effects responsible for the long-term evolution of galaxies. It has long been known that the relaxation and dynamical evolution of physical systems governed by long-range forces are determined chiefly by collective effects. The physical behavior of such systems is not completely reflected in the result of passive orbit calculations. Collective effects were known to enhance the speed of relaxation in many plasma systems (Kulsrud 1972; Balogh & Treumann 2013 and the references therein). Collective effects in the context of disk galaxies invariably involve local or global instabilities, and they operate by forcing a particle to systematically "collide with" or scatter off an aggregate of other particles collected together by the instabilities. The effective impact parameter in such a "collision" process is of the order of the size of the instability-feature formed. In effect, besides experiencing the smooth part of the potential, a particle which participates in a collective process also bounces off the local short-to-intermediate-range grainy scattering-potential. These collisions or scatterings are not random but rather are highly coordinated, and in the case of disk galaxies they enable the formation and stabilization of global density wave modes, as well as the long-term secular morphological evolution of their parent galaxies.

A collective dissipation mechanism responsible for the secular evolution of the disks of spiral galaxies was first proposed and analyzed in Zhang (1996), based on the more qualitative studies in Zhang (1992). It was shown that there exists a characteristic radial distribution of an azimuthal phase shift between the perturbation potential and density pairs of a spontaneously formed spiral or bar *mode*, such that the density leads the potential inside the corotation radius, and lags the potential outside the corotation. Due to the resulting secular torque between the density wave and the disk matter implied by the potential-density phase shift, the disk matter inside the corotation radius loses energy and angular momentum to the density wave and accretes inward, and the matter outside corotation gains energy and angular momentum from the wave and excretes. This happens for the entire disk surface, not just at the wave-particle resonances as predicted by older analyses. As a result, the disk surface density becomes more and more centrally concentrated, together with the build-up of an extended outer envelope, consistent with the direction of entropy evolution in self-gravitating systems. Furthermore, a local physical mechanism was also found to account for the collective secular dissipation as is revealed and required by the phase shift. This mechanism takes the form of a temporary local gravitational instability of the streaming disk material at the spiral arms, which makes spiral and bar modes akin to the collisionless shocks in plasma physics (Balogh & Treumann 2013). Subsequent work by Zhang (1998, 1999, 2008, 2016), Zhang & Buta (2007, 2012, 2015), and Buta & Zhang (2009) further established the relevance and importance of this collective dissipation process to the observed structure, kinematics, and secular morphological evolution of galaxies.

The conceptual support for these results also comes from the connection to theories of "dissipative structures" (Prigogine 1969, 1980; Glansdorff & Prigogine 1971; Nicolis & Prigogine 1977). Prigogine and coworkers have found that the large-scale coherent patterns present in complex, nonlinear, and far-from-equilibrium systems were not merely impressive veneers, they serve the important dynamical role of greatly accelerating the speed of entropy evolution of the underlying nonequilibrium systems. At the close-to-equilibrium regime, and under certain restrictive conditions, entropy production was found to be a minimum through transport processes (Prigogine 1969); whereas at the far-from-equilibrium regime, through the formation of dissipative structures, local entropy production is greatly accelerated, yet the global pattern is not destroyed in the process because the locally produced entropy is exported to the environment through the operation of the very same dissipative structures, such that the local entropy density can remain at a quasi-steady value. The accelerated entropy evolution refers to the system *plus* the environment, the paradox of entropy increase in the face of the formation of highly organized structures is thus solved. Nonequilibrium can become the source of order.

For self-gravitating systems, as we have commented before, the direction of entropy evolution is toward configurations of ever-increasing central concentration, together with an extended outer envelope (Antonov 1962; Lynden-Bell & Wood 1968). In the case of galaxies, this is the same as the direction of morphological evolution along the Hubble sequence from the late to the early Hubble types. Density wave patterns accelerate the local entropy production (the conversion of the organized orbital-motion energy into random motions of the stars and gas), but possess the ability to transport energy, angular momentum, and entropy to the environment through the coordinated interaction of the wave and the basic state.

1.3 Organization of the Material

The content of this monograph covers the analytical, numerical, and observational aspects of secular morphological evolution of galaxies driven by an internal dynamical process mediated by density wave modes. The choice of material reflects not only the author's past experience (as well as its unavoidable limitations), but also the fact that in the study of collective phenomena, a purely deductive analytical approach (or a "reductionist approach") is in principle not viable: the very essence of the self-organization process is such that the causes and effects become intermingled. This special feature in the collective global instabilities may be partly the reason that throughout the past centuries when classical mechanics made headway, only very slow progress had been made in the area of collective interactions within a complex system: because the necessary experimental, observational, and simulational techniques were not available. In cases where significant progress had been made before, such as Kolmogorov's derivation of the scaling laws for fully developed turbulence (Kolmogorov 1941a, 1941b, 1941c; Frisch 1995), a global, synthetic, balance-equation approach had been employed, rather than a deductive analytic approach[†]. Another example is the study of hydrodynamic shocks, where integral forms of the mass, energy and momentum balance-equations (i.e., the so-called Rankine-Hugoniot conditions) across the shock discontinuity need to be employed, together with the equations of state, to deal with the fact that a continuum (or differential) approach is no longer viable to aid with the task of crossing the shock front.

To properly employ a synthetic approach requires us to incorporate constraints from observational data of diverse range, as well as to use the results of numerical simulations to verify the products of logical reasoning. In a way, the approach adopted for writing this book mirrors the way progress had been made, and will continue to be made, in the study of complex and self-organized systems. Despite its employment of analytical equations and modern N-body simulations, in spirit this book has its kinship to Charles Darwin's *On the Origin of Species* (Darwin 1859) or Alfred Wegener's *The Origin of Continents and Oceans* (Wegener 1929), in that it is an explorer's account of how original discoveries were made, as well as a presentation of the diverse suite of evidence in support of a potential paradigm shift. The aim of the book is more to inspire a new generation of explorers to set foot onto this new frontier of galactic research, rather than to deliver the final words on a mature subject.

The organization of the chapters is as follows. In Chapter 2, we will present analytical derivation of the dynamical behaviors and emergent laws in disk galaxies containing spontaneously-formed density wave modes. Note that on a first reading, it is OK to skim through some of the more lengthy proofs in this chapter, as long as the central train-of-thoughts is followed. In Chapter 3, we will verify the theoretical

---

[†] In some sense the assumption of fully developed turbulence is like our assumption of the quasi-steady state of the wave mode, where the juggling act of energy injection and dissipation has been accomplished by nature, and the global self-consistency requirement is satisfied.

results through N-body simulations. Numerical techniques most relevant to the simulation of collective effects in disk galaxies will be highlighted. This will be followed by discussions of astrophysical implications and observational confirmations of the theoretical results in Chapter 4. In Chapter 5, we put together everything we have learned so far to infer further characteristics of the self-organized instabilities (including the relation of quasi-steady density wave modes to fully developed turbulence, as well as the origin of irreversible behavior in a nominally reversible dynamical system), as well as implications of this work on the cosmological evolution of galaxies. This will be followed by a brief Conclusions chapter (Chapter 6). The Appendix chapter (Chapter 7) discusses the link between the dynamical theory of nonequilibrium phase transition, to the relevant classical treatments in kinetic theory and fluid mechanics.

The current monograph focuses on the latest research results, and by itself does not serve as a complete course on galactic dynamics. The author encourages the reader to make full use of the classical texts (e. g., Rohlfs 1977; Gilmore et al. 1990; Shu 1992, especially Chapters 11 and 12, which together form a streamlined introduction to density wave theory; Binney & Tremaine 2008; Bertin 2014), in conjunction with the study of the current text, in order to arrive at a personal perspective. This was also the initial path that the author had taken in the late 1980s as a graduate student. Another piece of advice to the newcomers: take your own ideas and intuitions seriously, and follow them through with a lifetime of dedicated exploration.

2 Dynamical Drivers of Galaxy Evolution

We begin the formal presentation with a chapter devoted to the analytical formulation of the dynamical drivers for the secular morphological evolution of galaxies containing density wave modes. The formulation is built on the foundation of the density wave theory of galaxies as developed in the past few decades, which itself is built on the foundations of classical kinetic theory and fluid dynamics. As the presentation proceeds, however, we will point out that within this older theoretical framework, we cannot self-consistently accommodate a density wave mode of finite amplitude. The application of the continuum-based density wave theory leads either to neutral-wave solutions (as in the local WKBJ treatment), to linear and infinitely growing density wave modes (as in the fluid-disk treatment of global modes), or to non-self-consistent nonlinear solutions (as in the response of a gaseous disk to stellar density wave forcing). This sets the background for us to explore the new elements needed to construct a globally self-consistent theory that leads both to a self-sustained density wave modal-set in disk galaxies, as well as to the secular morphological evolution of its parent-disk mass distribution.

2.1 Motivation and Outline for the Theoretical Approach

In order to obtain sufficient radial mass flux to transform the Hubble types of galaxies over the age of the universe (i. e., within the so-called Hubble time, or 13.7–13.8 billion years), with the understanding that a significant fraction of galaxies may have formed much later, thus their ages are only a fraction of the Hubble time, common sources of microscopic viscosity were long since known to be drastically inadequate. From the analogy with other types of self-organized global patterns, such as the hexagonal convection cells in atmospheric flow (i. e., the Bénard problem), we naturally suspect that the strikingly beautiful density wave patterns in disk galaxies, such as spirals, bars, rings, and lenses, could perform a similar function of greatly accelerating the speed of entropy evolution and mass redistribution in their parent systems. After all, the whirlpool appearances of some of these galaxies (such as the namesake "Whirlpool Galaxy" M51) give us the impression that the orbiting stars and gas on the disks of these galaxies were spiraling into their central region. In fact, this very inkling was what had set the current author out on this strenuous quest in the late 1980s, when she was just beginning her astronomy graduate study at UC Berkeley[†].

[†] A popular account of the genesis of the spiral density wave theories, including the part that the author's work had played, can be found in an article by Jack Lucentini in September 2002 issue of the *Sky & Telescope* magazine.

The situation with the galaxies turns out to be infinitely more complicated than the flow of water in the whirlpool, which at first glance may seem to underlie what is going on in a spiral galaxy. The main difference is that the process in the whirlpool is not a *self-organized* process, it is instead what we would call a *driven* or a *passive* process. The gravitational potential difference between the surface and the drain of a typical whirlpool coerces the accumulated water to spiral down the drain. The stars and gas in a galaxy, however, had achieved their quasi-equilibrium configuration, that is, that of the mass and velocity distributions on the galactic disk, through negotiating a delicate balance between gravity and pressure forces[†]. Unless they are perturbed by external driving forces (such as the tidal forces of a companion), there is not an easy "path of least resistance" for the galactic mass to follow, in order to overcome the angular momentum barrier due to their orbital motion.

Here is where the analogy with other types of self-organized dissipative structures comes into play. If the spirals and bars in galaxies are indeed another instance of dissipative structures, they will generate emergent physical characteristics to speed up entropy production and export, and to allow the original diffusion-type angular momentum transport to be replace by a convective (or advective) type of angular momentum transport. Since the density wave patterns are perturbations on the basic state mass distribution, and since most of the galaxy mass resides in the basic state, the key to this wave-assisted advective evolution process is to be found in a set of physical mechanisms which are able to (1) load the angular momentum from the basic state onto the wave within the inner disk, (2) allow the wave to transport the angular momentum to the outer disk, and (3) unload the transported angular momentum by the wave back onto the basic state at the outer disk region. This way the wave can be thought of as a kind of "lorry," performing the task of secular angular momentum transport while itself remains quasi-steady.

In order to load the angular momentum from the basic state onto the wave in the inner disk and unload it from the wave to the basic state at the outer disk, there needs to be an irreversible/dissipative interaction between the basic state and the wave. For classical WKBJ type (tightly wound) density waves, past studies have shown that there is no wave and basic state interaction for steady-amplitude waves (Goldreich & Tremaine 1979; Goldreich & Nicholson 1989). Therefore, to enable

---

[†] We can regard the stellar circular motion, or the so-called angular momentum barrier, as a kind of generalized pressure-force as well, when considering the system in a virial-equilibrium configuration. Incidentally, in Toomre (1964)'s derivation of the disk-galaxy local stability condition, it is the pressure *force* (or the local velocity dispersion) that enters into the stability criterion. However, in dealing with the global stability of a galactic system, when considering the stability of each galactic annulus separately, it is rather the pressure *gradient*, as well as centrifugal force of the stellar orbital motion, that balance the inward gravitational pull. Furthermore, if the global stability problem is considered from the point of view of the virial equilibrium of the galaxy as a whole, then once again the pressure force itself enters the consideration, and the orbital angular momentum can be viewed as just another form of stellar motion.

wave/basic-state interaction, one must look to the open types of spiral waves (i. e., those that have nonzero pitch angles).

What is the new physical element added by the open waves? As we all know, gravitational interaction is long range, meaning that the potential of a mass distribution is obtained from the Poisson integral of the mass distribution, and the resulting potential does not have to look like the mass distribution itself. In the case of a mass distribution in the shape of a spiral, the Poisson integral in general produces a potential spiral that is *phase shifted* in azimuth from the density spiral that generates the potential field. This phase shift is of vital importance to the dynamical mechanism that enables the secular mass redistribution in galaxies since, as we will show below, in the disk geometry in order for a *quasi-steady* wave pattern to exchange angular momentum with the basic state mass distribution secularly, a phase shift between the perturbation potential and density distributions is the *necessary and sufficient condition*.

But would the radial distribution of the azimuthal phase shift be of the correct sense to allow angular momentum loading in the inner disk, and unloading in the outer disk? As it turns out, this is exactly the case for spontaneously formed density wave *modes*. The types of potential and density distributions for modes are such that the phase shift is positive (means density leads potential in the azimuthal direction of galactic rotation) inside the corotation radius, and negative outside. This pattern of phase shift distribution is equivalent to the torquing by the density wave on the basic state matter in each annular ring of the galactic disk in just the correct sense to produce a secular mass redistribution trend as dictated by the entropy law of self-gravitating systems. Therefore, nature seems to have engineered a mechanism in the precise fashion to allow a disk galaxy to achieve the goal of accelerated entropy evolution. The same phase shift distribution will be shown to be responsible also for the spontaneous emergence of the wave mode in the linear regime, and for its stabilization to a constant amplitude at the quasi-steady state.

With the mechanism for loading and unloading of the angular momentum now available, and with the mechanism for outward transport of the angular momentum previously found by Lynden-Bell & Kalnajs (1972)[†], we are in possession of a dynamical process for the secular evolution of the disk-galaxy mass distribution. However, much more subtle intricacies of the process still await discovery. The phase-shift/torque/angular-momentum-transport process is a global description. There is the need to explore the local dynamical processes that allow these global relations to be fulfilled. This is shown to be related to the local instability condition at the density

---

[†] Lynden-Bell and Kalnajs had advocated for a *constant* angular momentum flux for a steady wave train, due to their *a priori* assumption of no-wave-basic-state interaction. Whereas with the continuous angular momentum loading within the inner disk, and unloading in the outer disk, which we will demonstrated in this work, the angular momentum flux turns out to be of a characteristic bell shape, with the peak of the bell at the corotation radius.

wave crest of an open spiral (or skewed bar) mode. This instability condition, coupled with the demonstration of the transonic velocity jump across the spiral arm (see Section 3.3), shows that the density wave modes possess the essential features of plasma "collisionless shocks" (Zhang 1996; Balogh & Treumann 2013).

The collisionless shock at the density wave crest is a formal singularity of the solutions of the underlying fluid (or stellar dynamical) equations used to model the finite-amplitude density wave modes in galactic disks. We will show that within such formal singularities of the dissipationless equation sets, which are used to model the spontaneously formed density wave modes, the differential form of the Poisson equation is no longer valid (in fact, the continuum formulation itself is invalid), which is to be expected since a formal singularity of the solutions of the governing differential equation set implies precisely the breakdown of the differential forms of governing equations, or the breakdown of mean-field/continuum treatment. So we will need to seek a genuine particle treatment of the problem to repair this "breakdown," and to put the different pieces of the self-organization process back together.

As part of this process of putting together, as we will show in Chapter 3, when advancing articles on their trajectories in N-body simulations, we in fact are making use of velocities generated from the force field of the matter distribution that incorporated the multitude of *correlations* among disk particles. The kinematics and the particle mass distribution together enable the self-organization process, and together they encode particle correlations. This is why a proper treatment of the gravitational viscosity generated by self-organized density wave patterns requires a genuine N-body simulation, which naturally models particle-correlation-induced gravitational viscosity through the use of the integral form of the Poisson equation, whereas schemes such as the smoothed-particle-hydrodynamics (SPH, Monaghan (1992) and the references therein), which is based on the mean-field differential formulation, will necessarily require the specification of gravitational viscosity as an artificial input parameter.

We emphasize at the outset that a *proper basic state choice* that allows genuine unstable (or self-organized) modes to spontaneously emerge is crucial for the attainment of the correct secular evolution behavior of the basic state of the disk. Only such basic states that admit intrinsic unstable modes display *correlated fluctuations* that produce the kind of wave/basic-state interactions that lead to galaxy evolution along the Hubble sequence. Over-stable disks that allow only *transient wave trains* to emerge will not serve this purpose. We will address this point further in Section 5.1.2.

To follow the intricate details of the self-organization process, as we have mentioned before, a particle approach is required, which we will defer the discussion to Chapter 3. In this chapter, we present analytical derivations of the relevant secular evolution dynamics mostly from a global closure-relation point of view. We will make use of the conservation requirements of the various physical entities to derive new closure relations at the quasi-steady state of the wave mode, which turn out to be closely related to the potential-density phase shift we had mentioned earlier. The process through which the older closure relations in the angular momentum transport and exchange processes is replaced by emergent new closure relations is similar in spirit to the mechanism of spontaneous breaking of gauge symmetry in high-energy physics to arrive at new dynamical laws, when the energy scale of fundamental physical processes is traversed. In fact, nature appears to organize the hierarchies of physical laws in the different regimes of physical parameters through invoking just such symmetry-breaking processes, both in the high-energy and low-energy physics. We will come back to the discussion of this topic toward the end of this monograph.

2.2 Density Wave Crest as the Site of Gravitational Instability[†]

Much of the new dynamics responsible for enabling the secular morphological evolution of galaxies originates from the operation of collective effects due to the multitudes of mutually interacting stars and molecular complexes in the galactic disk, especially near the potential minimums of the density waves (Zhang 1996).

For collective effects to operate in a spiral or a barred galaxy, individual stars have to be *aware* of their "neighbors" directly, besides experiencing the smoothed axisymmetric plus the smoothed spiral potential. However, as is well known, binary encounters in disk galaxies are extremely rare compared to the age of a galaxy (Binney & Tremaine 2008). Under this circumstance, the scattering of a star off its neighboring stars can only happen when the disk is locally gravitationally unstable – even if just marginally so. Therefore, the first step in establishing that a spiral structure can induce collective dissipation is to show that a spiral structure can lead to local gravitational instability in an originally marginal stable disk.

We point out at the outset that in the derivations in this section, even though expressions for WKBJ waves were initially used, the final conclusions of collective dissipation in self-organized modes depend on the patterns being *open*, or have finite pitch angles. *The WKBJ formulation is used here only to illustrate its own inadequacy* – that is, we will arrive at the conclusion that only by going *beyond* the WKBJ approximation in the density wave theory, can we expect to arrive at the true source of local gravitational instability, as well as collective dissipation and secular morphological evolution of galaxies. Furthermore, even though we start with a generic density *wave* formulation, in later sections of this chapter we will show that the global-self-consistency requirement dictates that the kind of open waves that allow the secular and quasi-steady mass redistribution need to be *unstable wave modes* of the underlying basic state of the galactic disk.

[†] Portions of this section used material previously published in Zhang (1996), reproduced with modifications from The Astrophysical Journal @ AAS. Reproduced with permission.

2.2.1 Local Stability Condition at the Spiral Arm and Interarm Region

By considering the competing influence of pressure-force and rotational stabilization effects in a stellar disk configuration, Toomre (1964) arrived at the following well-known local stability condition against axisymmetric type of instabilities in a disk geometry:

$$Q \equiv \frac{\sigma_r \kappa}{3.36 G \Sigma} > 1, \tag{2.1}$$

where σ_r is the radial velocity dispersion, κ is the epicycle frequency, Σ is the surface density of the disk, G is gravitational constant and Q is Toomre's stability parameter. For a fluid disk, the factor 3.36 in the denominator is changed to π.

At the different azimuthal locations, the streaming motion of the disk material under the influence of a spiral perturbation potential changes the values of the radial velocity dispersion σ_r, the epicycle frequency κ, and the surface density Σ from their original values appropriate for an axisymmetric disk. In the following we derive the variations of these parameters with the phase of the spiral and calculate how these variations influence the value of the instability parameter Q at the spiral arm and the interarm region. We will first consider a linear and WKBJ (i. e., tightly wrapped) spiral wave, and then discuss what modifications we need to introduce when considering a more open type of wave in the nonlinear regime.

In the following we adopt the simpler Eulerian fluid formulation (rather than the stellar dynamical equation used in the original Lin–Shu theory) for the discussion of the stability condition in the spiral-arm region. For an m-armed spiral density wave of pattern speed Ω_p, the gravitational potential at the disk location (r, ϕ) and time t can be written as (Rohlfs 1977; Shu 1992)[†]

$$\mathcal{V}(r, \phi, t) = \mathcal{V}_0(r) + A(r) \exp\{i[m\Omega_p t - m\phi + \Phi(r)]\}, \tag{2.2}$$

where $|A| \ll |\mathcal{V}_0|$ and where $\Phi(r)$ is related to the pitch angle i and the wavenumber $k = \lambda/2\pi$ of the spiral through

$$\frac{d\Phi}{dr} = \frac{m}{r \tan i} = k, \tag{2.3}$$

with $k < 0$ corresponds to a trailing spiral. The WKBJ approximation further demands that

$$|kr| \gg 1, \tag{2.4}$$

[†] Despite using somewhat different notations, the formulations of Rohlfs (1977) and Shu (1992), when both using Eulerian fluid equation set, are completely equivalent. For example, the solution for the perturbation spiral density under the linear and WKBJ approximation, is represented by eq. (68) in Rohlfs (1977), and by eq. (11.44) in Shu (1992). These two solutions can be shown to be identical after straightforward variable substitutions.

or that the wavelength of the wave is much smaller than the system dimension under concern, so locally the WKBJ waves in the disk geometry approximate plane waves.

The solution for the azimuthal velocity of the streaming stars within the WKBJ approximation is

$$v(r, \phi, t) = v_c(r) + i \frac{kA}{2\Omega_0} \frac{1}{1 - v^2 + x} \exp\{i[m\Omega_p t - m\phi + \Phi(r)]\}, \tag{2.5}$$

where v is the normalized encounter frequency of streaming stars with respect to an m-armed spiral pattern, and

$$v = m(\Omega_p - \Omega_0)/\kappa_0, \tag{2.6}$$

$$x = k^2 \sigma_{r0}^2 / \kappa_0^2, \tag{2.7}$$

and where v_c, σ_{r0}, Ω_0, and κ_0 are the unperturbed circular velocity, radial velocity dispersion, angular frequency and epicyclic frequency, respectively. For convenience, in the following discussions we assume a flat rotation-curved galaxy, that is, $v_c(r) = v_c$ is a constant. Since we are considering WKBJ waves, which is a local plane-wave approximation of the density waves, the constant circular velocity assumption will not increase the level of error within the WKBJ approximation.

The corresponding density variation is

$$\Sigma(r, \phi, t) = \Sigma_0(r) - \Sigma_0(r) \frac{k^2 A}{\kappa_0^2} \frac{1}{1 - v^2 + x} \exp\{i[m\Omega_p t - m\phi + \Phi(r)]\}. \tag{2.8}$$

Since

$$\kappa^2 = 2r\Omega \frac{d\Omega}{dr} + 4\Omega^2, \tag{2.9}$$

in the following we first calculate the change in Ω and $d\Omega/dr$ due to the presence of a spiral.

The angular frequency at a location (r, ϕ) in the presence of spiral perturbation becomes

$$\Omega(r, \phi, t) = \frac{v(r, \phi, t)}{r} = \Omega_0(r) + i \frac{kA}{2\Omega_0 r} \frac{1}{1 - v^2 + x} \exp\{i[m\Omega_p t - m\phi + \Phi(r)]\}. \tag{2.10}$$

Therefore,

$$\frac{d\Omega}{dr}(r, \phi, t) = -\frac{v_c^2}{r^2} - \frac{k^2 A}{2\Omega_0 r} \frac{1}{1 - v^2 + x} \exp\{i[m\Omega_p t - m\phi + \Phi(r)]\}. \tag{2.11}$$

The effective κ^2 in the presence of the spiral potential can thus be calculated to be

$$\kappa^2(r, \phi, t) = \kappa_0^2 \left\{ 1 - \frac{k^2 A}{\kappa_0^2} \frac{1}{1 - v^2 + x} \exp\{i[m\Omega_p t - m\phi + \Phi(r)]\} \right\}, \tag{2.12}$$

where $\kappa_0^2 = 2\Omega_0^2$ for a flat rotation-curved galaxy, and where we have dropped a few terms of order $1/kr$ or higher compared to the dominant terms in deriving eq. (2.12).

On the other hand, from eq. (2.8) we know that

$$\frac{\Sigma}{\Sigma_0} = 1 - \frac{k^2 A}{\kappa_0^2} \frac{1}{1 - v^2 + x} \exp\{i[m\Omega_p t - m\phi + \Phi(r)]\}, \quad (2.13)$$

we have thus demonstrated that

$$\frac{\kappa}{\kappa_0} = \left(\frac{\Sigma}{\Sigma_0}\right)^{0.5} \quad (2.14)$$

for WKBJ waves.

The change in the velocity dispersion σ_r depends on the energy conversion process assumed. If, as in the linear density wave theory, we assume an adiabatic process for stars entering and leaving the spiral arms, so that the self-gravitating potential energy of the streaming stars is temporarily converted into the random velocities of these stars, we expect that

$$\frac{\sigma_r^2}{\sigma_{r0}^2} \approx \frac{\Sigma}{\Sigma_0}, \quad (2.15)$$

where we have assumed that the compression is one dimensional, as is appropriate for WKBJ waves.

From eqs (2.1), (2.14), and (2.15), we see that at the location of the spiral arm, which corresponds to the location of the density enhancement, the epicycle frequency κ increases so that the stabilizing effect of the Coriolis force is increased. The velocity dispersion increases too by an amount determined by the degree of density enhancement and also by the overall energy conversion process at the spiral arms. The final stability state at the spiral arm region will be determined by the competition of these different factors.

In the case when only orbit crowding but no energy loss (due to the energy exchange with the density wave) occurs, the effective Q will not change significantly from its unperturbed value for the basic state of the disk, that is, we have $Q_{arm} \approx Q_{interarm} \approx Q_0$ for a linear WKBJ wave, due to the fact that both κ and σ_r in eq. (2.1) scale as $\Sigma^{0.5}$, so together they cancel the Σ dependence in the denominator. This result is to be expected since WKBJ wave is not a true instability of the disk. It can be shown that the WKBJ waves have zero potential and density phase shift (i. e., comparing eqs 2.2 and 2.8), and the existence of phase shift is a key factor in the ability for density waves to exchange energy and angular momentum with the disk matter. This exchange is responsible both for the spontaneous growth of the wave and the secular dissipation effect it induces. Therefore, the WKBJ wave can neither grow spontaneous as a disk instability, nor induce secular changes (except in the case where a dissipative gas component is invoked in a non-self-consistent manner, as will be discussed later in the text).

However, for a more *open* spiral pattern with finite pitch angle and finite amplitude, the potential field at the spiral arm is generated not only by the local streaming mass, but also by the matter in the rest of the spiral pattern. The phase shift between the potential and density spirals of an open spiral pattern means that a patch of disk material, when entering the spiral arm, experiences an extra compression contributed by the rest of the disk matter, besides that contributed by its own self-gravity. Thus the originally marginal stable disk material becomes *temporarily unstable* when crossing the spiral arms. The same argument applies for streaming stars within skewed bar mode as well.

In some sense, the global open spiral wave itself *is* an instability structure on the originally marginal stable axisymmetric disk. The local instability at the spiral arms is the constituent of the global spiral instability.

2.2.2 Length Scale of Spiral Instability at the Solar Neighborhood

At the solar neighborhood, ~10–20% of the local surface density is contributed by the gas component, which has random velocities between 4 and 8 km^{-1} (Spitzer 1978, p. 231; Stark 1979; Liszt & Burton 1981). The presence of the low-velocity dispersion gaseous component generally makes the galactic disk less stable. The quantitative effect of the gas on the stability condition of the star/gas combined disk can be analyzed through the two-fluid dispersion relation of Jog & Solomon (1984). Additional analyses of the local instability conditions for galaxy disks containing multiple mass components can be found in Bertin & Romeo (1988), Romeo (1992, 1994a), Elmegreen (1995), Jog (1996), Rafikov (2001), Romeo & Wiegert (2011), Romeo & Falstad (2013), etc.

The distinctive features of the two-fluid instability as compared to the one fluid instability are, first, in the case where the stars and gas are separately stable, the combined two-fluid disk can be unstable; and second, the length scale of the most unstable instability feature is significantly reduced from that of the pure stellar case. Including the finite-disk-thickness correction, the two-fluid dispersion relation can be cast in the form (Jog & Solomon 1984)

$$\omega^2 = \frac{1}{2}\left\{(\alpha'_s + \alpha'_g) - [(\alpha'_s + \alpha'_g)^2 - 4(\alpha'_s\alpha'_g - \beta'_s\beta'_g)]^{1/2}\right\}, \tag{2.16}$$

where

$$\alpha'_s = \kappa^2 + k^2 v_s^2 - 2\pi Gk\Sigma_s\left\{[1 - \exp(-kh_s)]/kh_s\right\}, \tag{2.17}$$

$$\alpha'_g = \kappa^2 + k^2 v_g^2 - 2\pi Gk\Sigma_g\left\{[1 - \exp(-kh_g)]/kh_g\right\}, \tag{2.18}$$

$$\beta'_s = 2\pi Gk\Sigma_s\left\{[1 - \exp(-kh_s)]/kh_s\right\}, \tag{2.19}$$

$$\beta'_g = 2\pi Gk\Sigma_g\left\{[1 - \exp(-kh_g)]/kh_g\right\}, \tag{2.20}$$

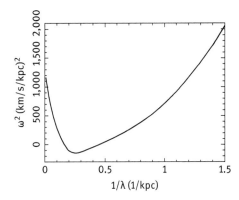

Figure 2.1: Growth rate of the two-fluid instability at the solar neighborhood (adapted from Zhang 1996). Parameters used: $\Sigma_{total} = 80$ $M_\odot$, 15% of the surface density is in the gas phase, $v_s = 30$ km s^{-1}, $v_g \approx 6$ km s^{-1}, $h_s = 175$ pc, $h_g = 75$ pc, $\kappa = 36$ km s^{-1} kpc^{-1} (references for the origins of parameters used in this figure were given in Zhang (1996)).

where ω is the angular frequency of the two-fluid instability, h_s and h_g are the scale heights, Σ_s and Σ_g are the surface densities, and v_s and v_g are the velocity dispersions of the stellar and gaseous fluids, respectively, and $k = 2\pi/\lambda$ is the wavenumber of the axisymmetric two-fluid instability under consideration.

Making use of the two-fluid dispersion relation, we plot in Figure 2.1 the ω^2 versus $1/\lambda$ curve for parameters appropriate for the solar circle. Since there is expected to be extra compression on the streaming disk matter due to the fact that the perturbation potential and density are phase-shifted with respect to each other, we have adopted the highest known total surface density at the solar neighborhood, $\Sigma_{total}(r = r_\odot) = 80$ $M_\odot$ pc^{-2} (Bahcall 1984) in the calculation of the instability length scale. From Figure 2.1, we see that the solar neighborhood is slightly below the stability threshold, and the most unstable mode (the one with the largest magnitude of negative ω^2) has wavelength

$$\lambda_{\text{most unstable}} \approx 3.6 \text{ kpc}. \tag{2.21}$$

The instability structure formed usually has an extent of $1/2\,\lambda$ for the region of density enhancement, thus a radius of $1/4\,\lambda$. Therefore, the magnitude of the radius of the instability structure is

$$r_c \sim 0.9 \text{ kpc}. \tag{2.22}$$

Compared to the Galactic orbital circumference at the solar neighborhood, which is ~ 50 kpc, the size of the instability structure can fit comfortably in the circumferential direction. This length scale is also very close to that of the size of the giant HI and molecular cloud complexes observed near the spiral arm region of many external galaxies (Elmegreen & Elmegreen 1983). It is likely that the appearance that these giant molecular cloud complexes are gravitationally bound is an indication that the underlying stellar disk is also unstable, since, on the scale of a kiloparsec, the gaseous material is not likely to be decoupled from the stars in its stability state.

The stability of the solar neighborhood thus is compatible with it being marginally unstable if it is near the spiral arm of a moderately open type.

2.3 Potential-Density Phase Shifts for Density Wave Modes[†]

The marginal gravitational instability condition at the spiral arms allows the streaming stars to scatter off their neighbors when crossing the arms, thus producing collective dissipation effect. What then would be a sure sign that such secular dissipation effect is indeed occurring? In the following we will show that for a spiral structure that is quasi-stationary on local dynamical time scales, the *only* possible way for the orbiting stars and the density wave pattern to exchange angular momentum secularly is to have a spiral density distribution that is phase-shifted in azimuth with respect to the spiral potential distribution. A self-sustained potential-density phase shift pattern is both the result and the enabler of the secular dissipation process mediated by the spiral density wave. This is what we mean by the global self-consistency requirement: in a disk geometry one cannot have secular energy and angular momentum exchange between the wave and the disk matter without it leaving a trace in the form of the potential-density phase shift; conversely this trace of secular dissipation cannot survive as a quasi-steady feature of the galaxy without the Poisson integral (as well as the equations of motion) making an allowance for it to be accommodated by a pair of potential and density spirals. The realization of this requirement for self-sustained phase shift pattern to exist through both the Poisson integral and the equations of motion in order to enable secular dissipation, was what had originally led the author to search for, and independently discovered its existence numerically, through the Poisson integral; and the demonstration by the author of the existence of the phase shift in the equations of motion of stars supporting a quasi-steady density wave mode, on the other hand, appears to be a genuine first derivation (Zhang 1996, 1998), both of which we will reproduce in the following subsections.

2.3.1 Definition of the Potential-Density Phase Shift

Consider a disk galaxy, with a total potential distribution of $\mathcal{V}$ and a total density distribution of Σ, each of which contains an axisymmetric part and a perturbation of the spiral form. For an annular ring located at radius r with width dr, the (z-component) torque applied by the total potential field on the material in this annular ring is

$$T(r) = rdr \cdot \int_0^{2\pi} -\Sigma \cdot (\vec{r} \times \nabla \mathcal{V})_z \, d\phi$$

$$= rdr \cdot \int_0^{2\pi} -\Sigma \cdot \left(\frac{\partial \mathcal{V}}{\partial \phi}\right) d\phi, \qquad (2.23)$$

[†] Portions of this section used material previously published in Zhang (1996, 1998), reproduced with modifications from The Astrophysical Journal @ AAS. Reproduced with Permission.

where we have used $\vec{r} = r\hat{r} + z\hat{z}$.

Equation (2.23) can also be written in the form of

$$T(r) = rdr \int_0^{2\pi} -\Sigma(r,\phi) \frac{\partial \mathcal{V}_1(r,\phi)}{\partial \phi} d\phi, \qquad (2.24)$$

where the subscript 1 on the potential denotes the spiral perturbation component, since the axisymmetric component of the potential gives zero ϕ derivative. Manifestly, eq. (2.24) describes the torque applied by the *spiral* part of the potential on the *total* disk surface density. Therefore, it also gives the amount of the angular momentum transport from the spiral wave to the disk matter in this annular ring per unit time.

Equation (2.24) can be further written in the form of

$$T(r) = rdr \int_0^{2\pi} -\Sigma_1(r,\phi) \frac{\partial \mathcal{V}_1(r,\phi)}{\partial \phi} d\phi, \qquad (2.25)$$

with the subscript "1" on both the potential and density variables. This is because the axisymmetric component of the density integrates to a null value in eq. (2.24), as long as the perturbation potential is periodic in ϕ (which it is in a disk-based geometry). Equation (2.25), however, should still be regarded as the torque applied by the spiral wave on the *total* disk matter in the annular ring.

Dividing the above expression by $2\pi rdr$, the area of the annular ring, we obtain that the azimuth-averaged torque density $\overline{\mathcal{T}}$ applied by the spiral potential field on the disk surface density, which is equal to the averaged rate of angular momentum flow per unit area from the density wave to the disk material, at a particular radius r is

$$\overline{\mathcal{T}}(r) = \overline{\frac{dL}{dt}}(r) = -\frac{1}{2\pi} \int_0^{2\pi} \Sigma_1(r,\phi) \frac{\partial \mathcal{V}_1(r,\phi)}{\partial \phi} d\phi. \qquad (2.26)$$

A similar expression was first written down by Kalnajs (1972) in analyzing the angular momentum exchange *between the stellar and gaseous waves* (i.e., not in the context of the *self-interaction* of the density wave modes).

The torque integral in eq. (2.26) vanishes for potential and density profiles which have *identical* waveforms (i.e., possessing zero phase shift), even if they are nonsinusoidal. This can be demonstrated by expanding these waveforms into their Fourier components, which would have the property that the Fourier coefficients of the same harmonics for the potential and density are proportional to each other. Using trigonometric identities, we can easily show that the torque integral in eq. (2.26) for such a pair of waveforms vanishes. Therefore, the azimuthal non coincidence of the potential and density spirals in the form of an equivalent phase shift is the *only* means for secular angular momentum transfer between the disk material and a quasi-stationary spiral density wave. This is what gives importance to the study of the phase shift.

From the above torque relation, we can define the equivalent phase shift ϕ_0 between two nonlinear wave forms as

$$\phi_0 \equiv \frac{1}{m} \sin^{-1}\left(\frac{1}{m} \frac{\int_0^{2\pi} \Sigma_1 \frac{\partial \mathcal{V}_1}{\partial \phi} d\phi}{\sqrt{\int_0^{2\pi} \mathcal{V}_1^2 d\phi} \sqrt{\int_0^{2\pi} \Sigma_1^2 d\phi}} \right), \tag{2.27}$$

where Σ_1 represents the perturbation density wave form and $\mathcal{V}_1$ the perturbation potential wave form, and m is the number of spiral arms. The sign convention of the equivalent phase shift is such that the positive sign of the phase shift corresponds to the perturbation density leads perturbation potential in the direction of galactic rotation.

The equivalent phase shift is the amount of phase shift which would be present between two sinusoidal wave forms if each is endowed with the same energy as the corresponding nonlinear wave form, and which would lead to the same value for the torque integral as would the nonlinear waveforms. Note that in the above expression the perturbation waveforms must have their azimuthal mean values subtracted.

In the following subsections, we will show that under the boundary conditions of basic state that admit density wave *modes*, a phase shift can be found for the potential and density spirals (or skewed bars) related through the Poisson integral; in the solution of the linearized Eulerian equations of motion; as well as in the orientation of the linear periodic orbit.

The Poisson integral gives a *constant* phase shift with radius for a spiral pattern which is infinite in radial extent and has a power law radial density profile. This constant phase shift is positive (which means that the density leads the potential) for power-law spirals with radial density falloff slower than $r^{-3/2}$ and is negative for faster radial density falloff.

The sign of the phase shift obtained from the Eulerian equations of motion and from the linear periodic orbit solution is such that for quasi-steady spiral or skewed bar *modes* the potential lags the density inside corotation, and vice versa outside corotation (see Figure 2.2 for an illustration). This shows that *in principle* a self-consistent spiral or bar density wave solution can be constructed with a phase shift between the potential and density perturbations if a proper basic-state disk surface density distribution is chosen. Later on we will show that indeed such a density distribution is naturally chosen by spontaneously formed density wave *modes*.

We want to comment briefly here on the meaning of the phrase "energy and angular momentum exchange between the disk material and the density wave." A question the author has frequently been asked is: "How could stars give energy and angular momentum to the wave (or vice versa)? Isn't the wave itself made of stars?" Yes, a self-sustained spiral wave is supported by the motion of many individual stars. However, a wave indeed has a separate existence other than the straightforward superposition of the individual stellar orbits. In the potential field of a relatively large amplitude spiral wave, it can be shown that the orbits themselves become chaotic (Section 3.8).

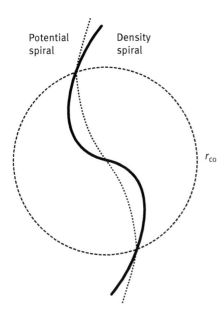

Figure 2.2: Schematic of potential-density phase shift for a self-sustained density wave mode in a galaxy disk with corotation radius r_{co}. The solid line indicates the locus of the peaks of the density spiral, and the dotted line indicates the locus of the troughs of the potential spiral. The circle indicates the corotation circle of the galaxy where the density wave pattern speed and the basic state matter's circular speed are the same (adapted from Zhang 2016).

After each crossing of the spiral arm a typical star loses all information (memory) of its previous orbital phase. Thus the information about the wave is not stored merely in the individual star's orbital orientation, as appears to be the case in Kalnajs' kinematic spiral representation (Kalnajs 1973), which is accurate only for infinitesimal wave amplitude (Kalnajs himself had in fact already stressed this point in his 1973 paper). Rather, the information about the wave is stored in the collective force field (or potential field) of the wave, which "collapses the chaos" inherent in the individual star's orbital motion. The local field in the spiral pattern is effectively contributed by all the stars in the disk due to the long-range nature of gravitational interaction. Thus, when we say that a star exchanges energy and angular momentum with the wave, this exchange occurs simultaneously with all the rest of the stars in the disk which contribute to the wave motion. This exchange is not random but is organized by the wave. So, for practical purposes it is infinitely more convenient to talk about the energy and angular momentum exchange between the wave and the stars than to talk about the energy and angular momentum exchange of one star with the rest of the stars in the disk in the manner constrained by the wave field.

2.3.2 A Phase Shift Given by the Poisson Integral

As we have commented before, due to the long-range nature of the gravitational interaction, the potential field generally has a different distribution from the mass density which generates it. In the case of a spiral wave, this difference appears in the form of a phase shift between the potential and density spirals related through the Poisson integral.

In the branch of mathematical physics dealing with potential theory, it has long since known that there exist a constant (i.e., independent of radius) phase shift between a self-consistent pairs of spiral potential and spiral density having power-law radial density profiles (Snow 1952). This so-called spiral transform has been applied successfully by Kalnajs (1965, 1971) to the study of galactic spiral structure. However, before the current work, this phase shift in the Poisson-integral spiral transform pair was mostly considered just a nuisance in the analytical calculations of the spiral modes. For example, in the SWING amplifier investigation of Toomre and Zang (Toomre 1981), they had deliberately chosen the kind of power-law profiles ($r^{-3/2}$ in density profile and $r^{-1/2}$ in potential profile), in order to make the phase shift come out exactly zero. No one had suspected the relevance of this phase shift to the collective dissipation and secular evolution of spiral galaxies. In fact, to fully appreciate the roles of phase shift to the spontaneous emergence as well the collective dissipation effect of a spiral pattern, one will need to deal with spiral or bar *modes* which have a specific radially-modulated density profiles that lead to the sign change of phase shift at the corotation radius (a result first discovered in this work), rather than waves – especially not those waves that have power-law density profiles previously studied in many over-stable disks used to explore transient density waves.

Here we present a brief derivation of the constant phase shift between a power-law potential and density spiral pair (Snow 1952; Kalnajs 1965).

Define $u = \ln r$; it follows that a power α of r, r^α, can be written as $e^{\alpha u}$.

From the potential theory, we know that a reduced spiral density

$$r^{3/2} \Sigma(r, \phi) = e^{i(\alpha u + m\phi)} \tag{2.28}$$

will produce a reduced spiral potential

$$r^{1/2} \mathcal{V}(r, \phi) = -2\pi G K(\alpha, m) e^{i(\alpha u + m\phi)}, \tag{2.29}$$

where K is the ratio of several gamma functions (Kalnajs 1971). Since $K(\alpha, m)$ is an analytic function of α, the above relation between the reduced potential and reduced density is still true for complex values of α, provided that the imaginary part is sufficiently small. Write

$$\alpha = \alpha_r + i\alpha_i, \tag{2.30}$$

we have

$$e^{i(\alpha_r + i\alpha_i)u} = e^{-\alpha_i u} \cdot e^{i\alpha_r u}, \tag{2.31}$$

with the $e^{-\alpha_i u}$ term contributes to the departure from the $r^{-3/2}$ radial density falloff.

The above expression, when substituted back into eqs (2.28) and (2.29), indicates that for an infinitely long density spiral with radial density falloff differing from $r^{-3/2}$, its corresponding potential spiral, although still has the property that its radial modulation function is r times the radial modulation function of the density spiral, is phase-shifted with respect to the density spiral because $K(\alpha, m)$ now becomes slightly complex. By expanding $K(\alpha, m)$ in a Taylor series around α_r, we have

$$K(\alpha, m) = K(\alpha_r + 0i) + \frac{\partial K}{\partial \alpha}|_{\alpha=\alpha_r}(i\alpha_i) + \text{higher order terms} \qquad (2.32)$$

For $\alpha_r > 0$ (which corresponds to a trailing spiral in our convention), it can be shown that the first derivative of $K(\alpha, m)$ is negative (Kalnajs 1971, Table 1) for small α. Therefore if $\alpha_i > 0$, which means that the density (potential) falloff is faster than $r^{-3/2}$ ($r^{-1/2}$), the potential spiral will lead the density spiral (i.e., the potential is in the form of $Ce^{i[\alpha_r \ln r + m(\phi - \phi_0)]}$, with $\phi_0 > 0$), and vice versa for $\alpha_i < 0$.

These results of course apply only to spiral density and potential pairs having a *constant exponent* for radial density falloff of the power-law shape. For realistic radial density profiles of the spiral or bar *modes*, global self-consistency requirement dictates that these profiles must be such that the potential-density phase shifts predicted by the equations of motion are simultaneously satisfied by that predicted by the Poisson integral. These dual requirements mandate that the density profile of an unstable mode cannot be arbitrary, but must be such that it leads to a kind of radial phase shift distribution which is conducive to the spontaneous emergence and long-term maintenance of the mode, as our analyses below will further demonstrate.

2.3.3 Phase Shifts in the Eulerian Equations of Motion and in the Linear Periodic Orbit Solution

In order to obtain a self-consistent spiral wave solution which admits a phase shift, there has to be a corresponding phase shift in the potential and density relation given by the equations of motion, besides that given by the Poisson equation. As is shown in equation (D12) of Lin & Lau (1979), the relation between the spiral potential and density, obtained from the higher-order (i.e., higher than the basic WKBJ) asymptotic solution of the linearized Eulerian equations of motion and the equation of continuity, for an open spiral mode, indeed contains a so-called out-of-phase term. Through careful examination, first performed by the current author in the mid-1990s, one can show that the sign of this term is such that for a trailing spiral the density leads in phase to the potential inside corotation, and vice versa outside corotation.

In the linear regime (before the formation of spiral collisionless shock), the surface density of a spiral disk can be considered to be composed of the superposition of periodic stellar orbits for the case of near-zero random velocity of stars (a cold disk). We thus expect that the relative phase shift of the potential and density spirals is also reflected in the orbital response of a star in a more open type of spiral potential, especially since in the linear regime there is an exact correspondence between the Eulerian and Lagrangian approaches, at least for the pressureless case (Lau & Bertin 1978). In what follows we demonstrate that there is indeed a phase offset in the orientation of the linear periodic orbit, which is obtained in the corotating frame of a spiral potential.

The linearized orbit equations in a frame that corotates with a rotating potential pattern of angular speed Ω_p are (Binney & Tremaine 2008, eqs (3.142a) and (3.142b))

$$\ddot{r}_1 + \left(\frac{d^2\Phi_0}{dr^2} - \Omega^2\right)_{r_0} r_1 - 2r_0\Omega_0\dot{\phi}_1 = -\left(\frac{\partial\Phi_1}{\partial r}\right)_{r_0}, \tag{2.33}$$

$$\ddot{\phi}_1 + 2\Omega_0\frac{\dot{r}_1}{r_0} = -\frac{1}{r_0^2}\left(\frac{\partial\Phi_1}{\partial\phi}\right)_{r_0}, \tag{2.34}$$

where r_1 and ϕ_1 are the perturbed orbital coordinates, Φ_0 is the axisymmetric potential, $\Omega = \sqrt{1/r d\Phi_0/dr}$ is the angular speed, $\Omega_0 \equiv \Omega(r_0)$, with r_0 the zeroth-order radius where the potential and the angular speed are evaluated, and, finally, Φ_1 is the perturbation potential, which we choose to be of the spiral form

$$\Phi_1(r, \phi) = F(r)\cos[f(r) + m\phi], \tag{2.35}$$

where $\phi = \phi(t) = \phi_1(t) + (\Omega_0 - \Omega_p)t$, which for non resonant stars can be approximated by $\phi \approx \phi_0(t) \equiv (\Omega_0 - \Omega_p)t$.

Integrating eq. (2.34), we obtain

$$\dot{\phi}_1 + 2\Omega_0\frac{r_1}{r_0} = -\frac{(\Phi_1)_{(r_0,\phi_0)}}{r_0^2}\frac{1}{(\Omega_0 - \Omega_p)} + \text{constant}. \tag{2.36}$$

Substituting eq. (2.36) into eq. (2.33) to eliminate $\dot{\phi}_1$, we obtain

$$\ddot{r}_1 + \kappa_0^2 r_1 = -\left(\frac{\partial\Phi_1}{\partial r}\right)_{(r_0,\phi_0)} - 2\Omega_0\frac{\Phi_1(r_0)}{r_0}\frac{1}{\Omega_0 - \Omega_p}, \tag{2.37}$$

where we have ignored the constant term for the same reason as given in Binney and Tremaine (2008), and where

$$\kappa_0 = \left(r\frac{d\Omega^2}{dr} + 4\Omega^2\right)_{r_0} = \left(\frac{d^2\Phi_0}{dr^2}\right)_{r_0} + 3\Omega_0^2 \tag{2.38}$$

is the epicycle frequency.

2.3 Potential-Density Phase Shifts for Density Wave Modes

Using the expression of the perturbation potential of eq. (2.35), the forcing terms on the right-hand side of eq. (2.37) can be found to be

$$rhs = \left[-F'(r_0) - 2\Omega_0 \frac{F(r_0)}{r_0} \frac{1}{(\Omega_0 - \Omega_p)} \right] \cos[f(r_0) + m\phi_0]$$
$$+ F(r_0)k(r_0) \sin[f(r_0) + m\phi_0], \tag{2.39}$$

where $\phi_0 = \phi_0(t) = (\Omega_0 - \Omega_p)t$, and $k(r_0) \equiv (df/dr)_{r_0}$.

We now see clearly that the forcing consists of two terms which are 90° out of phase with each other. We thus expect that the forced orbital response, or the particular solution of eq. (2.37), will also contain two similar terms. In fact, the particular solution of eq. (2.37) can be written as

$$r_1(t) = \frac{1}{\kappa_0^2 - m^2(\Omega_0 - \Omega_p)^2} \times \left\{ \left[-F'(r_0) - 2\Omega_0 \frac{F(r_0)}{r_0} \frac{1}{\Omega_0 - \Omega_p} \right] \cos[f(r_0) + m\phi_0] \right.$$
$$\left. + F(r_0)k(r_0) \sin[f(r_0) + m\phi_0] \right\}. \tag{2.40}$$

This solution for r_1 can be further written as

$$r_1(t) = C \sin(m\phi' + m\delta) \tag{2.41}$$

where $\phi' = f(r_0)/m + \phi_0$, and

$$C = \sqrt{A^2 + B^2}, \tag{2.42}$$
$$\delta = \frac{1}{m} \tan^{-1}(A/B), \tag{2.43}$$

with

$$A = \frac{1}{\kappa_0^2 - m^2(\Omega_0 - \Omega_p)^2} \left[-F'(r_0) - 2\Omega_0 \frac{F(r_0)}{r_0} \frac{1}{\Omega_0 - \Omega_p} \right], \tag{2.44}$$

and

$$B = \frac{1}{\kappa_0^2 - m^2(\Omega_0 - \Omega_p)^2} F(r_0)k(r_0). \tag{2.45}$$

Note that it is the sine form of the orbital response that should be compared with the negative cosine form of the forcing spiral potential (since we must take radial derivative to get from the potential to the radial force) in order to derive the relative phase shift. We have also assumed $F(r) < 0$ here.

Therefore, we have derived that the phase shift δ of the orbit with respect to the forcing potential can be expressed as

$$\delta = \frac{1}{m} \tan^{-1} \left[\frac{-F'(r_0) - \frac{2\Omega_0 F(r_0)}{r_0} \frac{1}{\Omega_0 - \Omega_p}}{F(r_0) k(r_0)} \right]. \quad (2.46)$$

Since the rate of amplitude variation of the density wave $F'(r_0)$ is expected to be small, eq. (2.46) can be further simplified to

$$\delta \approx \frac{1}{m} \tan^{-1} \left[-\frac{2\Omega_0}{\Omega_0 - \Omega_p} \frac{1}{k(r_0) r_0} \right], \quad (2.47)$$

which tells us that the phase shift δ is negative for a trailing wave ($k > 0$ in our current convention) inside corotation. A negative δ here means that the orbit leads in space (lags in time) with respect to the spiral potential; the opposite is true for orbits outside the corotation. This agrees with what we had inferred from the fluid equations of motion.

The discussion of the phase shift of linear-orbit orientation from the forcing potential is absent in the corresponding section (Section 3.3.3) of Binney and Tremaine (2008), which focused mainly on orbit resonances.

Note furthermore that the above orbit solution formally diverges at the corotation radius. This is to be expected since such divergence is what is behind the concept of "wave-particle resonances." This signifies that more sophisticated treatment of the response of orbits is needed at such resonances. In reality, the phase shift transitions through zero value at the corotation, as we will show in Section 2.4, in an alternative derivation of the existence of phase shift, as well as its sign change, at the corotation radius from the point of view of the spontaneous emergence of global density wave modes. From there it follows that the result of the sign change of the potential-density phase shift at corotation is a robust result originating from the distribution and sign change of the angular momentum density of the wave mode itself, and is not affected by the singular nature of the orbital phase at the corotation radius as we have shown above. This robust result will be used later in Chapter 4 as a practical method to locate the corotation resonances (CR) in observed galaxies.

2.3.4 Phase Shift in the Linear and Nonlinear Regimes of the Eulerian Solutions

Since both the Poisson integral and the linearized Eulerian equations of motion can admit a spiral potential and density pair which are phase-shifted from each other, a self-consistent linear global spiral mode can be constructed with a phase shift between the potential and density, as long as the radial density modulation of the spiral mode is such that it causes the phase shift given by the Poisson integral to change sign at the corotation radius.

Within the context of the linear calculation, however, the phase shift we have just demonstrated does not have a dissipation effect. This is because linear calculation aims to obtain a self-consistent flow solution, which is periodic in the case of a spiral galaxy, with neighboring streamlines (in the fluid approach) never crossing one another. Incidentally, if we calculate the net angular momentum gain (or loss) of a star when completing a full cycle on the *inclined* linear periodic orbit (so as to take into account the potential-density phase shift in the linear regime) through calculating the integral

$$-\int_0^{P_r} \frac{\partial \Phi}{\partial \phi} dt = -\int_0^{P_r} \left[\left(\frac{\partial^2 \Phi}{\partial r \partial \phi} \right)_{(r_0,\phi_0)} r_1 + \left(\frac{\partial^2 \Phi}{\partial \phi^2} \right)_{(r_0,\phi_0)} \phi_1 \right] dt, \qquad (2.48)$$

with $\phi_0 = \phi_0(t) = (\Omega_0 - \Omega_p)t$, and with the use of eq. (2.47) which relates $\sin\delta$ and $\cos\delta$, we find that it is zero. For orbit to display secular dissipation behavior, the graininess of the potential within the spiral-arm collisionless shock (its detailed nature will be demonstrated in Section 3.3) will need to be taken into account.

The dissipationless Eulerian equations are not capable of predicting all the physical effects in a spiral disk which admits collective dissipation processes, even though it is successful at predicting the morphology and kinematics of open density wave modes in the linear regime. The validity of the Eulerian approach in the nonlinear regime is compromised by the tendency of the global wave solution to steepen into a singularity when solved as an initial value problem in an iterative solution (Lubow, Balbus, & Cowie 1986).

The situation here is very similar to the steepening of acoustic waves in a nonlinear and dissipationless medium (see, e. g., Shu 1992, pp. 203 ff; Frisch 1995 and the references therein. See also Burgers 1948). Owing to the difference in effective propagation speed for waves of different amplitudes, when the nonlinear Eulerian equations are solved as an initial-value problem, the peak region of a finite-amplitude sound wave gradually catches up to the trough of the wave located originally at its front, and the solution first becomes singular and subsequently unphysical. The non physical solution in the nonlinear waves calculated using the dissipationless equations is usually remedied by the introduction of a hydrodynamic shock, as well as the associated shock viscosity and dissipation, in order to bring the solution back to the physically meaningful form. In the case of a spiral wave, the dissipation effect introduced by the collective instabilities at the spiral arms is similarly responsible for halting the wave steepening process indicated in the nonlinear Eulerian solutions.

In summary, an open spiral wave calculated by the linearized Eulerian equations can self-consistently sustain a phase shift without invoking dissipation at the expanse of continuous modal amplitude growth. A self-consistent nonlinear and open spiral solution obtained with the Eulerian equations, when solved as an initial value problem, cannot reach a time-steady solution but will always steepen with time. Another piece of supporting evidence is that the past non-self-consistent (Roberts 1969; Shu, Milione, & Roberts 1973) or partially self-consistent (Levinson & Roberts 1981) WKBJ calculations have invariably found large-scale spiral shock solutions for finite amplitude spiral forcing. A phase shift between the potential and density spirals introduces exactly this element of local (i. e., within the same annulus) non coincidence between the forcing spiral potential and the response spiral density, in a globally self-consistent, open spiral modal solution. A self-consistent nonlinear WKBJ wave, however, does not steepen because it does not introduce a potential-density phase shift, which is what is responsible for driving the nonlinear wave-steepening process. But such WKBJ plane-wave solutions can only be meaningful in a local sense under the disk geometry.

2.4 Linear Regime and Quasi-Steady State of the Wave Modes[†]

In the previous section, we have demonstrated the possibility for the existence of potential-density phase shift, through examining the Poisson integral and equations of motion separately. In this section, we will take a more integrated view and examine the close relation between radial distribution of potential-density phase shift and the spontaneous emergence and stabilization of density wave modes. For ease of presentation, we will refer these modes mostly as spirals, though the conclusions apply for skewed bar modes as well.

We will demonstrate that the kind of density wave patterns in disk galaxies which can remain quasi-stationary on the time scale of a Hubble time are spontaneously formed open, global modes, which acquired their dynamical equilibrium state through the competing processes of spontaneous growth due to a radial positive feedback loop between the central region of the galaxy and the corotation, and local dissipation due to the irreversible interaction between the perturbation wave mode with the basic state of the galactic disk. The existence of the phase shift is shown to be crucial both for the spontaneous emergence of the wave mode in the linear regime, as well as for the damping and stabilization of the mode to a finite amplitude in the nonlinear regime.

As an alternative (but equivalent) view of the wave-amplitude stabilization process, we show that the radial distribution of the total torque coupling integral (which is equal to the total radial flux of angular momentum) of a spontaneously-formed spiral or bar *mode* must be (and is found to be) of a characteristic bell shape, with the peak of the bell located at the corotation radius of the mode. This distribution of the total torque coupling is shown to be concordant with the radial distribution of the

---

[†] Portions of this section used material previously published in Zhang (1998), reproduced with modifications from The Astrophysical Journal @ AAS. Reproduced with Permission.

phase shift between the potential and density spirals. Whereas the existence and the sign of the total torque coupling imply that there is angular momentum being carried outward by a trailing spiral density wave mode, as have already been pointed out by Lynden-Bell & Kalnajs (1972), the bell-shaped radial distribution (which revises Lynden-Bell & Kalnajs' conclusion of a radially constant torque-couple/angular-momentum-flux distribution) further indicates that there is angular momentum being picked up from or deposited onto each annular ring of the galactic disk *en route* of this outward angular momentum transport.

The amount of angular momentum being deposited by the density wave mode is shown to be negative inside corotation (indicating an angular momentum pick-up) and is positive outside corotation (indicating a true deposition). Since a spiral mode has negative angular momentum density inside corotation and positive angular momentum density outside corotation, it follows that the bell-shaped torque coupling of an open spiral mode leads to its own spontaneous growth in the linear regime. In the modal growth process from the linear regime (or small amplitude) to the progressively nonlinear regimes, an increasingly larger fraction of the deposited angular momentum is channeled onto the basic state of the disk (which causes the dissipative secular evolution of the basic state) through a collisionless shock, and progressively lesser portion is used for the wave modal growth; and finally, in the fully nonlinear regime (the quasi-steady state of the wave mode), all of the angular momentum deposited by the trailing spiral mode is given to the basic state which leads to its secular evolution. The spiral mode can then remain quasi-stationary (the meaning of "quasi" will be defined later in the text) at the expense of a continuous dissipative basic state evolution (Zhang 1998).

2.4.1 The Process of Reaching the Quasi-Steady State

As we have commented before, the type of global structures formed spontaneously in nonequilibrium systems are not neutral structures. They are rather large-scale instabilities of the underlying systems, which are sustained by a continuous flux of energy and entropy, as well as by a local dissipation process which offsets their spontaneous growth tendency. This notion is reinforced by the well-known "fluctuation-dissipation" theorem for nonequilibrium thermodynamic systems (proven so far mostly for the close-to-equilibrium, or linear nonequilibrium thermodynamic, regimes; see de Groot & Mazur [1962] and the references therein), which implies that no finite amplitude fluctuation can be sustained without invoking a corresponding dissipation process.

The physics of the self-organized spiral mode appears to be an extension of the fluctuation-dissipation theorem into the far-from-equilibrium regime, that is, the kind of spiral modes capable of spontaneously emerging as large-scale fluctuations have built-in characteristics such that these fluctuations can be sustained as quasi-stationary patterns through the action of local dissipation. The capability of a trailing spiral to transport angular momentum outward (Lynden-Bell & Kalnajs 1972) and to deposit angular momentum in the course of this outward transport (Zhang 1998) allows a spiral mode to grow spontaneously. On the other hand, the local instability condition at the spiral arms and the presence of a phase shift between the potential and density spirals allow the deposited angular momentum to be channeled onto the basic state through a collisionless shock, at a finite amplitude of the wave. A quasi-steady state is obtained when the above two processes reach dynamical equilibrium.

An unstable mode is capable of growing spontaneously and homogeneously. Since a spiral mode has negative energy and angular momentum densities inside corotation and positive densities outside corotation, the spontaneous growth of the spiral mode requires that the radial derivative of its total torque coupling integral $C(r)$, which determines how much angular momentum is left onto unit-width each annular ring as a result of the torquing by the disk material on both sides of the annular ring, must allow negative angular momentum to be deposited onto each annular ring inside corotation, and positive angular momentum to be deposited outside corotation. In other words, a spiral perturbation will grow spontaneously if the shape of its $C(r)$ in the linear regime allows it to be self-enhancing.

The total torque coupling integral due to a spiral structure is the sum of the gravitational torque coupling integral C_g and the advective torque coupling integral C_a (Lynden-Bell & Kalnajs 1972), and we will present the details of this demonstration in Section 2.5.

The requirement of local angular momentum balance in an annular ring of unit width at radius r, during the outward angular momentum transport process by a trailing spiral wave, leads to

$$\frac{dC}{dr} = \frac{d(C_a + C_g)}{dr} = -2\pi r \frac{d\bar{L}}{dt}, \qquad (2.49)$$

where $\bar{L}$ is the azimuthally averaged angular momentum density at radius r, which we have defined and used in the previous sections, and where $C > 0$ for a trailing spiral mode.

Now $\bar{L}$ can be decomposed into $\bar{L}_{\text{basic state}} + \bar{L}_{\text{wave}}$. During the linear modal growth process, $\frac{d\bar{L}_{\text{basic state}}}{dt} = 0$ away from the resonances[†], and the angular momentum deposited by C is used entirely for the increase of the wave amplitude. So we have

$$\frac{dC}{dr} = \frac{d(C_a + C_g)}{dr} = -2\pi r \frac{d\bar{L}_1}{dt}, \qquad (2.50)$$

where $\bar{L}_1$ is the azimuthally averaged angular momentum density of the wave mode at radius r.

[†] This result comes about simply from the definition of the basic state, which is the steady azimuthally symmetric part of the disk on top of which linear perturbations of sinusoidal form are superimposed in the linear modal calculation.

Since inside corotation, $\bar{L}_1 < 0$, modal growth means that $d\bar{L}_1/dt < 0$ inside corotation. This demands further, through eq. (2.50), that $dC/dr > 0$ inside corotation. Similarly, modal growth demands $dC/dr < 0$ outside corotation. Therefore, we arrive at the conclusion that *for an unstable mode in the linear regime to grow spontaneously, the radial profile of its total torque coupling integral must be of a characteristic bell shape, with the top of the bell at the corotation radius.*

For *homogeneous* modal growth throughout the disk, we must further have

$$\frac{d\bar{L}_1}{dt} = 2\gamma_{g0}\bar{L}_1, \qquad (2.51)$$

where γ_{g0} is the (constant) amplitude growth rate of the mode in the linear regime (thus $2\gamma_{g0}$ for the angular momentum growth rate).

We thus have, from eqs (2.50), (2.51)

$$\gamma_{g0} = -\frac{dC/dr}{4\pi r \bar{L}_1} = \text{constant}, \qquad (2.52)$$

in other words, the magnitude of the linear growth rate γ_{g0} of an unstable mode is *determined* by the radial gradient dC/dr of the total torque coupling $C(r)$, as well as by the wave angular momentum density $\bar{L}_1$, with the ratio of the two being a constant independent of radius.

In the linear regime, the gravitational torque coupling C_g is related to the torque integral $\mathcal{T}$ through (we will demonstrate this in Section 2.5.2, and we also used eq. (2.26))

$$\begin{aligned}\frac{dC_g}{dr} &= r\int_0^{2\pi} \Sigma_1(r,\phi)\frac{\partial \mathscr{V}_1(r,\phi)}{\partial \phi}d\phi \\ &= -2\pi r\overline{\mathcal{T}}(r) \\ &\equiv T_1(r).\end{aligned} \qquad (2.53)$$

Since in the linear regime, the distribution of $C(r) = C_a(r) + C_g(r)$ for an unstable spiral mode is dominated by the distribution of $C_g(r)^\dagger$, we thus see that the bell-shaped $C(r)$ distribution is consistent with the existence and the sign of phase shift for a trailing spiral, which gives a positive phase shift and a positive $T_1(r)$ inside corotation,

---

† This is because that in the linear regime, in the hydrodynamic limit of $ka < 1$, where k is the wave number and a is the sound speed of the medium, C_a always opposes C_g in terms of the direction of angular momentum flux it carries (Lynden-Bell & Kalnajs 1972; Goldreich & Tremaine 1979). Since a trailing spiral mode can grow only by transporting angular momentum outward, and since C_g for a trailing spiral pattern is always positive (which leads to outward angular momentum transport), it follows that the kind of trailing spiral modes capable of spontaneous growth must have its C_g larger in magnitude than C_a in the linear regime. This conclusion is also borne out from analyzing the detailed expressions of C_a and C_g for the various wave branches of WKBJ waves (Goldreich & Tremaine 1979).

and negative ones outside corotation. In other words, *the existence and the sign of the phase shift between the potential and density spirals of a trailing spiral mode is what leads to the spontaneous growth of such a mode in the linear regime.*

Away from the linear regime, the exponential growth of the wave amplitude gradually saturates, and eventually the wave growth halts completely.

When the wave amplitude stops growing, the spiral pattern still has a very coherent appearance; and our N-body simulations in the next chapter will demonstrate that the total torque coupling due to the spiral pattern retains a bell shape throughout the modal growth process (as reflected in the two-humped distributions of both the potential-density phase shift, as well as the radial mass flow rate, of each mode). The typical distribution of the different torque coupling integrals at the quasi-steady state is shown in Figure 2.3, left frame, for the N-body spiral mode calculated in Zhang (1996). The total torque coupling integral in Figure 2.3 left frame clearly appears to have a bell shape, which is peaked at the corotation radius r_{co} (r_{co} = 30 for the mode in this simulation). Since the wave amplitude at the time step of the plotting is reasonably steady, it follows that such bell-shape for the torque couplings is intrinsic to the quasi-stationary spiral mode under concern, and not due to some spurious effect of the N-body simulation, such as the broadening of the resonances due to a fluctuating wave amplitude.

Note that in the various torque coupling curves shown in Figure 2.3 left frame there is also a second weak hump at the outer disk. This second hump is likely to be due to the reflection of the wave energy from the outer disk edge, which results in the formation of a secondary mode outside the outer Lindblad resonance of the dominant mode, which is located at r_{OLR} = 42 (Donner & Thomasson 1994).

The bell-shaped total torque-coupling integral such as shown in Figure 2.3 left frame indicates that the spiral pattern continues depositing energy and angular

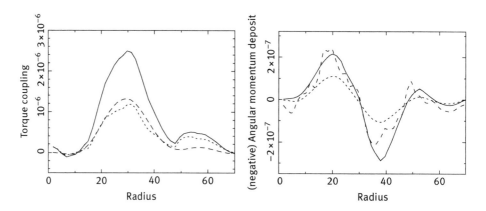

Figure 2.3: *Left*: Gravitational (dotted), advective (dashed), and total (solid) torque couples from the N-body simulations of Zhang (1996, 1998). *Right*: Gradient of gravitational (dotted) and total (dashed) torque couples, and the volume torque T_1 (solid), from the same N-body simulations (adapted from Zhang & Buta 2012).

momentum onto the disk throughout the modal growth process, as well as at the quasi-steady state of the wave mode. Since at the quasi-steady state the deposited energy and angular momentum no longer increase the amplitude of the wave, they can only be left onto the basic state. This is consistent with the observed radial migrations of the basic state stars seen throughout the same set of N-body simulation (Chapter 3).

This channeling of the deposited wave energy and angular momentum is achieved by the spiral gravitational shock at the spiral arms. When stars are crossing the spiral arms, the near-collision condition produced by the shock destroys the phase coherence of the orbits, so stars can no longer align themselves to that which corresponds to a kinematic spiral (Kalnajs 1973) in order to reinforce the wave amplitude. The sudden velocity deceleration at the shock, for a star inside corotation, also means that the star loses part of its orbital angular momentum to the wave (or the wave deposits negative angular momentum onto the basic state star), and therefore spirals inward. As the wave amplitude increases and therefore the shock gets narrower and stronger, a growing larger fraction of the deposited negative energy and angular momentum by the wave is used for the braking (deceleration) of the star's orbital motion, rather than for the recruitment of particles to contribute to the growth of the wave. This characteristic of the nonlinear regime is also partially reflected in the sign of C_a shown in Figure 2.3 left frame, which is the same as the sign of C_g, unlike what we should expect for such an open wave in the linear regime. The possibility for such sign change of C_a in the nonlinear regime of the wave has also been indicated in Lynden-Bell and Kalnajs (1972).

At the quasi-steady state of the wave mode, instead of having the linear-regime relation $dC_g/dr = T_1(r)$, the proper closure relation now becomes $dC/dr = dC_a/dr + dC_g/dr = T_1(r)$ (see the discussion in Section 2.5.3). All the angular momentum deposit indicated by dC/dr (or equivalently $T_1(r)$) now goes to the basic state, and none is used for wave growth. This is reinforced by Figure 2.3, right.

The quasi-steady state for a spontaneously-formed spiral mode is a dynamical equilibrium state, with the spontaneous growth tendency of the mode balancing the dissipative damping tendency. In some of the published works on the subject of wave damping mechanisms, it had often been speculated that the relevant damping mechanism must be nonlinear. In the following, however, we will show that any *reversible* nonlinear mechanism (which serves to deform the azimuthal profile of the wave) by itself is not capable of limiting the wave amplitude to a finite value in a finite time. The relevant wave stabilization mechanism must also involve a dissipative energy and angular momentum exchange process between the spiral density wave and the basic state of the disk.

The generic form of the amplitude evolution equation of a spontaneously-formed spiral mode is

$$\frac{dA}{dt} = \gamma_g(A)A. \tag{2.54}$$

The influence of nonlinearity on the growth of a spontaneous spiral mode is expected to be chiefly a reduction of the growth rate $\gamma_g(A)$. This is because as the wave amplitude increases as a result of over-reflection at the corotation and feedback, the azimuthal profile of the wave gradually becomes nonsinusoidal. The growth rate $\gamma_g(A)$ of the mode thus decreases since part of the wave energy now is channeled into the higher-order harmonics in azimuth, which cannot be directly amplified by the galactic resonant cavity. However, the growth rate will never actually reach zero through the deformation of the azimuthal wave profile alone. This is especially true for the spiral patterns in physical galaxies, where the departure of the azimuthal wave profiles from sinusoidal is never serious, and the $m = 2$ sinusoidal modal component is always dominating. Thus the amplitude of the mode will grow without bound in the presence of a nonlinear but dissipationless mechanism alone.

However, the dissipative channeling of the wave energy and angular momentum onto the basic state through the action of the spiral shock gives rise to an effective amplitude damping rate γ_d, with the net growth rate of the mode being $\gamma'_g = \gamma_g - \gamma_d$, which *replaces* the γ_g in eq. (2.54).

It is expected that γ_d increases with increasing wave amplitude. This is because γ_d, being the manifestation of a *local* dissipation process, depends mainly on the *steepness* of the shock, which is itself determined by local values of the wave amplitude A and potential/density phase shift ϕ_0. Through direct numerical calculations it can be shown that the phase shift itself is mainly determined by the pitch angle of the spiral and by its radial density envelope function and is not sensitively dependent on the azimuthal (nonlinear) profiles of the spiral pattern. In fact, for azimuthal density profiles as diverse as that shown in Figure 2.4, the calculated phase shift distribution as shown in Figure 2.5 is amazingly similar. Therefore, $\gamma_d[A, \phi_0(A)] \approx \gamma_d(A, \phi_0)$, and it tends to increase with increasing wave amplitude.

The trend of variation of $\gamma_g(A)$ and $\gamma_d(A)$ stated above means that the effective growth rate $\gamma'_g(A) = \gamma_g(A) - \gamma_d(A)$ will reach zero at a finite wave amplitude – this amplitude is the equilibrium amplitude of the wave mode, at which $dA/dt = 0$ according to eq. (2.54). Since the manner in which the growth rate decreases with the wave amplitude, as well as the manner in which the dissipation rate increases with the wave amplitude, are both determined by the basic state, as well as by the modal properties, with the later determined ultimately also by the basic state, we see that the equilibrium amplitude of a spiral mode depends only on the basic state characteristics.

2.4.2 Reconciliation of the Two-Wave Superposition and the Unstable-Mode Points of View

It is a well-known fact from wave theory that two oppositely propagating traveling waves of the same frequency and the same amplitude sum up to a *constant-amplitude*

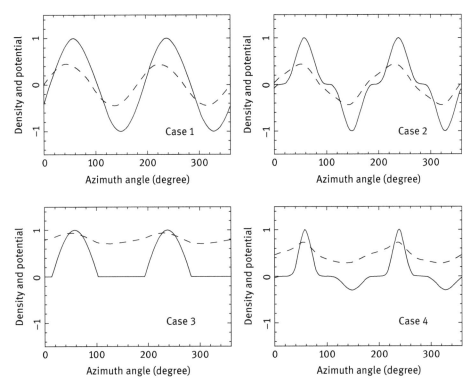

Figure 2.4: Azimuthal profiles of the spiral density (solid lines), as well as the spiral potential (dashed lines) calculated through the Poisson integral of the corresponding density, for several truncated spiral waves (adapted from Zhang 1998).

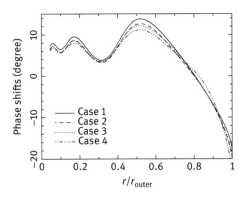

Figure 2.5: Radial dependence of the equivalent phase shifts for the truncated spiral waves of Figure 2.4. The ripples observed are due to the truncation of the computation domain at r_{outer}. The phase shifts should be a constant independent of radius for an infinitely long constant amplitude spiral wave (adapted from Zhang 1998).

standing wave, or a steady mode. However, the issue of what kind of characteristics the two traveling waves must have in order that their superposition leads to a *spontaneously growing unstable mode* is hardly ever addressed. We show below that in a simplified one-dimensional case an exponentially decaying profile for the wave action (we use the phrases "wave action" and "total torque coupling integral" interchangeably in the following) is needed along each wave's propagation path, which, together with the over-reflection mechanism at corotation and the feedback mechanism near the galactic center, leads to a spontaneously and homogeneously growing wave mode. This understanding also helps to bridge a connection between the current work and some of the previously published work, notably that of Lynden-Bell & Kalnajs (1972) and Goldreich & Tremaine (1979).

In their 1972 paper, Lynden-Bell & Kalnajs found that when the contributions from the gravitational and advective torques are summed together, the total torque-coupling is in the form of angular momentum density multiplied by the group velocity of the wave (this result had previously been obtained by Toomre 1969). Therefore, for a wave inside corotation, which has negative angular momentum density, the *net* group velocity of the two waves must be directed inward in order for the trailing wave to transport angular momentum outward. Fortunately, this condition is satisfied for most of the trailing spiral structures we observe, thanks to the over-reflection mechanism at corotation (Mark 1976; Toomre 1981), which makes the inward propagating trailing wave trains always more powerful (or have a bigger amplitude) than the outward propagating leading or trailing wave trains.

Therefore, the essence of Lynden-Bell & Kalnajs' above result for the angular momentum transport by the density wave can be rephrased simply as that of "a spiral wave carries its angular momentum density with it as it propagates radially," and it does not interact with the basic state of the galactic disk. During the linear modal growth stage, while the Lynden-Bell & Kalnajs' original derivation does not apply (since they had assumed a constant-amplitude wave train), the conclusion of no-wave-basic-state-interaction should still be valid since the linear theory by design assumes a immutable basic state. How could we then arrive at the growing modal amplitude without the change in basic state? In what follows we will show that this is achieved by a special type of radial distribution of the total torque coupling (or angular momentum flux). This radial distribution is such that the wave amplitude progressively decreases *spatially* along its path of propagation – thus the newly arrived wave parcel is always of higher amplitude than the one being replaced, and this allows the wave amplitude everywhere across the galaxy to grow with time, without the participation of the basic state. Therefore, in this latter scenario of the linear modal growth, the waves inside and outside corotation become the source and sink of each other (through the over-reflection mechanism at corotation), with essentially no wave interaction with the basic state apart from the fact that the growing number of stars which participate in the growing wave ultimately come from the disk – yet this is not what is considered as basic state change per se, but rather a change of first-order perturbation wave amplitude.

Since in this section we are mainly interested in the general trend of the distribution of the magnitude of $C(r)$, we assume, without loss of generality, a one-dimensional wave propagation and reflection scenario. The two-dimensional picture will be needed when considering the detailed phase relation of the different wave branches, the superposition of which gives rise to the normal modes of a given basic state (Lin & Lau 1979). The direction of wave propagation in our one-dimensional case would correspond to the radial direction in a physical galactic disk. We make the simplifying assumption that the wave is amplified only at corotation radius r_{co}, with an over-reflection factor Γ for the wave action. The over-reflection mechanism at corotation, besides connecting the two oppositely propagating waves inside r_{co}, also produces an outgoing wave outside r_{co} (Mark 1976). We also assume a constant group velocity c_g for all the wave trains, as well as an inner-turning point of the wave located close to the center of the galaxy.

The exponential profiles we assumed for the three waves are illustrated in Figure 2.6. For $r \leq r_{co}$, the spatial distribution of the inwardly propagating wave C_1 is

$$C_1(r) = C_0 \exp\left(\frac{r - r_{co}}{r_0}\right), \quad (2.55)$$

and the spatial distribution for the outwardly propagating wave C_2 is

$$C_2(r) = -C_0 \exp\left(\frac{-r - r_{co}}{r_0}\right), \quad (2.56)$$

where r_0 is a constant scale length, assumed to be the same for all wave branches. The overall intensity of the wave inside corotation is the linear superposition of C_1 and C_2, that is,

$$C_{12}(r) = C_1(r) + C_2(r)$$
$$= C_0 \exp\left(-\frac{r_{co}}{r_0}\right)\left[\exp\left(\frac{r}{r_0}\right) - \exp\left(-\frac{r}{r_0}\right)\right], \quad (2.57)$$

The radial gradient of C_{12} is now (for $r < r_{co}$)

$$\frac{dC_{12}(r)}{dr} = \frac{C_0}{r_0} \exp\left(-\frac{r_{co}}{r_0}\right)\left[\exp\left(\frac{r}{r_0}\right) + \exp\left(-\frac{r}{r_0}\right)\right], \quad (2.58)$$

which is positive for $r < r_{co}$.

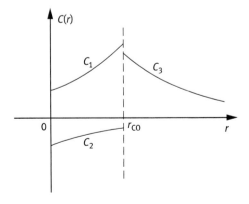

Figure 2.6: Superposition of wave trains to form an unstable mode (adapted from Zhang 1998).

To write also the expression for the wave branch outside corotation, we note that $C_1(r_{co}) = C_0$, and $C_2(r_{co}) = -C_0 \exp(-2\frac{r_{co}}{r_0})$, therefore the conservation of wave action across corotation requires that

$$C_3(r) = [C_1(r_{co}) + C_2(r_{co})] \exp\left(-\frac{r - r_{co}}{r_0}\right)$$

$$= C_0 \left[1 - \exp\left(-2\frac{r_{co}}{r_0}\right)\right] \exp\left(-\frac{r - r_{co}}{r_0}\right) \tag{2.59}$$

for $r > r_{co}$. The shape of C_{12} for $r < r_{co}$ plus C_3 for $r > r_{co}$ can be easily verified to be of a bell shape peaked at corotation[†].

If we put in the appropriate time dependence in the above three profiles to account for the fact that these are really traveling waves, that is, we change variable from r to $(r+c_g t)$ in C_1, and from r to $(r-c_g t)$ in C_2 and C_3, where c_g is the constant group velocity of the wave[‡], we immediately obtain

$$C_1(r,t) = C_0 \exp\left[\frac{(r + c_g t) - r_{co}}{r_0}\right], \tag{2.60}$$

$$C_2(r,t) = -C_0 \exp\left[\frac{-(r - c_g t) - r_{co}}{r_0}\right], \tag{2.61}$$

and

$$C_3(r,t) = C_0 \left[1 - \exp\left(-2\frac{r_{co}}{r_0}\right)\right] \exp\left[-\frac{(r - c_g t) - r_{co}}{r_0}\right], \tag{2.62}$$

which are the temporally exponentially growing waves (mode) at a given radius, with the constant growth rate being

$$2\gamma_g \equiv \frac{c_g}{r_0}, \tag{2.63}$$

where the factor of 2 is there because again we have chosen to use γ_g to represent the amplitude growth rate, hence $2\gamma_g$ for the wave action growth rate.

Furthermore, since the amplification of the wave is assumed to be restricted mainly to the corotation region, the amplitude growth rate of a wave mode γ_g should be related to the over-reflection factor Γ at corotation, where

$$\Gamma = \left|\frac{C_1(r_{co})}{C_2(r_{co})}\right| = e^{2\frac{r_{co}}{r_0}}, \tag{2.64}$$

[†] In fact, the peak has a cusp shape at corotation, at least under our multitude of simplifying assumptions. However, the N-body spirals always tend to smooth the cusp into a bell.

[‡] Once again this constant parameter is assumed for simplification of the presentation, without losing generality.

as well as the time it takes for the wave to travel from the corotation radius to the center of the galaxy τ through

$$\exp(2\gamma_g \cdot 2\tau) = \Gamma, \tag{2.65}$$

where

$$\tau = \frac{r_{co}}{c_g}. \tag{2.66}$$

It can be easily checked that with the relation of γ_g and r_0 as in (2.63), the total drop of wave action C along the wave propagation path, after a round trip on the C_1 and C_2 branches, is indeed $\exp(2\gamma_g \cdot 2\tau) = e^{2\frac{r_{co}}{r_0}} = \Gamma$, so the entire feedback loop is closed.

The past work of Lynden-Bell & Kalnajs (1972) and Goldreich & Tremaine (1979) had both shown that the total torque coupling is equal to the product of wave angular momentum density and the group velocity of the wave. In the linear regime, this product should be conserved along the path of wave propagation. We now see that the requirement of conservation of wave action along the path of wave propagation in the linear regime does not contradict our conclusion of a bell-shaped $C(r)$ function for a *spontaneously growing* open spiral wave mode, since this conservation is in the co-moving frame of each wave train (i. e., in a Lagrangian formulation), and is not across the galactic radius r at a given time, as Lynden-Bell & Kalnajs (1972) had previously assumed (which would be true for a steady wave rather than an unstable mode).

At the quasi-steady state of the wave mode, the distribution of $C(r)$ is expected to still have the same bell-shaped general profile, as is confirmed in the N-body simulations, and the distribution of each wave branch should also maintain similar radial profile from the linear to the nonlinear regime. The difference is that the deposited wave angular momentum is now no longer used for wave growth, but rather left onto the basic state. In the nonlinear regime, we do not simply have two propagating waves inside corotation, but rather propagating/decaying waves in the co-moving frame of each wave. Note that without decay, the propagation of the C_1 wave from $r + \Delta r$ to r would lead to the time increase of wave action at r; but the dissipative channeling of wave action onto the basic state makes the wave at the quasi-steady state to have the same action at all times when it arrives at r.

Finally, we mention that although we have shown that the spontaneous and homogeneous modal growth can be achieved if the traveling wave components of the mode have exponential spatial distribution[†], and that such a distribution is self-sustaining once it is established, we have not addressed the question of how such exponentially distributed wave action initially got established in the first place. The answer to this question is apparent when we realize that a spiral mode is an

---

[†] This conclusion is true, however, only in our simplified case of one-dimensional wave propagation, with a constant group velocity. In the two-dimensional disk, there will be further radial modulations of wave amplitude, which depends on the radial distribution of the basic state characteristics.

instability of the underlying disk. The spontaneous growth tendency of such an instability, in a particle disk, implies that, of all the possible *initial positions and velocities* of the N particles in the axisymmetric disk, there is a far greater portion of the $6N$-dimensional phase space which, when evolved with time, results in a spontaneously growing spiral mode (which can then be decomposed into two exponentially distributed waves inside corotation), than the portion of this phase space which does not result in such a spontaneous growth. So by randomly choosing the initial positions and velocities of the disk particles (while satisfying the large-scale equilibrium constraints) we are almost certain to land on an initial condition in phase space which leads to the spontaneous growth of spiral instability. The subsequent seemingly more ordered state of the disk with a spontaneously formed spiral mode is in fact of higher entropy than the state of the original smooth (but unstable) disk.

The fact of the homogeneous growth we almost always observe in a spontaneous N-body spiral is likely to be due to that for randomly distributed initial seed noise, there should *not* be a region of the disk which is favorably selected for an initial faster growth. Another aspect of the problem that supports homogeneous modal growth is that the dissipation mechanism is indiscriminating (i.e., it damps all Fourier components), but only the $m = 2$ global modal component (in a two-armed grand-design galaxy) gets amplified by the galactic resonant cavity, and the $m = 2$ modal component then gets channeled into nonlinear modal components at the spiral-arm collisionless shock through a process akin to turbulence cascade in fluid mechanics (see Chapter 5).

2.5 Torque Coupling and Angular Momentum Transport[†]

In this section, we give detailed derivation of the equations for torque coupling, angular momentum transport, as well as their relation to the time-variation of local disk angular momentum density (Zhang 1998). These relations will be used to further demonstrate the difference of their behaviors in the linear regime as well as in the quasi-steady state of the wave mode, which lead ultimately to a new closure relation at the quasi-steady state of the wave mode to serve as the fundamental dynamical relation for the secular morphological evolution of galaxies.

2.5.1 Torque Couplings due to Density Waves

The total torque coupling in a galaxy disk produced by a self-sustained galactic density wave pattern, which is equal (or equivalent) to the flux of angular momentum

---

[†] Portions of this section used material previously published in Zhang (1998, 1999), reproduced with modifications from The Astrophysical Journal @ AAS. Reproduced with permission.

transport across a particular galactic radius, can be divided into two parts (see, e.g. Lynden-Bell & Kalnajs 1972; Goldreich & Tremaine 1979). The first part is the (z-component of) the gravitational torque coupling C_g, owing to the mass distribution of the spiral, with its rate of the outward angular momentum transport across galactic radius r given by

$$C_g(r) = \frac{1}{4\pi G} r \int_{-\infty}^{\infty} \int_0^{2\pi} \frac{\partial \mathcal{V}}{\partial \phi} \frac{\partial \mathcal{V}}{\partial r} d\phi dz, \tag{2.67}$$

where $\mathcal{V}$ is the gravitational potential and G is the constant of gravitation.

The second part is the advective torque coupling, or coupling through Reynolds stress, owing to the large-scale velocity distribution produced by the spiral structure. In the Eulerian fluid formulation, the z-component of the rate of advective angular momentum transport across the galactic radius r is given by

$$C_a(r) = r^2 \int_0^{2\pi} \Sigma V_r V_\phi d\phi, \tag{2.68}$$

where Σ is the surface density, and where V_r and V_ϕ are the radial and azimuthal velocity perturbation relative to the circular velocity, respectively.

Using the continuum formulation, these two types of the torque couplings can be derived as follows (Lynden-Bell & Kalnajs 1972).

The differential form of the Poisson equation is (note the difference in sign convention from that used in Lynden-Bell and Kalnajs (1972), ours agree with that used in Binney & Tremaine (2008))

$$\nabla^2 \mathcal{V} = 4\pi G \rho. \tag{2.69}$$

Therefore, rearranging the constant factors and multiplying both sides of the equation by $\nabla \mathcal{V}$, we obtain that the gravitational force density $-\rho \nabla \mathcal{V}$ is

$$-\rho \nabla \mathcal{V} = -\frac{1}{4\pi G} \nabla^2 \mathcal{V} \nabla \mathcal{V}. \tag{2.70}$$

Using the divergence theorem, this can be further written as

$$-\rho \nabla \mathcal{V} = -\frac{1}{4\pi G} [\nabla \cdot (\nabla \mathcal{V} \nabla \mathcal{V}) - (\nabla \mathcal{V} \cdot \nabla)(\nabla \mathcal{V})] \tag{2.71}$$
$$= -\frac{1}{4\pi G} \nabla \cdot [\nabla \mathcal{V} \nabla \mathcal{V} - 0.5 \bar{\bar{I}} \cdot (\nabla \mathcal{V} \nabla \mathcal{V})],$$

where $\bar{\bar{I}} = \delta_{i,j}$ is the unit tensor.

Define $\vec{g} = -\nabla \mathcal{V}$ as the gravitational force vector, the above expression can be written as

$$-\rho \nabla \mathcal{V} = -\nabla \cdot \left[\frac{\vec{g} \vec{g}}{4\pi G} - \frac{g^2}{8\pi G} \bar{\bar{I}} \right], \tag{2.72}$$

therefore

$$-\rho\nabla\mathscr{V} = -\nabla \cdot \overline{\overline{T}}_g, \tag{2.73}$$

where

$$\overline{\overline{T}}_g \equiv \frac{\vec{g}\vec{g}}{4\pi G} - \frac{g^2}{8\pi G}\overline{\overline{I}} \tag{2.74}$$

is gravitational stress tensor.

In the cylindrical coordinates, across a cylinder located at r, the torque applied by the inner mass distribution on the outer mass distribution due to the above gravitational stress tensor can be expressed as

$$\vec{C}_g = \int (\vec{r} \times \overline{\overline{T}}_g) \cdot d\vec{S}. \tag{2.75}$$

Using the above expression for $\overline{\overline{T}}_g$, and taking into account that the isotropic part of the gravitational stress tensor in eq. (2.74) does not produce gravitational torque coupling, we arrive at the expression for z-component gravitational torque coupling of eq. (2.67).

To derive the z-component of the advective torque coupling expression (2.68), following Lynden-Bell & Kalnajs (1972) we define the stress tensor $\overline{\overline{S}}^*$ due to the stellar motion in terms of the distribution function f as

$$\overline{\overline{S}}^* \equiv \int f \vec{v}\vec{v} d^3 v, \tag{2.76}$$

then the advective torque couple across a cylinder located at radius R is

$$\vec{C}_a = \int \vec{r} \times \overline{\overline{S}}^* \cdot d\vec{S}, \tag{2.77}$$

which reduces to expression (2.68) for the z-component of the advective torque coupling under the Eulerian fluid approximation.

2.5.2 The Relation Between the Torque Couplings and the Rate of Angular Momentum Change in the Disk

We now demonstrate that the rate of change (the time rate of decrease) of angular momentum within a galactic radius r is given by the *sum* of $C_g(r)$ and $C_a(r)$. This demonstration, originally due to an anonymous referee of Zhang (1998), shows that the sum of these two fluxes is *solely responsible* for causing the angular momentum change within a galactic disk boundary, *when there is no additional external source or*

2.5 Torque Coupling and Angular Momentum Transport — 49

sink of angular momentum introduced within the disk. We begin with the use of stellar dynamical equation set, the same as used in the original Lynden-Bell & Kalnajs' (1972) derivation. At the end of the derivation we will take the fluid-dynamical limit in order to compare with the C_a and C_g expressions used earlier.

The rate of decrease of total angular momentum J contained within galactic radius r is given by

$$-\frac{dJ}{dt}(r) = -\frac{d}{dt}\int_{|\vec{r}'|\leq|\vec{r}|}\int_v \left(\vec{r}'\times\vec{v}\right) f\left(\vec{r}',\vec{v},t\right) d^3v d^3r' \quad (2.78)$$

$$= -\int_{|\vec{r}'|\leq|\vec{r}|}\int_v \left(\vec{r}'\times\vec{v}\right) \frac{\partial f}{\partial t} d^3v d^3r'.$$

By the continuity equation,

$$\frac{\partial f}{\partial t} + \frac{\partial}{\partial \vec{r}}(f\vec{v}) - \frac{\partial}{\partial \vec{v}}\left(f\frac{\partial \mathcal{V}}{\partial \vec{r}}\right) = 0, \quad (2.79)$$

hence we have

$$-\frac{dJ}{dt}(r) = \int_{|\vec{r}'|\leq|\vec{r}|}\int_v (\vec{r}'\times\vec{v})\left[\frac{\partial}{\partial \vec{r}'}(f\vec{v}) - \frac{\partial}{\partial \vec{v}}\left(f\frac{\partial \mathcal{V}}{\partial \vec{r}'}\right)\right] d^3v d^3r'. \quad (2.80)$$

The final term in the integrand is

$$-\frac{\partial}{\partial \vec{v}}\left[(\vec{r}'\times\vec{v})f\frac{\partial \mathcal{V}}{\partial \vec{r}'}\right] + \vec{r}'\times\frac{\partial \mathcal{V}}{\partial \vec{r}'}f, \quad (2.81)$$

so integrating in v and remembering that as $|\vec{v}|\to\infty$, $f\to 0$ and that $\int f d^3v = \rho$, we have

$$-\frac{dJ}{dt}(r) = \int_{|\vec{r}'|\leq|\vec{r}|}\int_v (\vec{r}'\times\vec{v})\frac{\partial}{\partial \vec{r}'}(f\vec{v}) d^3v d^3r' + \int_{|\vec{r}'|\leq|\vec{r}|}\vec{r}'\times\frac{\partial \mathcal{V}}{\partial \vec{r}'}\rho d^3r', \quad (2.82)$$

with the z-component of the final term reduces to the gravitational torque coupling $C_g(r)$ in (2.67) under the disk geometry; and the integrand of the first term can be written as

$$\frac{\partial}{\partial \vec{r}'}\left[f\vec{v}(\vec{r}'\times\vec{v})\right] - f\left(\vec{v}\cdot\frac{\partial}{\partial \vec{r}'}\right)(\vec{r}'\times\vec{v}), \quad (2.83)$$

here the second term is equal to zero because $\vec{v}\cdot I\times\vec{v} = 0$, where I is the identity matrix.

The first term in (2.83) when integrated gives

$$\int_{|\vec{r}'|=|\vec{r}|}\vec{r}'\times\int_v f\vec{v}\vec{v}d^3v \vec{ds} \equiv \int_{|\vec{r}'|=|\vec{r}|}\vec{r}'\times\overline{\overline{S}}^*\cdot\vec{ds}, \quad (2.84)$$

where $\int_v f\vec{v}\vec{v}d^3v$ is the total stress tensor $\overline{\overline{S}}^*$ of Lynden-Bell & Kalnajs' (1972).

Under the disk geometry, the z-component of the integral of (2.84) further reduces to

$$\int_\phi \int_r \int_v f v_r v_\phi d^3v d\phi dz. \tag{2.85}$$

In the fluid limit, we can define the averaged velocity dispersions V_r and V_ϕ such that

$$V_r V_\phi \equiv \frac{\int_v f v_r v_\phi d^3v}{\int_v f d^3v} \tag{2.86}$$

$$= \frac{\int_v f v_r v_\phi d^3v}{\rho},$$

Therefore, the first term in the z-component of $-d\vec{j}/dt$ becomes

$$\int_z \int_\phi r^2 V_r V_\phi \rho d\phi dz, \tag{2.87}$$

which reduces to the fluid expression of $C_a(r)$ (eq. 2.68) after the z integration. Thus we have proved

$$-\left[\frac{d\vec{j}}{dt}(r)\right]_z = C_a(r) + C_g(r), \tag{2.88}$$

that is, the decrease of total z-component angular momentum within a galactic radius is equal to the sum of gravitational and advective torque couplings through the boundary at the same radius. Taking the radial derivative of the above equation, we then obtained the results used in the main text,

$$-2\pi r \frac{d\bar{L}}{dt}(r) = \frac{d(C_a(r) + C_g(r))}{dr}, \tag{2.89}$$

where $\bar{L}$ is the azimuth-averaged angular momentum *density* at a particular radius.

Previously, we have showed that $C(r)$ is a bell-shaped curve for an N-body spiral mode, during *both* the linear modal growth stage and at the quasi-steady state of the spiral structure. Since $\frac{d\bar{L}}{dt}(r) \propto \frac{dC}{dr}(r)$ is thus non-zero, eq. (2.89) shows we can never have a true steady state as long as there is density wave perturbations present. However, when a skewed bar or spiral wave mode is present, these nonlinear dissipative structures have mean states that secularly change as a result of dissipation. Thus eq. (2.89) is to be interpreted as the rate of deposition of angular momentum into the mean background state to cause the secular mass redistribution at the quasi-steady state of the wave mode. Another way to look at the situation is that we can regard the interaction of the wave with the basic state as additional source/sink terms in the angular momentum balance equation for the wave, and the angular momentum transported by the wave in or out of a region can be regarded to come from the basic state (at the quasi-steady state of the wave mode), thus the wave mode can remain quasi steady despite its outward transport of angular momentum.

2.5.3 The Relation Between the Volume-Torque Integral $\overline{\mathscr{T}}$ and the Torque-Coupling Integrals C_g and C_a

The torque integral/phase shift relation we have derived earlier,

$$\overline{\mathscr{T}}(r) = -\frac{1}{2\pi}\int \Sigma_1 \frac{\partial \mathscr{V}_1}{\partial \phi} d\phi \equiv -\frac{1}{2\pi r} T_1(r), \qquad (2.90)$$

describes the averaged torque density of the spiral potential field on the disk material for an annular ring of unit width at disk radius r at the quasi-steady state of the wave mode. Therefore, it also indicates the rate of angular momentum exchange between the density wave and disk matter at the quasi-steady state. We now explore the relation between the torque integral $T_1(r)$ and the torque coupling integral $C(r)$.

Differentiating the expression of C_g in (2.67) with respect to r, we have

$$\frac{dC_g}{dr} = \frac{1}{4\pi G}\int dz \int d\phi \left[\frac{\partial \mathscr{V}}{\partial \phi}\frac{\partial}{\partial r}r\frac{\partial \mathscr{V}}{\partial r} + r\frac{\partial \mathscr{V}}{\partial r}\frac{\partial^2 \mathscr{V}}{\partial \phi \partial r}\right]. \qquad (2.91)$$

But the second term, which can be written as $\frac{r}{2}\frac{\partial}{\partial \phi}\left(\frac{\partial \mathscr{V}}{\partial r}\right)^2$, integrates to zero over ϕ.
Now the differential form of the Poisson equation in the cylindrical coordinates is

$$\frac{1}{r}\frac{\partial}{\partial r}r\frac{\partial \mathscr{V}}{\partial r} + \frac{1}{r^2}\frac{\partial^2 \mathscr{V}}{\partial \phi^2} + \frac{\partial^2 \mathscr{V}}{\partial z^2} = 4\pi G\rho, \qquad (2.92)$$

where ρ is the volume density of disk matter. Therefore,

$$\frac{dC_g}{dr} = \frac{1}{4\pi G}\int dz \int d\phi \frac{\partial \mathscr{V}}{\partial \phi}\left[4\pi G\rho - \frac{1}{r}\frac{\partial^2 \mathscr{V}}{\partial \phi^2} - r\frac{\partial^2 \mathscr{V}}{\partial z^2}\right]. \qquad (2.93)$$

The second term gives

$$\int dz \int d\phi \frac{\partial \mathscr{V}}{\partial \phi}\frac{\partial^2 \mathscr{V}}{\partial \phi^2} = \frac{1}{2}\int dz \int d\phi \frac{\partial}{\partial \phi}\left(\frac{\partial \mathscr{V}}{\partial \phi}\right)^2 = 0. \qquad (2.94)$$

The third term gives

$$\int dz \int d\phi \frac{\partial \mathscr{V}}{\partial \phi}\frac{\partial^2 \mathscr{V}}{\partial z^2} = -\int dz \int d\phi \frac{\partial \mathscr{V}}{\partial z}\frac{\partial^2 \mathscr{V}}{\partial \phi \partial z}$$

$$= -\frac{1}{2}\int dz \int d\phi \frac{\partial}{\partial \phi}\left(\frac{\partial \mathscr{V}}{\partial z}\right)^2 = 0. \qquad (2.95)$$

Therefore,

$$\begin{aligned}
\frac{dC_g}{dr} &= r \int dz \int d\phi \rho \frac{\partial \mathscr{V}}{\partial \phi} \\
&= r \int d\phi \Sigma \frac{\partial \mathscr{V}}{\partial \phi} = r \int d\phi \Sigma_1 \frac{\partial \mathscr{V}_1}{\partial \phi} = -2\pi r \overline{\mathscr{T}}(r) = T_1(r)
\end{aligned} \qquad (2.96)$$

in the linear regime (Tremaine 1995, private communication).

The relation we demonstrated above, that is, $dC_g/dr = T_1(r)$, is obtained under the standard mean-field/continuum approximation, which guarantees that the differential forms of the relations (such as the differential form of the Poisson equation) are valid. However, the mean-field approximation is not expected to hold away from the linear regime, after the spiral shock forms. This is because the local gravitational instability condition at the spiral arms, which is associated with the formation of spiral shock, makes the differential form of the Poisson equation (2.92) no longer valid between the gravitational potential at an arbitrary position, and the averaged volume density of the disk at the corresponding location. This breakdown of the differential form of the Poisson equation (as well as the differential form of the Eulerian equation of motion, as a matter fact, but this latter point does not affect our current discussion of the relation between $T_1(r)$ and dC_g/dr) is natural to expect since the collisionless shock is a formal singularity of the solutions of continuum formulation, and a continuum formulation cannot be carried within the shock singularity.

The necessity for the break down of the differential form of the Poisson equation can alternatively be understood as follows. If the differential form of the Poisson equation actually is valid inside a spiral instability, with the density ρ obtained through averaging the local grainy particle distribution, we can then integrate the differential form of the Poisson equation around an annular ring, and then throughout the disk, and obtain a corresponding spiral potential distribution which, lacking the particle graininess effect, in the case of steady wave amplitude will lead to no secular orbital change (as guaranteed by the conservation of the Jacobi under such conditions), a result obviously contradicting the results of N-body simulations (see next chapter). The differential form of the Poisson equation in effect erases the interparticle correlations by averaging on the mass distribution, and the corresponding Eulerian fluid equation set or the stellar dynamical kinetic equation further complete the continuum description and together they suppress the correlated fluctuations on the particle level. The most obvious manifestation of this suppression is the fact that in the SPH-type of algorithms for modeling disk galaxies, artificial viscosity will have to be introduced, since the continuum approach could not admit the natural gravitational viscosity of the collective instabilities.

Thus another way of saying that the differential form of the equation set (be it the Eulerian equation set or the stellar dynamical equation set) breaks down inside a collective spiral instability is that, *using the continuum formulation we can never* uncover all the physics, including secular matter migration, which could have existed in a particle spiral disk admitting self-organized collective instabilities. The crucial role of interparticle correlation, which is responsible for the self-organization and collective dissipation processes, is ignored in any continuum/mean-field theories (see the Appendix of this monograph for further discussion). Using such dissipationless continuum formulation, we would instead observe the steepening of the spiral shock until the singularity forms at the shock front, which invalidates the differential formulation, and which further requires the introduction of the viscous effect in the spiral shock region. Such viscous effect, however, cannot be properly modeled by introducing a uniform viscosity parameter into the Eulerian equation set to make it into a Navier–Stokes equation set, since the operation of the spiral shock viscosity is not homogeneous (it is most prominent at the spiral arms). So the particle disk formulation is the only one which incorporates naturally and self-consistently all the physics in a realistic spiral disk.

At the quasi-steady state of the wave mode, the result of the N-body simulations indicates that the torque integral $T_1(r)$ is close in both magnitude and shape to the sum of dC_a/dr and dC_g/dr. This forms a new closure relation for the quasi-steady state of the wave mode, which we will derive next in Section 2.5.4.

2.5.4 A Closure Relation for the Quasi-Steady State

In Figure 2.3, we had shown the result of an N-body simulation from Zhang (1996, 1998). The left frame, which displays the various torque couplings, shows curves which have the characteristic bell shapes that we have commented above. The peaks of the bell curves are near the CR of the dominant spiral mode in this simulation, at $r = 30$. In the right frame, we showed the calculated *gradient* of the gravitational and total torque couples, and compare them with $T_1(r)$, using results from the same set of N-body simulation. It is clear that $T_1(r) \neq dC_g/dr$, and rather $T_1(r)$ is closer to dC/dr, though the equality is not yet exact because this particular simulated N-body mode never achieved true steady state. We will now demonstrate the expected equality of dC/dr and $T_1(r)$ as forming the new closure relation at the quasi-steady state of the wave mode.

Previously, we have shown that the rate of angular momentum change in an annular ring of the galactic disk is related to the total torque coupling integral $C(r)$ through

$$-2\pi r \frac{d\bar{L}}{dt}(r) = \frac{dC(r)}{dr} = \frac{d(C_g(r) + C_a(r))}{dr}, \qquad (2.97)$$

where $\bar{L}$ is the azimuth-averaged angular momentum density at a particular radius, C_g is the gravitational torque coupling, and C_a is the advective torque coupling, or coupling through Reynolds stress.

Previously we have also introduced an alternative volume-type torque integral

$$T_1(r) = r \int_0^{2\pi} \Sigma_1 \frac{\partial \mathcal{V}_1}{\partial \phi} d\phi, \qquad (2.98)$$

which expresses the total torque experienced by the basic state stars in an annular ring due to the torquing of the perturbation spiral potential. We want to know how are $T_1(r)$ and $C = C_a + C_g$ related in the different regimes of the wave mode.

We have argued that although in the linear regime $T_1(r) = dC_g/dr$, this equality does not hold away from the linear regime, after the spiral shock forms, due to the fact that one of the relations needed in the proof of this equality, that is, the differential form of the Poisson equation, is no longer valid between the gravitational potential at an arbitrary position, and the averaged volume density of the disk at the corresponding location.

The results of N-body simulations (Zhang 1998, see also the discussion in the next chapter) indicate that at the quasi-steady state of the wave mode, the appropriate closure relation governing the wave and basic state energy and angular momentum exchange is in fact

$$dC/dr = d(C_a + C_g)/dr = T_1(r), \qquad (2.99)$$

we now give a more rigorous proof of this relation at the quasi-steady state.

We begin by observing that in the case of a particle disk the Reynolds stress indicated by C_a is also ultimately gravitational in nature, since in a formally collisionless disk the momentum transfer contained in the Reynolds stress expression is mediated through the short-to-intermediate range gravitational field of the particles constituting the spiral gravitational instability. So if we take into account also the particle discreteness effect and its contribution to the potential, the net torque experienced by particles in an annular ring (due to other particles external to the ring) should be purely gravitational in nature, and there is not a separate advective contribution.

Thus we can write that the total torque experienced by particles in a unit-width annular ring, which is equal to the rate of the angular momentum change, as

$$2\pi r \frac{d\bar{L}}{dt}(r) = \sum_i \vec{r}_i \times \vec{F}^i$$

$$= \int \sum_i \vec{r}_i \times \delta^2(\vec{r} - \vec{r}_i) m^i \left(-\frac{\partial \tilde{\psi}_i}{\partial \vec{r}} \right) d^2r = \int \vec{r} \times \Sigma(\vec{r}) \left(-\frac{\partial \tilde{\psi}_i}{\partial \vec{r}} \right) d^2r, \qquad (2.100)$$

where the summation i is over all particles in the unit-width annulus (thus $d^2r = rd\phi$), $\tilde{\psi}_i$ is the potential at the position $\vec{r}_i$ of the ith particle due to the rest of the particles in the disk, and the integral is over all $\vec{r} = (x, y, 0)$ in the annulus; also where $\Sigma \equiv \sum_i m^i \delta^2(\vec{r} - \vec{r}_i)$.

Taking the continuum limit and it follows that

$$2\pi r \frac{d\bar{L}}{dt}(r) = \sum_i \vec{r}_i \times \vec{F}^i = -r \int \Sigma_1 \frac{\partial \mathscr{V}_1}{\partial \phi} d\phi \equiv -T_1(r), \qquad (2.101)$$

that is, $-T_1(r)$ gives the total torque that the particles in a unit width annular ring experiences.

At the quasi-steady state, since this interaction is irreversible in nature, all the orbital angular momentum lost by this group of particles is permanently lost. So in terms of what the basic state stars experience at the quasi-steady state, it is much more natural for us to adopt a Lagrangian point of view, that is, $T_1(r)$ is the total loss/gain of *basic state* angular momentum of *this group of stars* in the annulus at a particular instant, and as a result of the action of $T_1(r)$ the stars will stream out of this annulus at the next instant, if the annulus is thin enough; but for the wave, on the other hand, since it is usually consisted of different particles and since it is at a quasi-steady state, it is more natural for us to adopt a Eulerian point of view, that is, we consider how much angular momentum the wave leaves onto each annular ring by calculating the gradient of $C_a(r) + C_g(r)$. At the quasi-steady state, by definition all the angular momentum deposited by the wave is left onto the basic state matter, so we come to the inevitable conclusion that $\frac{d(C_a(r)+C_g(r))}{dr} = T_1(r)$ at the quasi-steady state.

After arriving at this new closure relation, which is one of the central results of this work, we will now step back and take a closer look at the all assumptions that went into its derivation. The chief assumptions we made are that the density wave patterns are unstable modes of the underlying galactic disk, and these modes are self-sustained and have reached quasi-steady state through the competition of the spontaneous growth tendency of the global instability, and the irreversible collective interaction with the basic state of the galactic disk. Therefore, there is global self-consistency in the whole system, and the various processes that occur balance one another in a way that the quasi-steady state can be maintained at the expense of a slow secular evolution of the basic state.

This situation is in fact analogous to the fully developed turbulence studied by Kolmogorov (1941a, 1941b, 1941c), and his famous scaling laws are similarly derived through the assumption of quasi-steady state energy cascade and dissipation (Frisch 1995). Thus, the global energy balance approach seems to be a common one in deriving emergent laws governing the workings of the emergent structures.

Taking a closer look at both sides of the new closure relation, we note that the potential $\mathscr{V}$ in $T_1(r)$ is meant to be calculated from the integral form of the Poisson equation, which allows the particle correlations to be correctly incorporated in the fluctuations of the potential and mass distributions. In fact the potential used in the expression of $T_1(r)$ is in its gradient form, which means that in fact it is the gravitational force that enters into this volume torque expression. Therefore, because of Newton's third law, the mutual interaction and torquing of particles within the annular ring cancel each other out in the volume torque, what is left as represented by $T_1(r)$ is in fact the torquing from all the matter that is not situated in the unit-width annular ring in the definition of $T_1(r)$. This torquing therefore is *nonlocal*, and irreversible and dissipative at the quasi-steady state of the wave mode. The discrete particle effect of the basic state has to be taken into account in order for the physical process as represented by $T_1(r)$ to make sense.

The surface torque coupling as represented by C_a and C_g, on the other hand, represents *local* coupling of interactions. So the wave as an independent entity is transporting angular momentum outward as if through a series of interfaces (in galactic radius). The continuum assumption used for deriving the relation between the total C and the local rate of angular momentum change, which we have shown in Section 2.5.2, is still expected to hold for the wave in the linear as well as nonlinear regimes, since the wave itself is supported by the continuum of stellar fluid even though underlying it there is secular dissipation due to the discrete particle effect (i. e., the wave can be considered as a coarse-grained entity). This is the dichotomy in the organizations of galactic density wave modes: The continuum approach can be used for the calculation of the wave propagation properties (after all, the wave itself is a continuum concept); whereas the interaction with the basic state must be treated in a particle (or many-body) context. This is another example of the hierarchical organization of nature as well as its governing laws. We will return to this topic in Section 5.2.1 and in the Appendix of this monograph.

2.6 Rates of Secular Evolution[†]

From a direct numerical integration of nonlinear orbits in a spiral potential, it can be shown that the original conclusion of Lynden-Bell and Kalnajs of of no secular angular momentum exchange between a steady wave and a stellar orbit *is* the correct conclusion to all perturbation orders in a nonlinear calculation, *if* the star only experiences a smooth axisymmetric plus spiral potential but no collective effect. This conclusion is also supported by the fact that the Jacobi integral is a constant for a stellar orbit in a smooth spiral potential. In order to conserve the Jacobi integral, the amount of energy loss and the amount of angular momentum loss of a star in its interaction with the wave have to have the ratio Ω_p, or the pattern speed of the wave. An average stellar orbit, on the other hand, has its mean energy and angular momentum in the ratio of Ω, corresponding to the rotation curve at a given location. Therefore, the *secular* exchange of energy and angular momentum of a single star with a quasi-steady wave is prohibited by the constancy of the Jacobi integral at non-corotation radii, if there is no collective dissipation mechanism which converts part of the stellar orbital energy into heat (i. e., epicycle motion), whereby the phase of the non circular component of the stellar orbital velocity is decorrelated from the phase of the spiral wave.

[†] Portions of this section used material previously published in Zhang (1998, 1999), reproduced with modifications from The Astrophysical Journal @ AAS. Reproduced with Permission.

In fact, without collective dissipation, a finite amplitude spiral wave could not even obtain its coherent organization and form a self-consistent pattern. Only by dissipating part of their orbital energy at each crossing of the spiral arm potential do stars participate in wave motion. This is a view of the maintenance of the spiral pattern which is quite different from that offered by the "kinematic spiral" mechanism, where the orbits are the true "building blocks" of a global pattern. In essence, the single-orbit response in an applied spiral potential does not tell the whole story of a self-consistent spiral mode since it does not incorporate collective effects. All the secular evolution effects we derive in this section result from the collective effects (Zhang 1998, 1999; Zhang & Buta 2007, 2015).

2.6.1 Secular Change in the Mean Stellar Orbital Radius

We will derive the mean orbital radius change for a star in a galaxy disk harboring a quasi-steady density wave mode. Since we are interested in analytical expressions for our result, and since unfortunately most of the available analytical results for density waves have been obtained in the WKBJ regime only, in the following we will make use some of these expressions on open waves that strictly speaking are valid only for WKBJ waves. If, on the other hand, we are expressing the result in terms of the torque integral (2.26) itself, then the recourse to the WKBJ expressions can be avoided – that is, the secular dissipation effect expressed in terms of torque integral (rather than in terms of wave amplitudes, pitch angle, and phase shift) *does not need to involve WKBJ expressions at all.*

As we have shown before, at the quasi-steady state of the wave mode, the rate of angular momentum exchange between an open spiral pattern and the basic state of the disk, per unit area, is given by

$$\overline{\frac{dL}{dt}}(r) = -\frac{1}{2\pi}\int_0^{2\pi} \Sigma_1(r,\phi) \frac{\partial \mathcal{V}_1(r,\phi)}{\partial \phi} d\phi, \qquad (2.102)$$

which, for two sinusoidal waveforms, is given by

$$\overline{\frac{dL}{dt}}(r) = (m/2) A_\Sigma A_{\mathcal{V}} \sin(m\phi_0), \qquad (2.103)$$

where A_Σ and $A_{\mathcal{V}}$ are the amplitudes of the density and potential waves, respectively, and ϕ_0 is the phase shift between these two waveforms. We have assumed that ϕ_0 is greater than zero when the density spiral leads the potential spiral.

Furthermore, the fractional amplitude $F_{\mathcal{V}}$ of the potential wave is related to the potential wave amplitude $A_{\mathcal{V}}$ itself through

$$F_{\mathcal{V}} = \frac{|k| A_{\mathcal{V}}}{r\Omega^2}, \qquad (2.104)$$

where k is the wave number. This relation can be derived as follows: assume a perturbation potential wave of the form

$$\mathscr{V}_1(r, \phi) = A_{\mathscr{V}} \cos\left(m\phi + m\frac{\ln r}{\tan i}\right), \quad (2.105)$$

where i is the pitch angle, and m is the number of arms of the spiral pattern. If the potential wave amplitude $A_{\mathscr{V}}$ does not change with radius rapidly (a result certainly valid for WKBJ waves), we have that

$$\frac{\partial \mathscr{V}_1}{\partial r} = -A_{\mathscr{V}} \frac{1}{r} \frac{m}{\tan i} \sin\left(m\phi + m\frac{\ln r}{\tan i}\right). \quad (2.106)$$

Since the radial derivative for the axisymmetric component of the potential is

$$\frac{\partial \mathscr{V}_0}{\partial r} = \Omega^2 r, \quad (2.107)$$

it follows that

$$F_{\mathscr{V}} \equiv \frac{|\partial \mathscr{V}_1/\partial r|_{\max}}{|\partial \mathscr{V}_0/\partial r|} = \frac{|kA_{\mathscr{V}}|}{\Omega^2 r}, \quad (2.108)$$

where we have used $\tan i = m/|k|r$. Thus we have proved eq. (2.104).

We therefore have

$$\overline{\frac{dL}{dt}}(r) = (m/2)A_\Sigma A_{\mathscr{V}} \sin(m\phi_0)$$

$$= \frac{1}{2}F^2 v_c^2 \tan i \sin(m\phi_0)\Sigma_0, \quad (2.109)$$

where we have used $A_\Sigma = F_\Sigma \Sigma_0$, with Σ_0 being the surface density of the disk; $F^2 \equiv F_\Sigma F_{\mathscr{V}}$; $v_c = \Omega r$; as well as $m/|k|r = \tan i$.

For a single average star of mass M, the rate of its angular momentum loss is therefore

$$\overline{\frac{dL^*}{dt}}(r) = \frac{1}{2}F^2 v_c^2 \tan i \sin(m\phi_0)M. \quad (2.110)$$

Since $\overline{dL^*/dt}$ is also equal to (for a nearly flat-rotation-curved galaxy)

$$\overline{\frac{dL^*}{dt}}(r) = -v_c M \frac{dr_*}{dt}, \quad (2.111)$$

where r_* is the mean radius of the star under consideration, we have

$$\frac{dr_*}{dt} = -\frac{1}{2}F^2 v_c \tan i \sin(m\phi_0). \quad (2.112)$$

We have thus obtained the rate of secular change of the mean radius of a star due to the spiral-induced wave/basic state interaction. Since the value of ϕ_0 is positive inside corotation and negative outside, the resulting orbital change leads to the change of disk surface density toward configurations of ever-increasing central concentration, together with an extended outer envelope. This expression for the mean orbital radius change, which we have derived from the angular momentum closure relation at the quasi-steady state, will be confirmed in the N-body simulations in the next chapter.

2.6.2 Secular Heating of the Galactic Disk

Besides inducing the stellar mean orbital radius change, the wave/basic state interaction also leads to a secular increase in the stellar epicycle radius. This is due to the difference in the ratio of stellar energy and angular momentum loss (gain), compared to the ratio of the wave energy and angular momentum gain (loss), at a location other than corotation. Specifically, we have that the rates of loss of orbital energy and angular momentum for basic state stars are related through

$$\frac{dE_{\text{basic state}}}{dt} = \Omega \frac{dL_{\text{basic state}}}{dt}, \tag{2.113}$$

and the rates of gain of energy and angular momentum by the wave are related through

$$\frac{dE_{\text{wave}}}{dt} = \Omega_p \frac{dL_{\text{wave}}}{dt}. \tag{2.114}$$

Since $dL_{\text{basic state}}/dt$ is equal in magnitude to dL_{wave}/dt, we have that the rate of random energy gain (per unit area) of the disk stars is related to the angular momentum exchange rate (per unit area) through

$$\frac{d\Delta E}{dt} \equiv \frac{d(E_{\text{basic state}} - E_{\text{wave}})}{dt} = (\Omega - \Omega_p) \frac{dL_{\text{wave}}}{dt}, \tag{2.115}$$

where L_{wave} is the angular momentum density of the wave. Note that this expression is true (i.e., the random energy gain of the orbiting star has a positive sign) both inside and outside corotation, since both $\Omega - \Omega_p$ and dL_{wave}/dt change sign across corotation.

Now dL_{wave}/dt is equal in magnitude to $\overline{dL/dt}$ of eq. (2.103), we can thus write

$$\frac{d\Delta E}{dt} = \frac{1}{2}(\Omega - \Omega_p)F^2 v_c^2 \tan i \sin(m\phi_0)\Sigma_0. \tag{2.116}$$

Equation (2.116) gives the rate of random energy increase of stars per unit time and per unit area. The rate of random energy increase for a single star of mass M is then given by

$$\frac{d\Delta E^*}{dt} = \frac{1}{2}(\Omega - \Omega_p)F^2 v_c^2 \tan i \sin(m\phi_0)M, \tag{2.117}$$

where M is the mass of the star.

Since we also have

$$\frac{d\Delta E^*}{dt} = \frac{d(\frac{1}{2}M\sigma^2)}{dt}, \tag{2.118}$$

where σ is the (three-dimensional) velocity dispersion of stars, we finally obtain

$$\frac{d\sigma^2}{dt} \equiv D^{(3d)} = (\Omega - \Omega_p)F^2 v_c^2 \tan i \sin(m\phi_0), \tag{2.119}$$

where D^{3d} is the diffusion constant of the space velocity of a star. This rate of secular heating of the mean stellar orbit (or the diffusion of the stellar orbit in phase space) will also be confirmed in the N-body simulations of Chapter 3. In Chapter 4, we will derive further implications of the density-wave-induced secular heating effect on generating the age-velocity dispersion relation of the solar-neighborhood stars, as well as on generating the size-line-width relation of the Galactic molecular clouds, or the so-called Larson Law (Larson 1981).

2.6.3 Secular Mass Flow Rate Determination

We will now derive the equation for global radial mass flow rate through the orbital decay rate of an average star under the action of a density wave mode, which we had derived previously in Section 2.6.1. The (inward) radial mass accretion rate at a galactic radius r is related to the mean orbital decay rate $-dr/dt$ of an average star through

$$\frac{dM(r)}{dt} = -\frac{dr}{dt} 2\pi r \Sigma_0(r), \tag{2.120}$$

where $\Sigma_0(r)$ is the mean surface density of the basic state of the disk at radius r. The sign convention used here is such that for $-dr/dt > 0$ (or orbital decay), the enclosed mass within r is increasing.

We also know that the mean orbital decay rate of a single star is related to its angular momentum loss rate $-dL^*/dt$ through

$$\frac{-dL^*}{dt} = -v_c M_* \frac{dr}{dt}, \tag{2.121}$$

where v_c is the mean circular velocity at radius r, and M_* the mass of the relevant star. Here the sign convention is such that for a star that has a net loss of orbital angular momentum ($dL^*/dt < 0$, or $-dL^*/dt > 0$), the orbital decay rate $-dr/dt > 0$.

Now we have also

$$\frac{dL^*}{dt} = \overline{\frac{dL}{dt}}(r)\frac{M_*}{\Sigma_0}, \qquad (2.122)$$

where $\overline{\frac{dL}{dt}}(r)$ is the angular momentum loss rate of the basic state disk matter per unit area at radius r. Since

$$\overline{\frac{dL}{dt}}(r) = -\frac{1}{2\pi}\int_0^{2\pi} \Sigma_1 \frac{\partial \mathcal{V}_1}{\partial \phi} d\phi, \qquad (2.123)$$

at the quasi-steady state of the wave mode, we have finally

$$\frac{dM(r)}{dt} = \frac{r}{v_c}\int_0^{2\pi} \Sigma_1 \frac{\partial \mathcal{V}_1}{\partial \phi} d\phi, \qquad (2.124)$$

where the subscript 1 denotes the perturbed variables. The sign convention used here (consistent with that used in eq. (2.27)) is such that for a positive Σ_1 perturbation, the corresponding $\mathcal{V}_1$ perturbation is negative, and if the potential-density phase shift is positive (or if the density leads spiral) then $dM/dt > 0$, or that we have mass inflow.

The above mass flow rate equation, even though derived through the stellar orbital decay rate, is in fact general and can be applied to the mass accretion rate of both stars and gas as long as the relevant perturbation surface density is used. And we note that the potential perturbation to be used for the calculation of either the stellar or gaseous accretion needs to be that of the total potential of all the mass components, since the accretion mass cannot separate the forcing field component and responds only to the total forcing potential.

In the above mass flow rate equation, an equivalent volume-torque T_1 was effectively used in the determination of the total mass flow rate (i. e., in the derivation of eq. (2.124)), which included the contributions from both the gravitational and advective torque couples. The past calculations of the secular angular momentum redistribution rate (i. e., Gnedin, Goodman, & Frei 1995; Foyle et al. 2010) considered only the contribution from gravitational torque couple and ignored the contribution of the advective torque couple (the contribution from the advective torque couple cannot be directly estimated using the observational data – the main reason these authors have omitted its calculation in the above references – except through our round-about way of estimating the total torque using the volume type of torque integral $T_1(r)$, and then subtracting from it the contribution of the gravitational torque to arrive at the contribution of the advective torque). Past calculations of gas mass accretion near the central region of galaxies (e. g., Haan et al. 2009) are likely to have significantly underestimated the gas mass flow rate for the same reason.

In Chapter 4, we will see that for the kind of density wave amplitudes usually encountered in observed galaxies, the advective contribution to the total torque in fact is several times larger than the contribution from the gravitational torque. We will also show that in the nonlinear regime the advective torque couple is of the same sense of angular momentum transport as the gravitational torque couple – another characteristic unique to the nonlinear mode.

The density wave modes in observed galaxies show predominantly a kind of two-humped shape for the volume torque distribution, with zero crossing of the two humps (from the positive hump to the negative hump) located at the CR. This two-humped distribution of the volume torque is equivalent to the two-humped distribution of the phase shift because the phase shift is defined through the volume torque. Zhang & Buta (2007) and Buta and Zhang (2009) have found that galaxies often possess nested modes of varying pattern speeds, and each modal region has its own well-defined two-humped phase shift distribution. The constant pattern speed regime is delineated between the negative-to-positive (N/P) crossings of the adjacent modes, with the positive-to-negative (P/N) crossings (denoting the location of the CRs) sandwiched in between. The establishment of these nested resonances requires both inflow and outflow of matter, but, as we will show in the next chapter, apart from the dominant mode which occupies the major area of the galaxy, the central nested resonances often have their individual inflow/outflow patterns used to establish the modal shape be swamped by an overall mass inflow flux from the dominant mode, possibly aided by diffusive processes. Thus the galaxy on the whole has a mass flow pattern conducive to the trend indicated by the reverse Hubble sequence as well as the direction of entropy increase.

2.6.4 Secular Change of Disk Surface Density

In this subsection we will derive some general trends of the secular change of disk surface density, using well-known approximate forms of the disk surface density and velocity distributions, as well as the torque integral.

We have shown that

$$-\frac{1}{2\pi r}\frac{dC(r)}{dr} = \frac{d\bar{L}}{dt}, \qquad (2.125)$$

where $\bar{L}$ is the azimuthally averaged local angular momentum density and C is the total torque coupling integral. In the quasi-steady state of the wave, the (positive or negative) angular momentum deposited by the density wave is left entirely onto the basic state stars, which leads to the secular change of the mean stellar orbit,

$$\frac{d\bar{L}}{dt} = v_c \Sigma \frac{dr_*}{dt}, \qquad (2.126)$$

where v_c is the circular velocity, Σ the surface density of the basic state of the disk, and dr_*/dt the rate of change of mean stellar radius.

Using (2.125) in (2.126), we have for the rate of orbital change

$$\frac{dr_*}{dt} = -\frac{1}{2\pi r v_c \Sigma}\frac{dC}{dr}. \qquad (2.127)$$

Now the equation of continuity gives

$$2\pi r\frac{\partial \Sigma}{\partial t} = -\frac{\partial}{\partial r}\left(\frac{dr_*}{dt}\Sigma 2\pi r\right). \qquad (2.128)$$

Using eq. (2.127) in eq. (2.128), we obtain

$$\frac{\partial \Sigma}{\partial t} = \frac{1}{2\pi r}\frac{d}{dr}\left(\frac{1}{v_c}\frac{dC}{dr}\right). \qquad (2.129)$$

For nearly constant v_c with r, eq. (2.129) is solved by

$$\Sigma(r,t) = \frac{v_c}{2\pi r}\frac{d^2 C(r)}{dr^2}t + \Sigma_0(r). \qquad (2.130)$$

We have shown that the total torque coupling C is generally of an asymmetric bell shape, with the top of the bell located at the corotation radius of the relevant spiral mode. Such a shape can be approximated by the following composite Gaussian functions:

$$C(r) = \begin{cases} C_0 e^{-\frac{(r-r_{co})^2}{r_1^2}} & \text{for } r \leq r_{co} \\ C_0 e^{-\frac{(r-r_{co})^2}{r_2^2}} & \text{for } r \geq r_{co}, \end{cases} \qquad (2.131)$$

where r_{co} is the corotation radius.
We therefore have

$$\frac{d^2 C}{dr^2} = \begin{cases} \left[\frac{-2C_0}{r_1^2} + \frac{4C_0(r-r_{co})^2}{r_1^4}\right]e^{-\frac{(r-r_{co})^2}{r_1^2}} & \text{for } r \leq r_{co} \\ \left[\frac{-2C_0}{r_2^2} + \frac{4C_0(r-r_{co})^2}{r_2^4}\right]e^{-\frac{(r-r_{co})^2}{r_2^2}} & \text{for } r \geq r_{co}. \end{cases} \qquad (2.132)$$

We thus see that d^2C/dr is positive for $r < r_{co} - r_1/\sqrt{2}$ and $r > r_{co} + r_2/\sqrt{2}$, and is negative for $r_{co} - r_1/\sqrt{2} < r < r_{co} + r_2/\sqrt{2}$. Referring back to eq. (2.130), we see that this sign distribution of d^2C/dr^2 leads to the increase of Σ with time for $r < r_{co} - r_1/\sqrt{2}$ and $r > r_{co} + r_2/\sqrt{2}$, the decrease of Σ with time for $r_{co} - r_1/\sqrt{2} < r < r_{co} + r_2/\sqrt{2}$. This indicates that both the central region and the outer skirt of the disk will accumulate matter with time, and the surface density of the region near the corotation radius will decrease with time. These are consistent with N-body simulatioan results (Figure 3.18 later).

The above derivation assumed that C_0 is a constant, as well as r_{co}, r_1, and r_2. Over a longer time span, the evolution of the basic state is expected to also alter the kind of spiral mode present, therefore we expect that all of these "constants" will change slowly with time.

For a galaxy with flat rotation curve and with $\Sigma(r) \sim 1/r$, we have that $dr_*/dt \propto dC/dr$, from eq. (2.127). For certain radial range of the inner disk region, we can approximately write $dC/dr \sim$ constant, which leads to $dr_*/dt \sim$ constant independent of radius. Now since we also have

$$\frac{dr_*}{dt} = -\frac{1}{2}F^2 v_c \tan i \sin(m\phi_0), \tag{2.133}$$

we see that for such a galaxy the product of $F^2 v_c^2 \tan i \sin(m\phi_0)$ is also approximately constant, a result we will use in the following discussion.

The surface brightness distribution of the galactic disks is often found to be well-fitted by an exponential function of the form $I(r) = I_D e^{-r/r_D}$ (de Vaucouleurs 1959; Freeman 1970), where I_D and r_D are constants representing the extrapolated central brightness and the scale length of the exponential function, respectively. For a constant mass-to-light ratio, this leads to an exponential distribution for the surface density of the disk as well. In recent years, it becomes clear that most galactic disks have surface density distributions which are better described by a $1/r$ function, which approximates very well the exponential function over a wide range of intermediate galactic radii (Seiden, Schulman, & Elmegreen 1984). We now analyze the consequence of the radial mass accretion process to the evolution of the surface density profile of the galactic disks.

In general, the evolution of the disk surface density can be calculated by solving the two conservation equations for viscous accretion disks, that is, the angular momentum conservation equation

$$\frac{\partial C}{\partial r} = 2\pi r \frac{\partial}{\partial t}(\Sigma r^2 \Omega) + 2\pi \frac{\partial}{\partial r}(r \Sigma v_r r^2 \Omega), \tag{2.134}$$

and the mass conservation equation

$$r\frac{\partial \Sigma}{\partial t} + \frac{\partial}{\partial r}(r \Sigma v_r) = 0, \tag{2.135}$$

for $v_r(r,t)$ and $\Sigma(r,t)$, if we know the distribution of $\partial C(r,t)/\partial r$.

In the case of accretion induced by a galactic spiral structure, if the wave remain quasi-stationary (which translates to that most of the angular momentum deposited by the wave goes to the basic state), we further have that

$$\frac{\partial C}{\partial r} = \pi F^2 v_c^2 \sin(m\phi_0) \tan(i) \Sigma r, \tag{2.136}$$

here the wave amplitude F, pitch angle i, and phase shift ϕ_0 are themselves functions of both r and $\Sigma(r, t)$, as well as the distribution of the stellar velocity dispersion. So it is clear that we have a fully nested nonlinear and global problem, the solution of which could be obtained by an iterative approach.

However, an added level of complication to the above picture is introduced by the fact that orbital decay rate $dr/dt = v_r$ given by the quasi-steady state constraint *for the wave mode*, which leads to eq. (2.112) in the case of a flat rotation curved galaxy, does not always agree with that derived from the equation set (2.134)–(2.136), which governs the smooth change of *basic state properties* due to the presence of the wave mode. In cases where such a mutual consistency is lacking, we would not have a true quasi-steady state for the wave mode, that is, the wave mode will be co-evolving with the basic state characteristics during the accretion process.

It is easily verified that at least in one case we can obtain a consistent solution which satisfy all four equations. This is the case of a flat rotation-curved galaxy with $1/r$ surface density distribution, and with the spiral mode leading to a constant orbital migration rate v_r (which in reality can only hold for limited range in the inner galaxy, since v_r ought to change sign at corotation. In this case the accretion flux is continuous throughout the galactic radii under concern, and the only place the surface density is increasing is the central region of the galaxy.

In physical galaxies, we expect that the wave amplitude is nearly steady after a long period of evolution. We also observe that most galaxies do have flat rotation curves and $1/r$ surface density profiles over the intermediate radial range. This indicates that the requirement of global self-consistency during the mass accretion process due to a quasi-stationary spiral structure may be responsible for the creation and the maintenance of the approximate $1/r$ surface density profile of the disk under a flat rotation curve galactic potential.

2.6.5 Viscous Accretion Disk Analogy

Many types of astrophysical accretion disks suffer from the problem of the lack of a clearly identified source of microscopic viscosity which will allow the disks to accrete sufficient matter on astronomically significant time scales to fuel the growth of central objects (see Shu 1992, Chapter 7). An ad hoc assumption of "anomalous viscosity" is often brought in so the formal analysis can continue. Here, for the case of galactic disks, after we have identified the source of collective dissipation and quantify its magnitude, we can write out explicit expressions for the anomalous viscosity, The viscous accretion disk analogy for the galaxy disks undergoing secular mass redistribution will allow us to put our discussion in the general context of the viscous accretion disk theory, and to facilitate the comparison with other types of astrophysical accretion disks (Zhang 1997).

In the general viscous accretion disk theory (Pringle 1981), let $C(r, t)$ be the total z torque applied by the inner disk on the outer disk at radius r and time t (the C here is the equivalence of G used by Pringle). The torque is related to the viscosity parameters through

$$\frac{\partial C}{\partial r} = \frac{\partial}{\partial r}(2\pi r\nu\Sigma A_{\text{shear}}r), \tag{2.137}$$

where $A_{\text{shear}} = r\frac{d\Omega}{dr}$ is the shear rate, $\nu \approx l v_T$ is the kinematic viscosity, and here l is the mean free path, and v_T is the thermal velocity of the disk particles.

To calculate the effective viscosity due to the presence of a quasi-stationary spiral structure, we note that the torque gradient $\frac{\partial C}{\partial r}$ is related to our torque integral $T_1(r) \equiv r\int_0^{2\pi} \Sigma_1 \frac{\partial V_1}{\partial \phi} d\phi$ and phase shift ϕ_0 through

$$-\frac{\partial C}{\partial r} = 2\pi r\frac{\overline{dL}}{dt} = -T_1(r) = -r\int_0^{2\pi} \Sigma_1 \frac{\partial V_1}{\partial \phi} d\phi$$

$$= \pi F^2 v_c^2 \sin(m\phi_0) \tan i\, \Sigma_0 r, \tag{2.138}$$

here C is the sum of gravitational and advective torque couplings.

If we compare the two expressions (2.137), (2.138) for $\frac{\partial C}{\partial r}$, we arrive at

$$-\frac{\partial}{\partial r}(2\pi\nu_{\text{eff}}\Sigma r^3\Omega') = \pi F^2 v_c^2 \tan i \sin(m\phi_0)\Sigma r, \tag{2.139}$$

where $\Omega = v_c/r$ and $\Omega' = d\Omega/dr$.

We now try to obtain the expression of ν_{eff} for a particular type of disk. If the basic state of the disk has flat rotation curve and $1/r$ surface density distribution, we easily obtain from above

$$\nu_{\text{eff}} = \frac{1}{2}F^2 v_c \tan i \sin(m\phi_0) r. \tag{2.140}$$

It is obvious from the above expression of ν_{eff} that it changes sign at r_{co}, where ϕ_0 itself changes sign. Therefore, the effective viscosity due to a quasi-stationary spiral mode is positive inside corotation, and negative outside corotation, which is consistent with the fact that the disk matter accretes inside corotation, and excretes outside corotation.

For a 20% amplitude and 20° pitch angled spiral, the effective α parameter produced by the spiral-induced radial mass accretion process is about 0.01 (Zhang 1997). Therefore, the effective viscosity produced by non-axisymmetric instabilities is likely to be an important candidate for the long-sought-after "anomalous viscosity" in other types of astrophysical accretion disks as well, since these disks are expected to have similar amplitude density wave patterns, though because of the difference in basic state, different modal patterns from that in galaxies may be present (i. e., in Keplerian stellar accretion disks, one-armed spiral modes are known to dominate).

2.7 Relation to "Broadening of Resonances"[†]

As we have seen so far, correlated interactions of particle orbits in a galaxy possessing spontaneously formed wave modes will lead to "graininess" of the total potential, which differs from the passive galactic potential – even if the latter includes the smooth spiral potential. The passive calculation predicts the conservation of the Jacobi integral or the so-called adiabatic condition (Binney & Tremaine 2008, p. 179).

In the course of this work, the author has received inquiries of whether the collective dissipation effects are related to the resonant interactions of particle orbits with the wave, through the so-called broadening of resonances effect due to a time-varying density wave. This issue is especially important because some researchers have claimed that the Lynden-Bell & Kalnajs' (1972) theory already provided the complete foundation for secular evolution, and the fact that their theory proclaimed the immutability of basic state is only because they treated the wave as steady in amplitude, thus the resonances are localized. These researchers further speculate that once the waves are allowed to become transient the resonances themselves will become broadened. For extremely fast transients, the resonances are so broadened that they can effectively cover the entire galaxy disk and cause galaxy-wide secular evolution (see Carlberg & Sellwood 1985; Carlberg 1987).

We argue below that the collective effects we address in this work is of a totally different nature from the "broadening of resonances" effect due to the fluctuating wave amplitudes, mainly due to the fact that the resonance effect refers to the single-orbit's *passive* response to an *applied*, smooth, large-scale perturbation potential, it does not deal with the lateral interactions of all the particles participating in a collective instability.

Whereas the transient waves in over-stable disks *can* produce certain types of secular effect through resonant interactions, especially the secular heating of disk stars, the coordinated secular mass redistribution process for unstable modes we discuss in this work is not a direct result of the time-fluctuating nature of the density wave: in fact, the mass accretion rate expression becomes exact precisely at the quasi-steady state of the wave mode, whereas to produce any significant amount of secular change in the mean stellar orbits through transient mechanisms on an over-stable disk, the time scale of the variation of transient forcing has to be as fast as a local dynamical time scale (Carlberg & Sellwood 1985, see especially their abstract).

There are indeed random fluctuations of the overall potential in any N-particle disk harboring an unstable density wave mode, in addition to the correlated fluctuations. But these *random* fluctuations are not an integral part of the collective dissipation: as we will show in the next chapter, in N-body simulations one could

---

[†] Portions of this section used material previously published in Buta & Zhang (2009), reproduced with modifications from The Astrophysical Journal Supplement Series @ AAS. Reproduced with permission.

use different numbers of particles, different initial conditions, etc., and still obtain essentially the same mode and same secular mass flow rate, as long as the basic state is chosen to be the same and the force law (including the softening parameter) is chosen to be the same. The main change with the use of different particle numbers is the resulted heating rate: but even this rate contains two parts. The collective dissipation part is *independent* of the number of particles used, and only the *excess heating* has the kind of particle-number scaling behavior common to Poisson random noise. Significantly, the mass accretion rate and angular momentum exchange rate remain the same independent of the particle numbers used (Zhang 1998; see also Chapter 3 of this book). Zhang (1999) has also shown in the N-body simulation of quasi-steady modes that the inner Lindblad resonance in the simulation is very localized as shown by an isolated peak in the heating curve (see Fig. 4 of Zhang 1999), whereas mass redistribution is observed to occur over the entire surface of the disk.

In the resonance-broadening scenario, the mass flow directions predicted by it near the inner and outer Lindblad resonances were found by Carlberg and Sellwood (1985, p. 88) to be "independent of the winding sense of the spiral (Lynden-Bell & Kalnajs 1972)." This is expected since resonant interaction is a local effect, so the winding sense of the pattern does not matter. In contrast, for the collective effect we are dealing with here, the sense of the phase shift is critically dependent on whether the wave is leading or trailing, since the geometrical phase shift is determined by the Poisson integral (which does not care about the motional state of the underlying matter), whereas the physical sense of whether the potential is torquing the density forward or backward depends on which direction the pattern is rotating. This fact alone tells us that the collective dissipation effect in spiral galaxies cannot be viewed as a broadening-of-resonance effect.

In addition, the resonance-broadening effect is a kind of top-down control effect – it is a single orbit's response to the applied potential (Carlberg & Sellwood 1985, p. 81, mentioned the equivalence of their transient-wave resonance-broadening approach with Lynden-Bell & Kalnajs' passive single orbit approach, even though one is Eulerian and the other is Lagrangian). In a self-organized instability it is the "sideways" interactions among stars themselves that leads to pattern coherence and collective dissipation. This latter effect cannot be modeled entirely as a top-down control hierarchy due to an *applied* spiral or bar potential. For particles to be aware of one another as in a collective instability, the formal singularity condition created by the collisionless shocks at the density wave crest is needed. The correlated interaction among particles in a collective instability is ignored in the derivation of the Boltzmann equation, as we will show later in the Appendix; whereas the resonant-interaction formalism is derived from Boltzmann's kinetic equation approach. Ultimately, the continuum approach needs to be abandoned if one seeks to derive collective dissipation from first principle. We will address this point further in Chapter 5.

2.8 In a Nutshell

To summarize, in order to properly account for the secular evolution dynamics of the unstable density wave modes, we need to go beyond the classical approach which describes the behavior of individual orbit's passive response under an applied potential, even if such an applied potential includes a spiral component. The root cause of the failure of the classical approach in producing the collective secular evolution behavior lies in its ignoring the correlations in the fluctuations of the participating ensemble of particles, and these correlated fluctuations are responsible both for the self-organization behavior during the growth of the modes as well as for their collective dissipative interaction with the basic state of the galactic disk.

The Lynden-Bell & Kalnajs (1972) theory solved part of the puzzle for the nature and origin of galactic density waves, yet it is not a globally self-consistent theory that can serve as the foundation of the dynamics of secular evolution of galaxies[†] – the claim that it can is an inference made by latecomers, and was not intended at the time by Lynden-Bell & Kalnajs themselves. To quote the Lynden-Bell & Kalnajs theory in a hand-waving fashion out of the context of its original intention will only lead to paradoxes and self-contradictions (such as the conclusion of outward angular momentum flow without the evolution of the basic state mass distribution). It will also prevent us from arriving at an accurate assessment as to the actual role of secular evolution of galaxies in the bigger picture of the cosmological evolution of galaxies and structures.

The advancement of our understanding of the fundamental structures of physical theories, and the exploration of the interconnections of the different scientific disciplines, have always been among the chief motivations for carrying out research in the physical sciences. Understanding the deep roots of galactic secular evolution phenomena in the broader context of nonequilibrium dynamics will provide us with a better appreciation of nature's wonderful and delicate hand at shaping the hierarchies of physical processes. The insight into this deep and hidden beauty of the theoretical structure of science is the privilege offered to explorers of the more abstract aspects of the universe's wonders.

[†] In general, no passive theory based on the top-down interactions can serve as the foundation for nonequilibrium phase transitions, which rely heavily on the lateral interactions of the many degrees of freedom of the system. Certain model-based theories, such as the Landau theory of phase transitions, can capture the phenomenological aspects of these transition processes, without going into details of the actual self-organization process (Stanley 1971).

3 *N*-Body Simulations of Galaxy Evolution

In the previous chapter, we have established the theoretical framework underlying the dynamically driven secular morphological evolution of galaxies. In this chapter, we will demonstrate through *N*-body simulations the quantitative validity of the theoretical formulation, while leaving further inference of the astrophysical consequences, as well as observational verifications of the theory, to Chapter 4.

The relevance of *N*-body simulation of self-gravitating galactic systems originates not only from its demonstrated effectiveness in reproducing the striking appearance of spiral and barred galaxies, but also from its innate ability to model the multitudes of interparticle correlations, as well as the amplification of fluctuations, throughout the nonequilibrium phase transition process. The analytical derivations in the last chapter depended on the (empirically well-supported) assumption that a quasi-steady state for the density wave modes can be attained, as well as on the (logically well-founded) assumption of global self-consistency of the set of governing laws as applied to the global density wave modes. With these in mind, we realize at the same time that the derivation (especially the procedure for obtaining the torque/angular-momentum-exchange closure relation) was not performed in a thoroughly deductive fashion. *N*-body simulations offer us the rare opportunity to peek into the inner workings of the very processes where new patterns and new dynamics emerge as an interwoven whole. Because of the breakdown of the older differential dynamics at the formal singularities of the continuum formulation, we simply would not have been able to cross the threshold of the phase transition in a totally deductive manner: a singularity is a formal barrier to analytical deduction. Logical global self-consistency arguments, coupled with *N*-body simulations, and supplemented by experimental/observational evidence, are our best aid for ensuring a safe voyage through the tumultuous waters of collective instabilities. Most of the quantitative results of this chapter had previous appeared in Zhang (1996, 1998, 1999, 2016).

3.1 Overview of *N*-Body Simulations of Disk Galaxies

Gravitational *N*-body simulations of galactic systems began with the early works of Miller & Prendergast (1968) and Hohl & Hockney (1969). Miller (1976) in particular was responsible for originating the two-dimensional polar grid approach used for the simulation of disk galaxies, and Miller's original code was the forerunner of the polar grid code developed by the current author, used in the simulations described in this text. The grid (or particle-mesh) approach, compared with the pair-wise direct force calculation approach, results in significant savings of computation time (Hockney & Eastwood 1988 Section 1-5-2), though with the accompanying algorithm complexity and the introduction of grid noise (see further the discussion in Section 3.9.4).

Sellwood (1987) presented an extensive review of the relevant N-body techniques that had been explored for the simulation of different types of gravitational N-body systems, for works published up to the mid-1980s.

The current work builds on the previous explorations but emphasizes the unique emergent processes due to collective instabilities in galactic disk systems. The simulation results described below are mostly drawn from two generations of numerical experiments: The first-generation tests were conducted during the 1990s and published in Zhang (1996, 1998, 1999); and the second-generation tests were conducted in the recent years and published in Zhang (2016). The second-generation tests had a much higher numerical resolution than the first-generation tests, reflecting the improvement in computational power which occurred during the past two decades. The importance of the proper treatment of the gravity softening parameter to the correct representation of the magnitude of collective effects was also only recently realized. We will note the numerical resolution and softening parameter used in each set of the numerical experiments described below.

3.2 Simulation Codes and Basic State Specifications

The N-body code used for the simulation of spiral disks presented in this work is a two-dimensional polar code, written by the author using the algorithm described in Thomasson (1989), which was a refined version based on Miller's (1976) original algorithm. The validity of the new code was checked against the examples presented in Thomasson (1989) and in Donner & Thomasson (1994). When running the same set of basic state parameters, good agreements on spiral modal properties were found between the results obtained using the current code and that presented in Donner & Thomasson (1994). Small differences do exist and are attributed to the slight difference in implementation of the two sets of codes. The slight difference in the details of the spiral mode formed can also be influenced by the different random-number generators used for the initial position and velocity assignments for the disk stars.

For the spiral mode calculated in Zhang (1996), the same basic state parameters as those used in Donner & Thomasson (1994) were chosen, at the time mainly for the purpose of not having to recalculate and replot many of the modal characteristics already discussed in the Donner and Thomasson paper, since the main focus of Zhang (1996) was on establishing the collisionless-shock nature of the wave mode, and on introducing the phase-shift concept. Zhang (1998, 1999) once again used nearly identical basic state parameters as that of Donner & Thomasson (1994), with only slight alteration of the disk-to-spheroidal ratio. For the sake of completeness, in what follows we will plot additional basic state and modal characteristics not given in Zhang (1996, 1998, 1999).

3.2.1 Choice of Basic State Parameters

We have chosen to use normalized units in the N-body calculations for this work, similar to that used in Donner & Thomasson (1994). In these units, the total mass of the disk, including the active disk mass m_d, the inert halo mass m_h, and the inert bulge mass m_b, sums to 1, that is,

$$m_{total} = m_d + m_h + m_b = 1. \tag{3.1}$$

Zhang (1996) used the set of mass ratios $m_d : m_h : m_b = 0.5 : 0.4 : 0.1$, the same as used in Donner and Thomasson, whereas Zhang (1998, 1999) used an alternative set of $m_d : m_h : m_b = 0.4 : 0.5 : 0.1$.

The disk surface density used in the simulations is in the form of a modified exponential

$$\Sigma_d(r) = \Sigma_{d0}(e^{-r/R_d} - e^{-2r/R_d}), \tag{3.2}$$

where we have chosen $R_d = 10$ for the disk scale length in the normalized unit, and Σ_{d0} is a constant chosen to make the total disk mass of the correct amount m_d in the normalized unit.

Figure 3.1 plots the initial radial variation of the surface density of the disk mass used in Zhang (1998, 1999). The chosen disk surface density is seen to have a central depression for the small-radius range, which will be offset by the mass of the rigid bulge specified below. This choice of the disk surface density avoids too drastic an initial mass inflow into the central void of the computation domain (which will have to be specially handled using algorithms different from the normal polar-code handling). It also avoids the corresponding large initial random velocities associated with a large central disk surface density. This choice of the disk surface density was initially made in Donner & Thomasson (1994), so Zhang (1996, 1998, 1999) conformed to that for ease of comparison. Subsequently tests by the author showed that changing the

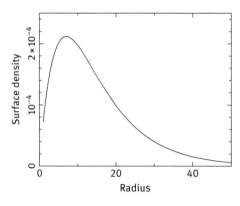

Figure 3.1: Initial disk surface density distribution for the simulations of Zhang (1998, 1999).

modified exponential (i. e., one with the central depression of surface density) to true exponential did not in fact change the qualitative characteristics of the modes formed.

An inactive bulge and a rigid halo (i. e., used purely for their gravitational effect on disk particles, while their own mass distributions are not recomputed as a response to the disk evolution) are also used, which are assumed to be of regular exponential shape:

$$\Sigma_h(r) = \Sigma_{h0} e^{-r/R_h}, \tag{3.3}$$

$$\Sigma_b(r) = \Sigma_{b0} e^{-r/R_b}, \tag{3.4}$$

The scale lengths are $R_h = 5$ and $R_b = 1$ for the rigid halo and bulge component, respectively.

Figure 3.2 plots the initial total rotation curve (due to the contributions of the disk, halo and bulge to the galactic potential) for the simulations of Zhang (1998, 1999). The rotation curve used in Zhang (1996) has a similar shape but roughly a factor of two smaller in numerical value because of the factor of 2 difference in grid resolution, as well as the slight difference in surface density specification of the disk and the halo.

Figure 3.3 plots the initial radial velocity dispersion for the above galaxy model. This corresponds to an initial stability parameter Q (c.f. eq. (2.1)) of 1.1 specified throughout the disk. The drop in velocity dispersion in both the central and the outer regions is due to the lower surface density there. The velocity dispersion will naturally increase in the central region as the computation proceeds.

This particular set of choices of the basic state parameters allows a resonant cavity to be set up between the corotation region (CR) (around $r = 30$ in the normalized unit) and the inner bulge region (which serves as a Q-barrier to reflect the inwardly propagating wave train to complete the feedback cycle), so a normal mode of the galactic disk can spontaneously emerge, as was first shown in Donner and Thomasson

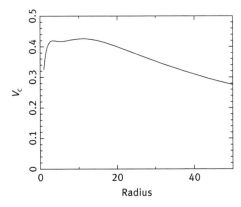

Figure 3.2: Initial total rotation curve for the simulations of Zhang (1998, 1999).

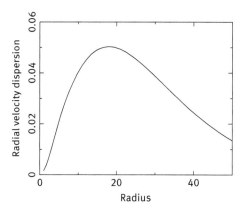

Figure 3.3: Initial radial velocity dispersion distribution for the simulations of Zhang (1998, 1999). This corresponds to an initial instability parameter $Q = 1.1$ throughout the disk.

(1994). In the following, we will present the results of modal characteristic calculations, but the emphases will be placed on the signature of collisionless shock and the associated collective dissipation effect in the disk, as well as on the demonstration of the presence of potential-density phase shift that accompanies the formation of the unstable spiral mode.

Incidentally, if one feels uncomfortable about the renormalized units used in the following calculations, one can always mentally multiply the length unit by kpc, and the mass unit by $10^{11} M_\odot$, and then if one interprets the computation results at the different time steps in terms of the rotation period at a given radius, and mass flow effect in terms of either the surface density or enclosed mass evolution, one will be able to get more of a physical sense of the magnitude of the secular evolution effect.

3.2.2 Choice of Simulation Parameters (First-Generation Tests)

The simulation grid for the two-dimensional disk geometry used in the galaxy simulations of this work is a polar grid similar to the original one used by Miller (1976), with radial grid lines distributed in an exponential fashion, and the azimuthal grid lines distributed in a uniform fashion, that is,

$$r = L e^{\alpha u}, \tag{3.5}$$

$$\phi = \alpha v, \tag{3.6}$$

where L is the constant length scale factor, usually chosen to be 1 in the normalized unit, and α is another constant usually chosen to make the total number of azimuthal v grid cells n_v a power of two, so as to allow the use of the FFT algorithm in the azimuthal direction. The total number of u grid cells in the radial direction n_u is chosen to ensure the disk surface density has decreased to negligible value at the outer region of the computational domain.

Donner & Thomasson (1994) selected a grid of $n_u = 50$ radial grid cells, and $n_v = 64$ azimuthal grid cells, together with 50,000 active disk particles. These parameters have been changed in the simulations of Zhang (1996, 1998, 1999), with Zhang (1996) used $n_u = 110$, $n_v = 128$ and the total number of particles $N = 200,000$, and Zhang (1998, 1999) used $n_u = 55$, $n_v = 64$, and $N = 100,000$.

The disk particles were initially distributed according to the surface density specification in an axisymmetric fashion, and the velocity assignment scheme ensures dynamical equilibrium and local stability. The cloud-in-cell (CIC) method is then employed for grid mass assignment from the disk-particle mass distribution. Fast Fourier transform (FFT) approach was employed for grid potential calculation based on grid mass, and forces on grid points were calculated using the difference of the grid potential. The forces on individual disk particles are then interplated once more through the CIC method from the grid force, and these disk particles are subsequently moved around through a time-centered leap-frog scheme. Sufficient time step resolution is chosen to ensure stability and accuracy.

The length of the time step is chosen through a parameter called the number of time steps per crossing time S_{stpcrt} at a reference radius (usually chosen to be $R_d = 20$, which is within the corotation radius $r_{co} = 25-30$). This number is defined as the time steps it takes for a particle to cross the specified disk radius R_d, that is,

$$\Delta t \cdot S_{stpcrt} = \frac{R_d}{V_c(R_d)}, \qquad (3.7)$$

where $\Delta t = 1$ is the normalized unit of time step, and V_c is the rotation curve value. In the original Donner and Thomasson (1994) simulation, S_{stpcrt} was chosen to be 50, which corresponds to 314 time steps per rotation period at radius 20. The same value was chosen also in Zhang (1998, 1999), and Zhang (1996) chose $S_{stpcrt} = 100$ to correspond to the factor-of-two increase in both the time and the spatial resolution. These S_{stpcrt} choices were made in order to guarantee the numerical stability of the computation algorithm.

After S_{stpcrt} is determined, a renormalized gravitational constant G_{new} is computed and is used to replace the G_{old} in the usual Newtonian equations. The new and old gravitational constants are related through

$$G_{new} = \frac{\frac{R_d^2}{\Delta t^2 \cdot S_{stpcrt}^2}}{m_d \frac{V_d^2(R_d)}{G_{old}} + m_h \frac{V_h^2(R_d)}{G_{old}} + m_b \frac{V_b^2(R_b)}{G_{old}}}, \qquad (3.8)$$

where V_d, V_h, V_b are the rotation velocity contribution of the different mass components to the total rotation curve of the galaxy, that is,

$$m_d V_d^2 + m_h V_h^2 + m_b V_b^2 = m_{total} V_c^2 = V_c^2, \qquad (3.9)$$

since $m_{total} = 1$ in the normalized units (see Freeman (1970) for the expression of rotation curve for an exponential disk). With these choices of Δt and G_{new} the desired S_{stpcrt} will be achieved.

A gravity softening length of 1.5 times the mesh length unit was used in the initial calculations of Zhang (1996, 1998, 1999). This choice of the softening parameter has been reexamined in the most recent work by the author (Zhang 2016) and these second-generation results will be described in Section 3.9.

Figure 3.4 shows the numerical relaxation time (due to the artificial noise of the smaller number of particles employed compared to those present in actual observed galaxies) calculated using the specification of (Rybicki 1971; White 1988):

$$t_{relax} = \frac{\sigma_r^3 a_{soft}^{eff} N}{5G^2 M_d \Sigma_0}, \qquad (3.10)$$

where σ_r is the radial velocity dispersion, a_{soft}^{eff} is the effective gravity softening length, here chosen as $a_{soft}^{eff} = \sqrt{s^2 + a_{soft}^2}$, where s is the local grid size, a_{soft} is the constant particle-softening length (to be discussed in more details in Section 3.9), and N is the total number of simulation particles, G is the gravitational constant, Σ_0 is the disk surface density, and M_d is the total disk mass, for the computation parameter set of Zhang (1998, 1999). Note that this plot is made using the basic state parameter set at the start of the simulation. As the basic state evolves, the relaxation behavior is expected to change as well. From this initial plot, we see that most of the mid-range of the disk will be unaffected by artificial relaxation throughout the entire run of the computation time (8000 steps); and in the very inner disk the relaxation time should improve as the computation proceeds since the velocity dispersion increases there the fastest; the outer disk, however, will suffer some artificial relaxation effect, as indeed we observe from the dilution of the pattern there as computation proceeds. Zhang (1996) carried out the calculation for effectively only half of the length of time compared to

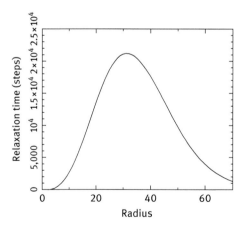

Figure 3.4: Numerical relaxation time for the simulation parameter choice used in Zhang (1998, 1999).

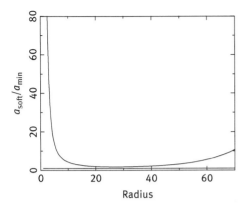

Figure 3.5: Ratio of actual softening length used in the $a_{soft} = 1.5$ simulation of Zhang (1998, 1999) to the lower limit of softening needed for local stability. The straight line in the lower part of the figure indicates a ratio of 1.

Zhang (1998, 1999) due to the twice-finer time step size and linear grid size, and the same number of steps, so the relaxation effect is even less of a concern.

Figure 3.5 plots the effective softening length to the minimum softening needed (Miller 1971) to ensure local stability for the Zhang (1998, 1999) set of simulation parameters. It is seen that over the entire range of the computational domain the minimum softening requirement is satisfied by the set of simulation parameters.

3.3 Signature of Collisionless Shock in *N*-Body Spirals[†]

The condition of mild local gravitational instability at the spiral arms, which we had demonstrated in the previous chapter, indicates that the streaming disk material experiences something akin to true collisions during a spiral-arm crossing. Therefore, the analogy of the stellar random velocity to the sound speed of a gas is much better established at the location of spiral arms. In plasma physics, the term "collisionless shock" has been used to refer to the kind of shock which is associated with the particle-scattering process inside an instability front (Krall & Trivelpiece 1973; Balogh & Treumann 2013), in contrast to the kind of hydrodynamic shock where true particle collisions are happening. The spiral gravitational shock we will explore in this section is in essence akin to the plasma "collisionless shock."

As is well known, over most of the galactic radii the entry speed of the disk material into the spiral arms is supersonic[†]; and for spiral forcing strength greater than ~3%, the periodic orbits of stars at neighboring galactic radii are found to intersect

---

[†] Portions of this section used material previously published in Zhang (1996), reproduced with modifications from The Astrophysical Journal @ AAS. Reproduced with permission.

[†] Here we need to distinguish the effective sonic speed of stars (as well as gas clouds treated as particles), which is their mean velocity dispersion; and of gas when considering the atomic and molecular level of motion.

(Wielen 1975). Furthermore, the presence of the phase shift between the potential and density distributions means that the local state of the disk matter and the wave has to be in one of the two situations. First, if the wave pattern has reached a quasi-steady state, then the phase shift indicates a secular dissipation process as given by the torque integral in eq. (2.26). Second, if the local dissipation rate due to the collective instability is insufficient compared to that corresponds to the value of the phase shift, the wave has to change in shape (amplitude and/or profile). A natural way for the wave to change shape is for it to steepen into a nonlinear shock-like profile, so that shock dissipation can relieve some of the "stress" applied or required by the phase shift. These factors, together with the analogy to the steepening of nonlinear acoustic waves into shock waves, point to the formation of galactic-scale spiral shocks in the *stellar* medium, in addition to the known galactic-scale gaseous density wave shocks.

The calculation of galactic-scale spiral shocks, originally thought to exist *only* in the gaseous component, began with the work of Fujimoto (1968). The possibility of quasi-stationary gaseous spiral shocks of galactic scale was first demonstrated by Roberts (1969) in a non-self-consistent calculation, within the framework of WKBJ theory and with the self-gravity of the gas ignored. Shu et al. (1973) demonstrated that at least in the non-self-consistent case (i.e., with the self-gravity of the gas ignored from the forcing spiral potential, and with the gas modeled passively), the equilibrium flow solution *always* contains shocks as long as the forcing is more than a few percent. A finite-amplitude spiral potential modifies the velocity field of the galactic flow from that of entirely supersonic (or entirely subsonic) to that which contains sonic transitions, and a shock forms near the location where the supersonic flow velocity changes to subsonic velocity in the form of a sudden jump. The dissipation in the galactic shock solutions leads to a phase shift between the stellar and gaseous WKBJ solutions, which is only possible because these solutions are not self-consistent: since the WKBJ solutions cannot self-consistently admit a potential-density phase shift.

The time-dependent calculation of the formation and steepening of spiral shocks was first carried out by Woodward (1973, 1975), again ignoring the self-gravity of the gas and again using the WKBJ approximation. Later work on the nonlinear development of spiral structures, although still employing the WKBJ approximation, had incorporated the self-gravity of the stars (Shu, Yuan, & Lissauer 1985) and gas (Lubow et al. 1986; Lubow 1988). It is found that the self-gravity of the disk material generally makes the density peaks that formed near the spiral potential minimum more symmetric than that of the forced hydrodynamic shock (Shu et al. 1985; Lubow et al. 1986; Lubow 1988). It is also found that for fully self-gravitating nonlinear WKBJ waves, the streamlines of the fluid never cross one another (Shu et al. 1985), so presumably no dissipative shocks can form. This result is also hinted in the earlier nonlinear analysis of the self-consistent WKBJ waves by Vandervoort (1971). Nonlinearity by itself thus is insufficient to support a global potential-density phase shift.

From the above-mentioned results, we expect that shock formation could be a general characteristic of the nonlinear development of the galactic spiral waves even in stellar disks *as long as* the wave is somewhat *open* in morphology: since the tightly wound self-consistent WKBJ waves were shown not to contain galactic shocks. The presence of a phase shift between the potential and density of an open spiral pattern indicates that the streaming matter is locally only partially self-gravitating before entering the spiral arm, and this provides the possibility for the material to "unexpectedly" shock onto the potential wave as it crosses the spiral potential minimum. The phase shift allowed by the Poisson integral for *open* spiral structures also leads to the possibility for a *globally self-consistent galactic shock solution* since shock dissipation is now allowed in the global formulation of the entire dynamical equation set.

One of the possible ways to demonstrate the steepening of nonlinear spiral wave modes into spiral shocks is by the iterative solution of the set of nonlinear Eulerian fluid equations, as well as the equation of continuity and the Poisson integral, starting from a known linear spiral modal distribution for a given basic state, together with the proper inner and outer boundary treatment. In the following, however, we will adopt another route by employing the well-developed technique of N-body calculations. There is an added advantage for adopting the N-body approach, in that the viscosity due to the graininess of the particles is naturally incorporated into the simulation, closely resembling the situation in real stellar disks.

Zhang (1996) used a computation mesh (110 radial divisions, 128 azimuthal divisions) about twice as fine and a total number of particles (200,000) about four times as many as used in Donner & Thomasson (1994) in order to better resolve the shock structure. The time resolution is also increased correspondingly through a choice of $S_{stpcrt} = 100$.

In Figure 3.6 we plot the calculated density wave morphological evolution for a spiral mode first obtained in Zhang (1996), but re-run for this book with slightly different random number seed for initial condition setup, which appears to have had little effect on modal characteristics. Twelve selected time steps are shown, and 20,000 randomly selected particles are plotted in each frame. The time-step choice in this simulation is twice finer than that used in the simulations presented in Donner and Thomasson (1994), or in Zhang (1998, 1999), which we will describe later.

In Figure 3.7 we plot a close-up view of the spiral morphology at time step 3200 from the original Zhang (1996) simulation. Figure 3.8 plots the calculated potential-density phase shift versus radius at this time step. Here we observe that the phase shift is mostly positive in the inner disk, albeit with oscillations, in Figure 3.7[†], and is negative in the outer disk. It is reassuring to see that the main transition between the positive and negative phase shifts happens near the corotation radius of $r = 30$. To observe this kind of coherent phase-shift distribution, the spiral mode used would need to have achieved a high degree of organization throughout the disk. At time steps

---

[†] These oscillations are partly due to the truncation of the spiral at the outer disk, as we had demonstrated in Section 2.4.1, and partly due to the non-steady nature of the mode formed.

80 — 3 *N*-Body Simulations of Galaxy Evolution

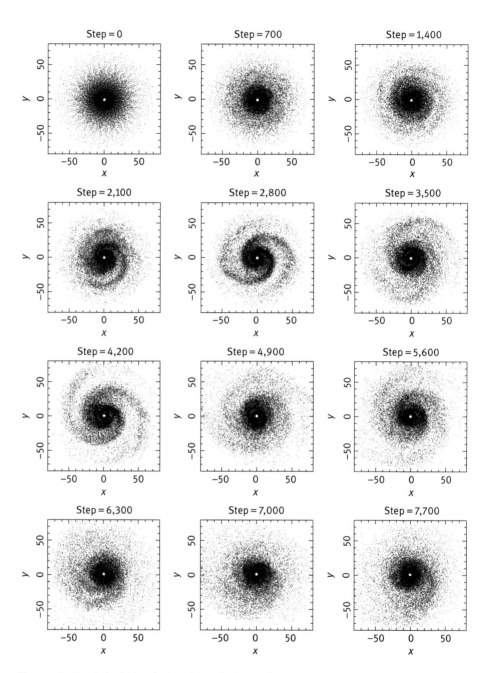

Figure 3.6: Morphological evolution of an *N*-body spiral mode first studied in Zhang (1996). Pattern speed is about 0.006 radian/step.

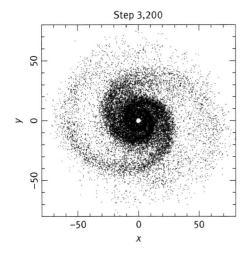

Figure 3.7: Close-up view of the spiral mode calculated in Zhang (1996) at time step 3200. Pattern speed is about 0.006 radian per step. Adapted from Zhang (1996).

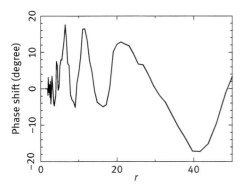

Figure 3.8: Phase shift versus radius plot for the spiral mode of Figure 3.7 at time step 3200. For this particular mode, the inner Lindblad resonance is near $r = 10$, the corotation radius is near $r = 30$, and the outer Lindblad resonance is near $r = 42$ (adapted from Zhang 1996).

other than 3200, as the spiral mode loses its coherence, the phase-shift distribution also becomes less regular.

We now focus on the main issue of our concern here, which is the formation and steepening of spiral gravitational shocks. In Figure 3.7, we could already discern some evidence for the presence of spiral shocks. Sharp density maxima are seen to be present at the leading edge of the spiral pattern in the inner disk (inside corotation), reminiscent of the narrow dust lanes found at the leading edges of the spiral arms of real galaxies, which are thought to represent the location of large-scale gaseous shocks. However, this simulation is purely particle simulation with no dissipative gas component. It is obvious that a pure particle disk can indeed display behavior normally thought to be only present in dissipative gaseous disks,

In Figure 3.9(a–f), we further plot the azimuthal distributions of surface density and negative potential, radial velocity dispersion, epicycle frequency κ, Toomre's Q parameter, and the velocity components parallel and perpendicular to the spiral arms. Here the parallel and perpendicular velocity components are related to our grid

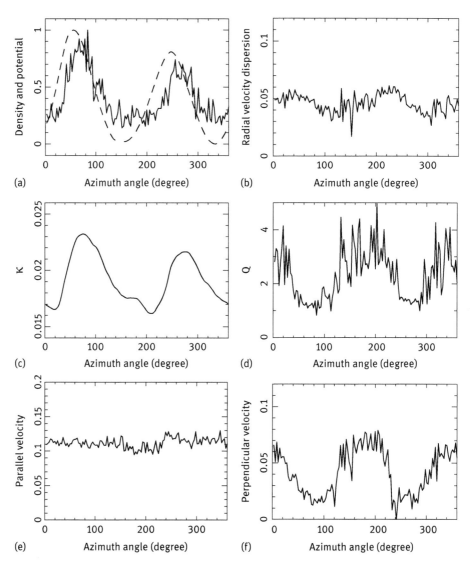

Figure 3.9: Spiral gravitational shock. Different frames show the azimuthal distributions of the following parameters: (a) surface density (solid line) and negative potential (dashed line). Here the density is normalized to have a maximum of 1. The potential has an arbitrary scale and is shifted in the vertical direction to be displayed on the same frame as the density. (b) Radial velocity dispersion. (c) Epicyclic frequency κ. (d) Toomre's Q parameter. (e) Velocity component parallel to the spiral arm. (f) Velocity component perpendicular to the spiral arm. The centers of the bins where these properties are averaged have a radius of 14.5. The time step is 3800 (adapted from Zhang 1996).

velocities in the radial and azimuthal direction v_r and v_ϕ, as well as to the spiral pattern speed Ω_p and pitch angle i through (Roberts 1969)

$$v_\perp = v_r \cos i + v_\phi \sin i - \Omega_p r \sin i, \qquad (3.11)$$

and

$$v_\parallel = -v_r \sin i + v_\phi \cos i - \Omega_p r \cos i. \qquad (3.12)$$

The pattern speed for this spiral mode is about 0.006 radians per time step (again the numerical value is about one half of that obtained in Donner & Thomasson 1994 because our time step is half as fine as theirs), and the pitch angle is ~16.8°. The distributions in Figure 3.9(a–f) are obtained by averaging the relevant characteristic for each individual computational spatial bin in an annulus centered around $r = 14.5$, at time step 3800.

From Figure 3.9(a), we see that the density profile is of the nonlinear shape similar to that found in the N-body simulations of gaseous spiral shocks of Levinson and Roberts (1981), with a maximum arm-interarm density contrast of 3 to 1 (for comparison, Levinson and Roberts obtained a density contrast of 2 to 1 in their gaseous simulations). Owing to the self-gravity of the disk material, the spiral shock here acquires a more symmetric shape, as was also found in the previous gaseous shock simulations. The potential profile is seen to be phase shifted from the density profile in the correct sense for a radial location inside corotation.

Figure 3.9(b–d) present the variations of σ_r, κ, and Q versus the spiral phase as expected. The radial velocity dispersion σ_r starts to decrease not long after entering the negative potential region (or positive half-cycle of the potential curve on this plot). The spiral phase where σ_r starts to decrease coincides with the phase where $v_\perp$ suffers a sharp downward jump (Figure 3.9(f)), with the value of $v_\perp$ going from supersonic (note that the equivalent sound speed in this case is around 0.04, as is indicated by Figure 3.9(b)) before the jump to subsonic after the jump. This clearly indicates the presence of a shock[†]. It has been verified that the shock actually caused the radial velocity component v_r itself to change sign near the location of the shock, which is partially responsible for the minimum of σ_r observed there. Thus we see that because of shock dissipation, the velocity dispersion σ_r at the spiral arm region is smaller than that given by the linear WKBJ theory. This causes Toomre's Q parameter to suffer a drastic decrease in the spiral arm region, to a value very close to (and sometimes smaller than) 1. This confirms our prediction that there should be a temporary local gravitational instability at the spiral arms. Since the presence of this instability is a result of the nonlocal (long-range) nature of the gravitational interaction, and is brought about by the relative phase shift of potential and density in an open spiral pattern, we expect the strength of the gravitational instability to correlate with the value of the phase shift, in a quasi-steady state. Similar behavior of Σ, σ_r, κ, and Q is also found for other radial locations inside corotation.

[†] In any realistic physical systems which contain dissipation, the transition from supersonic to subsonic flow can *only* be accomplished by a shock (see, e. g., Woodward 1975; Shu 1992, p. 77).

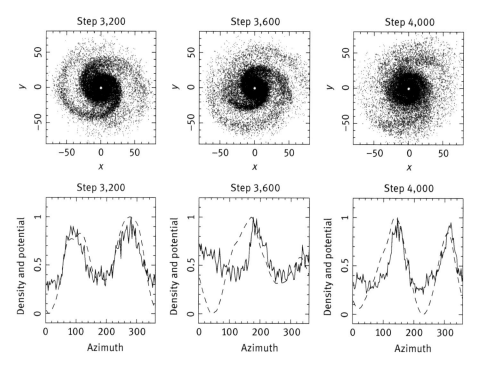

Figure 3.10: Steepening of spiral shock with time. Top three frames: surface morphology evolution. Bottom three frames: azimuthal density (solid line) and negative potential (dashed line) profiles at radius $r = 15$. The circle in the first frame indicates the radial location where the bottom azimuthal profiles are plotted. For the bottom three frames, the density is normalized to have a maximum of 1, the potential has arbitrary scale, and is shifted in the vertical direction (adapted from Zhang 1996).

In Figure 3.10 we plot the disk morphology at three different time steps, as well as the azimuthal profiles of the grid density and potential at radius $r = 15$ (the location of the circle in the first frame) for the corresponding time steps, to illustrate the shock steepening process and its relation to the local phase-shift value. The density profile is seen to steepen with time from a more sinusoidal profile to a shock-like nonlinear profile. It is also seen from Figure 3.10 that the shock strength increases as the local phase shift increases, although there seems to be a time lag between the moment when the phase shift is largest (bottom middle frame) and the moment when the phase shift succeeded in compressing the density distribution into the narrowest nonlinear profile (bottom right frame). The shape of the density distribution in the middle frame, however, has the best resemblance to the classical hydrodynamic shock, where there is a very steep rising edge followed by a much more gradual falloff.

The time lag observed in Figure 3.10 between the moment of the largest phase shift and the moment of narrowest density distribution is not unexpected. Only when a (nonlinear) spiral mode reaches a quasi-stationary stage can we expect an exact correspondence between the local phase-shift value and the local shock strength, whereas the state of the art of N-body simulations is still such that the nonlinear profile of a spiral mode cannot yet be made quasi-stationary, even though the underlying $m = 2$ mode is indeed found to have achieved a quasi-stationary amplitude in both the Donner & Thomasson (1994) simulation and in our current simulation. Despite the inadequacy of the current N-body simulations in producing long lasting large-scale spiral shock distribution, we do continue to believe that real galaxies can do so for much longer than the local dynamical time scale, just from the number of observed spirals which have well-defined grand design spiral patterns.

Due to the presence of the local gravitational instability at the spiral arms, the width of the spiral shocks is effectively the size of the instability structure formed, that is, on the order of 1 kpc for a galaxy like our own, instead of the free particle mean free path. In other words, whenever there is a local gravitational instability, the range of influence of the gravitational scattering potential is no longer restricted to that around the individual particle itself but rather around the collection of particles that form the instability. Such views have already been expressed in Kulsrud (1972). The observed narrow width of the dust lanes in the leading edge of many physical spirals, on the other hand, may very well reflect the size of the gas cloud mean free path in the spiral instability, which could be smaller than the stellar mean free path there if the two fluids are not completely coupled.

The signature of spiral shocks in the stellar component has also been found in observations (Elmegreen & Elmegreen 1989), where it is noticed that for many spiral galaxies the leading edge of the spiral arm is generally sharper than the trailing edge, and the shape of the wave in both the blue and the near-infrared band has the appearance of a water wave on the verge of breaking up.

3.4 Modal Nature of a Spontaneously Formed Pattern[†]

During the past few decades' work on the density wave theory of spiral structure, one of the central issues of interest (and of controversy) has been the longevity of spiral patterns in disk galaxies (see, e.g., Binney & Tremaine 2008). Although the statistical studies of grand design spiral galaxies in groups show evidence that spiral patterns are long-lived (Elmegreen & Elmegreen 1983, 1989), the mechanism through which this longevity is achieved has not been firmly established. In recent decades, it has gradually become apparent that the long-lived spiral patterns in disk galaxies are normal modes of the underlying disks (see, e.g., Bertin et al. 1989a, 1989b and the references therein). However, there remains the question of how to offset the violent growth tendency of a large fraction these spiral modes to obtain modes with

---

[†] Portions of this section used material previously published in Zhang (1998), reproduced with modifications from The Astrophysical Journal @ AAS. Reproduced with permission.

amplitudes that are approximately steady. In the last chapter, we have presented a dynamical mechanism whereby the wave modes can remain quasi-steady as a result of the competition of the opposing dynamical processes of spontaneous growth of the wave mode, and irreversible channeling of the wave energy and angular momentum onto the basic state of the galactic disk. We will demonstrate the modal longevity that can be achieved through this new dynamical mechanism in the following simulations. Later on, in Chapter 5, we will also describe a new criterion and practical approach for judging the quasi-steady state of the wave mode, that turn out to be closely related to Kolmogorov's theory of fully developed turbulence in the inertial range.

In the N-body simulations described in this section, which were first performed in Zhang (1998) but here re-run with a slightly different random-number seed for initial-condition assignment (which should only affect each individual particle's orbit but not the modal characteristics), we have used a polar grid of 55 radial divisions and 64 azimuthal divisions (i. e., coarser than used in Zhang (1996)). The time step used corresponds to 314 time steps per galactic rotation period at radius 20. About 100,000 equal-massed particles are used to represent the active disk component. A gravity softening length of 1.5 times the grid length unit is used.

The basic state specification is similar to that used in the last section, with a modified exponential disk component, and a rigid bulge and rigid halo components. The initial value for the stability parameter Q is set to 1.1. However, in the current section, the mass ratio of disk:halo:bulge is 0.4:0.5:0.1 (as in Zhang (1998, 1999)), instead of 0.5:0.4:0.1 as that used in the last section or in Zhang (1996). This reduced active disk mass avoided the early central bar formation in the previous simulation, and this allowed the spiral mode to survive for a much longer time.

In the different frames of Figure 3.11, we plot the evolution of the disk morphology at various time step. Note that for this run a time-step range of 2000 would correspond to a time-step range of 4000 in Figure 3.6 because of the factor of 2 coarser time resolution here. About 20,000 randomly chosen particles are plotted in each frame. The spiral pattern is seen to have kept the global coherence throughout the period from time step 1000 to 4500, which is about six pattern revolutions (the pattern speed is about 0.011 radian per time step, as compared to the 0.006 radian per time step in the Figure 3.6 simulation mainly because of the factor-of-2 difference in spatial and temporal resolution since the difference in basic state specification is minor). The inner spiral pattern inside the corotation radius is seen to have maintained the longest period of pattern coherence.

At larger time steps, the nonlinear spiral pattern gradually fades away due to the heating of the disk stars. However, Fourier analysis shows that the underlying $m = 2$ mode retains nearly constant amplitude, after the initial exponential growth, until the very end of the simulation run (time step 8000), which is about 14 pattern revolutions (Figure 3.12). In Section 3.6.2, we will show that the heating of the disk stars in the N-body simulation is much exaggerated due to the small number of particles used compared to that which is present in a physical galaxy. The amount of this *excess heating* due to the limitations of the current generation of N-body simulations can be

3.4 Modal Nature of a Spontaneously Formed Pattern — 87

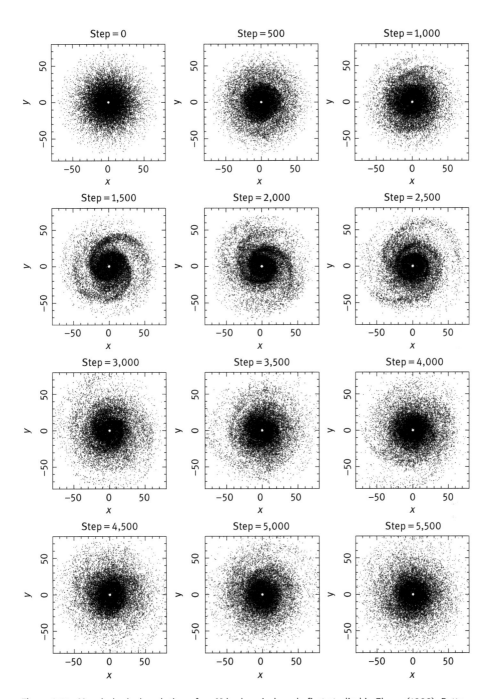

Figure 3.11: Morphological evolution of an *N*-body spiral mode first studied in Zhang (1998). Pattern speed is about 0.011 radian/step.

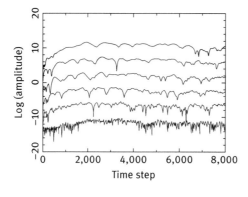

Figure 3.12: Amplitude evolution of the $m = 2$ component of the spiral mode in Figure 3.11. From bottom to top, the different curves are for radius 10, 20, 30, 40, 50, and 60, respectively. The curves are shifted in the vertical scale by multiples of 5 with respect to the $r = 10$ curve (adapted from Zhang 1998).

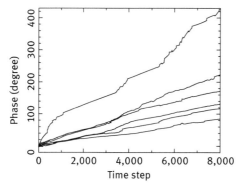

Figure 3.13: Phase evolution of the $m = 2$ component of the spiral mode in Figure 3.11. From top to bottom, the different curves are for radius 10, 20, 30, 40, 50, and 60, respectively.

calculated, and true heating due to the spiral pattern alone can be obtained after a simple rescaling of the N-body results. We will then see that for many observed disk galaxies the nonlinear spiral shock patterns can survive on the order of a Hubble time.

The pattern speed of the $m = 2$ component is also found to be nearly constant throughout the simulation run. In Figure 3.13, we show the phase evolution of the $m = 2$ component at six different radii in the same set of simulation as we had shown for the amplitude. We see that in the middle range of the radius values $r = 20$–50 (recall that the corotation radius for this mode is located at $r = 30$ and the outer Lindblad resonance is at $r = 42$, whereas the inner Lindblad resonance is close to $r = 10$), the phase-versus-time curves for the different radii have nearly the same constant slope, indicating a nearly rigidly rotating underlying $m = 2$ modal pattern, confirming what we had found through the morphological study before. The topmost curve for $r = 10$ appears to correspond a nested inner mode that rotates faster than the outer dominant mode. The degree of the non-parallelness in the lower curves indicates a mild winding tendency for the dominant mode.

In Figure 3.14, we plot the phase versus radius of the $m = 2$ component for the above spiral mode at six different time steps. It is seen that in the intermediate radial range ($r = 20$–50) this mode retains slowly varying pitch angle throughout the simulation run, as reflected in the nearly constant slopes of the curves in the intermediate radial range, confirming our visual impression of the modal morphology. The

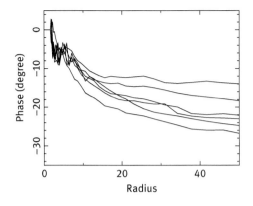

Figure 3.14: Phase versus radius of the $m = 2$ component of the spiral mode in Figure 3.11, for time steps 1000, 2200, 3400, 4600, 5800, and 7000, respectively.

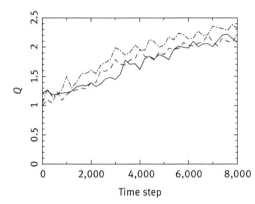

Figure 3.15: Evolution of the Q-parameter at radius $r = 20$ (solid), $r = 30$ (dash), and $r = 40$ (dash-dot), respectively, with time for the N-body mode of Figure 3.11 (adapted from Zhang 1998).

inner region though ($r \leq 10$) appears to have a nested inner mode having a more tightly wrapped pattern (steeper slope of the phase versus r), conforming our observation from the phase-versus-time signature that we have seen previously in Figure 3.13 top curve.

In Figure 3.15, we plot the evolution of Toomre's Q parameter (Toomre 1964). It can be seen that the Q parameter is monotonically increasing during the entire simulation run, and the stabilization of the wave amplitude after time step 1200, observed in Figures 3.11 and 3.12, has no observable correlation with the increase of the Q parameter, which measures the degree of disk heating in this case. In fact, at step 1200 when the wave amplitude saturates, the disk is still rather cool.

We can thus make the following observations about the amplitude saturation process of this N-body spiral mode:

- The amplitude stabilization is not due to the dissipation in the gas component since no such component is included in the current simulation.
- The amplitude stabilization is not due to the excess heating of the disk stars either, since the equilibrium amplitude is reached when the disk is still rather cool, and the subsequent continued heating of the disk does not significantly change the equilibrium amplitude.

90 — 3 N-Body Simulations of Galaxy Evolution

- The amplitude stabilization is not due to the channeling of wave energy into multi armed spiral modes, since the spiral pattern remains predominantly two-armed throughout the simulation run.
- The presence of a rigid halo component did not stop the initial exponential growth of the wave mode, and the subsequent wave amplitude saturation could not be due solely to the presence of the halo component.

It is thus most natural to explain the amplitude stabilization process through the nonlinear damping mechanism we had presented in the last chapter.

3.5 Qualitative Signature of Secular Mass Redistribution[†]

In virtually all of the N-body experiments which formed spiral patterns, there is a tendency for stars to migrate in their mean radius. In cases where a spiral mode is

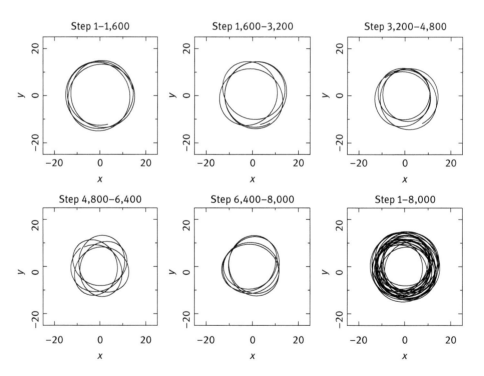

Figure 3.16: Evolution of orbit trajectory for a typical star inside corotation, for the N-body mode of Figure 3.6. Pattern speed is 0.006 radian/step.

[†] Portions of this section used material previously published in Zhang (1996), reproduced with modifications from The Astrophysical Journal @ AAS. Reproduced with permission.

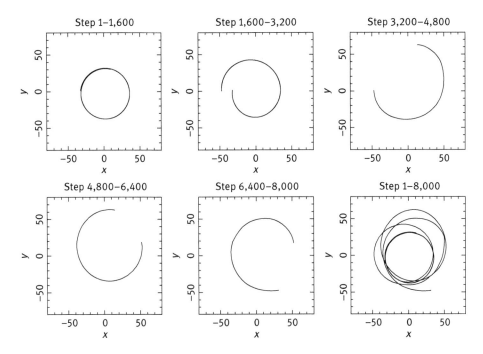

Figure 3.17: Evolution of orbit trajectory for a typical star outside corotation for the N-body mode of Figure 3.6. Pattern speed is 0.006 radian/step.

formed, the particles are found to accrete inward inside corotation and excrete outside corotation (Donner & Thomasson 1994). This same trend is also observed in our own N-body results.

In Figures 3.16 and 3.17, we have plotted the different segments of a typical orbit inside corotation and a typical orbit outside corotation, respectively, both obtained from the re-run of the N-body simulation of Zhang (1996) with a different random number seed. Note that the force interpolation scheme used in the N-body simulation would smear out any small scale "kinks" in the orbit that could be produced by the local gravitational instability at the spiral arms. These kinks are expected to be very gentle in any case for real galaxies since they are produced by small-angle scatterings, with the particle mean free path for scattering on the order of 1 kpc. The cumulative effect of these small-angle scatterings nevertheless survives, as is reflected in the often sharp change of the orbital orientation, observable especially in Figure 3.16, which is reminiscent of the sharp-pointed oval-shaped streamlines in a gaseous spiral shock (Roberts 1969, Figure 4). The dissipation effect of the spiral shock is also revealed through the secular decrease (or increase) in the mean orbital radius, inside and outside of the corotation radius, respectively.

In Figure 3.18, the corresponding frames for the disk surface density are plotted. The figure clearly demonstrates that there is a secular increase in the disk surface density in the inner disk region, together with a slight density increase in the outer disk, consistent with the trend shown in Figures 3.16 and 3.17 for the mean orbit

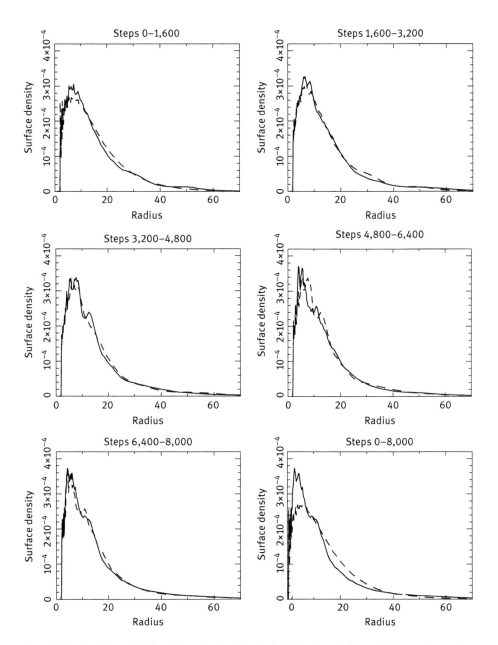

Figure 3.18: Evolution of disk surface density. The dashed line in each frame indicates the surface density at the earlier time step, and the solid line the later time step (adapted from Zhang 1996).

evolution. The surface density at the corotation radius $r_{co} = 30$ is also decreasing with time because of both the inflow of matter at smaller r and outward of matter at larger r away from corotation. Note also that the second density peak on some of the frames in Figure 3.18 near $r = 10$ is due to the temporary accumulation of disk matter near the inner Lindblad resonance.

The amount of surface density displayed in Figure 3.18 is not sufficient to transform the Hubble types of a galaxy significantly over the time span of the simulation (about seven pattern rotation periods). This will later be shown to be related to the choice of the softening parameter. We will address this issue further in Section 3.9 when discussing the second-generation N-body tests.

3.6 Longevity of the Spiral Modes[†]

We have argued in the previous chapter that a quasi-stationary spiral mode can be sustained by the dynamical balance of its spontaneous growth tendency and a local dissipation process, as well as by a flux of matter, energy and angular momentum through the system. Therefore in principle, a spiral pattern can remain quasi-stationary as long as this flux of matter and energy can be maintained. However, we have also shown that the source and sink of this flux lie in the basic state of the galactic disk, so the price to pay for maintaining a quasi-stationary spiral structure is the continuous evolution of the basic state. The question of "how long the spiral patterns in disk galaxies can remain quasi-stationary" thus becomes how long it takes the basic state to change significantly so that it will no longer be able to support a given spiral pattern.

In the following, we first calculate the expected mass inflow rates using analytical expressions derived in the last chapter, and apply these to the stellar N-body spiral mode we have calculated in Section 3.4. We will then compare this expected mass flow rate against the actually measured basic state evolution rates in the same N-body calculations. A confirmation of the evolution rates serves to validate both our analytical expressions and the N-body approach itself. Once such confirmation is made, we could use the N-body results to infer the lifetime of spiral patterns in physical disk galaxies.

3.6.1 Faithfulness of the *N*-Body Simulations

In this subsection we will use the results of N-body simulations to confirm the analytical results pertaining to the secular evolution rates. For this purpose, we first need to address the issue of how faithfully the N-body technique can model the secular evolution of the disks of spiral galaxies.

Since the number of particles used in N-body simulations is usually many orders of magnitude less than what is actually present in physical galaxies, much attention has been paid to the issue of whether there is exaggerated rate of relaxation due to the small number-of-particle effect. In a two-dimensional disk, it is found that even

---

[†] Portions of this section used material previously published in Zhang (1998), reproduced with modifications from The Astrophysical Journal @ AAS. Reproduced with permission.

though the two-body relaxation time scale is comparable to the local dynamical time scale for a razor-thin disk (Rybicki 1971), when properly chosen softened gravity and mesh-grid are used, the two-body relaxation becomes negligible over the time scale of interests, at least in cases where the disk is stable against large-scale gravitational instabilities (Hohl 1973; Romeo 1990, 1994a).

However, as has been cautioned by various authors (e. g., Sellwood 1987; Weinberg 1993), the emergence of large-scale gravitational instabilities could significantly alter the picture. In the case of the two-dimensional modeling of spiral galaxies conducted in the 1990s (when the first-generation N-body tests described here were conducted), which generally employed 50,000–100,000 particles, significant heating associated with the emergence of spiral structure can always be observed, and the spiral patterns formed quickly decay as a result of this self-induced heating. Since the calculated two-body relaxation time scale in such cases is always longer than the time span of the simulation run, this heating effect in the N-body simulations has been extrapolated by many as indicating that spiral patterns in physical galaxies cannot be long-lived since their self-induced heating always leads to their self-destruction.

In the following, we will demonstrate that the rate of heating observed in the N-body spiral disks is much larger than in physical galaxies, and this rapid heating is produced partly by the particle scattering process at the spiral arms. Since this scattering process is an integral part of the spontaneous formation of spiral pattern, the N-body simulations of unstable spiral modes (as opposed to tidal spirals in originally stable disks) can *never* be made truly collisionless. This seems to suggest that we are out of luck in trying to use a small number of particles to simulate the evolution of a physical spiral disk since in the presence of particle scatterings the number of particles used *will* matter. We show in the following that the situation is in fact a lot more optimistic. It turns out that the N-body simulations using the presently employed number of particles can accurately model the angular momentum evolution (or the mass distribution evolution) of the basic state. It is also shown that indeed the small number of particles used exaggerates the heating of the disk. But through a proper rescaling of the excess heating rate, using the dependence of the Poisson noise on the number of particles, we can also obtain an estimate of the actual heating rate in the physical disk galaxies.

Another caution often raised by the previous practitioners of the N-body technique is the large errors often found in the individual particle's trajectories. A calculated orbit generally diverges exponentially from the true orbit in the presence of small random error (Miller 1967). Furthermore, the kind of N-body system which admits a spontaneous global instability structure, such as a spiral pattern or a bar, is known to show even stronger level of such dynamical instability, which, as we have stated before, is due to the collective dissipation process associated with these global structures. Collisions or close encounters are known to increase the dynamical instability of an N-body calculation (Miller 1967).

In the face of such dynamical instability, one wonders how and in what sense our N-body calculation could simulate a realistic physical system. We point out here that if our aim is to model the global behavior of an instability structure, the sensitivity of individual orbits to random errors need *not* worry us at all. This is because the very same collective dissipation process associated with such a global instability makes such nominally Hamiltonian systems *effectively* a dissipative system. As is well known, the large-scale structures formed in dissipative systems possess asymptotic stability, and are *insensitive* to the details of small random errors. This asymptotic stability is responsible for the faithfulness of the N-body simulations in modeling the macroscopic properties of large-scale instability structures in nonequilibrium systems.

3.6.2 Three Test Runs Using Different Numbers of Particles

In this subsection, we present the result of the N-body simulations of a spontaneously formed spiral mode, first presented in Zhang (1998). This simulation uses the same basic state as that in the earlier stellar simulations presented in Section 3.4, but employ 100,000, 200,000, and 300,000 particles, respectively, to represent the active disk component, for the three simulation runs. It was confirmed that the global morphologies of the spiral mode formed are very similar in all three cases, even though the spiral pattern in the smaller number-of-particle case emerges a little earlier and fades a little sooner. Fourier analysis shows that the pattern speeds and equilibrium amplitudes of the $m = 2$ spiral mode formed are also similar for the three cases.

When plotting out the evolution of the basic state parameters, we found that the circular velocity v_c and the surface density Σ have very similar evolution behavior for all three cases. The values of radial velocity dispersion v_r, however, show marked difference. In particular the stars in the smaller number-of-particle run are found to heat up significantly more than those in the larger number-of-particle run. Another observation is that the heating effect is the smallest near the corotation radius for all the cases, consistent with the behavior of spiral modal heating as derived in Section 2.6.2.

In order to quantify the above observations, we now calculate the expected radial mass accretion rate and heating rate at a particular radius inside corotation, for the spiral patterns obtained in these simulations, and compare them to that actually observed in each case.

We begin by choosing an arbitrary radius of $r \approx 13.5$ (in the normalized units of Thomasson 1989) to verify the radial mass accretion eq. (2.112). We found that at this radius $v_c = 0.4$, $F = 0.1^\dagger$, $i = 17°$, and $\phi_0 = 5°$. Therefore we obtain dr_*/dt (theoretical) $= -0.0001$ (per time step), or a $\Delta r = -0.3$ in 3000 time steps, from eq. (2.112). This is obviously a rather small rate of radial drift due to the small equilibrium

---

† The value of F is calculated by first calculating the fractional potential wave amplitude F_ψ and density wave amplitude F_Σ, and demanding that $F^2 = F_\psi \cdot F_\Sigma$. In the current case, at $r = 13.5$, we found that $F_\psi = \frac{kA_\psi}{\Omega^2 r} = 0.04$, where $A_\psi \equiv \frac{\sqrt{\int_0^{2\pi}(\psi - \overline{\psi})^2 d\phi}}{\sqrt{2\pi}}$; and $F_\Sigma = \frac{A_\Sigma}{\Sigma_0} = 0.25$, where $A_\Sigma \equiv \frac{\sqrt{\int_0^{2\pi}(\Sigma - \overline{\Sigma})^2 d\phi}}{\sqrt{2\pi}}$. Small typos in Zhang (1998) for the A_ψ and A_Σ expressions are here corrected.

amplitude for this particular spiral mode. This makes the direct comparison to the actually observed orbital decay rate in the simulations rather difficult. So instead, we calculate the expected rate of increase of total mass inside the radius $r = 13.5$, as a result of the stellar orbital decay,

$$\frac{dM}{dt}(r \le 13.5)_{\text{(theoretical)}} = \left(-\frac{dr_*}{dt}\right)_{\text{(theoretical)}} \cdot 2\pi r \Sigma(r)|_{r=13.5}. \quad (3.13)$$

Now the local surface density at $r = 13.5$ is about 1.5×10^{-4}. Therefore we obtain that the expected mass increase rate is $\frac{dM}{dt}(r \le 13.5)_{\text{(theoretical)}} \approx 1.2 \times 10^{-6}$.

We then integrate the actual surface density obtained in the N-body simulations from $r = 0$ to $r = 13.5$, and obtain the $M(t)_{\text{(measured)}}$ for the three runs with different numbers of particles, which we plot in Figure 3.19. Immediately we see that the long-term trends of variation of the surface density in these three runs are very similar. From the slope of these curves, we obtain $dM/dt(r \le 13.5)_{\text{(measured)}} \approx 10^{-6}$ between time step 3000 and 6000, which is quite close to the theoretical prediction (3.13). This gives us confidence that the theoretical radial orbital decay rate (2.112) gives a good description of the actual mass accretion observed in the N-body simulations. The close agreement of the three curves in Figure 3.19 also shows that N-body simulations using small number of particles can quite accurately model the angular momentum evolution of the basic state induced by a galactic spiral structure.

Next, we look into the issue of disk heating in the N-body experiment. Again we consider the disk region near radius $r = 13.5$. We have that $\Omega \approx 0.029$, $\Omega_p \approx 0.01$, as well as $F \approx 0.1$, $i \approx 17°$, $\phi_0(r = 13.5) \approx 5°$, $v_c(r = 13.5) = 0.4$ as before. This gives us $D^{2d}(\text{theoretical}) \approx 1.6 \cdot 10^{-6}$ according to eq. (2.119)[†]. Therefore $D^{1d}_{\text{(theoretical)}} = 0.8 \cdot 10^{-6}$.

In Figure 3.20, we plot the actual evolution of the mean-square radial velocity for a group of stars near $r = 13.5$[††]. The first thing we notice is that the three curves

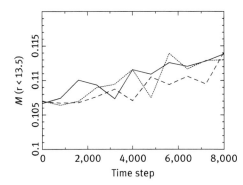

Figure 3.19: Evolution of the total mass within $r = 13.5$. Solid: $N = 100,000$ run; Dashed: $N = 200,000$ run; Dotted: $N = 300,000$ run (adapted from Zhang 1998).

[†] Note that since we are here performing a 2D simulation, the overall energy input from the spiral is distributed in two dimensions, rather than in three dimensions as is the case for physical galaxies.

[††] In the actual N-body calculation, since the particle numbers are assigned sequentially starting from the inner disk, the particles used for calculating the mean-square radial velocity are those with serial numbers between $N/4$ and $N/4 + N/200$.

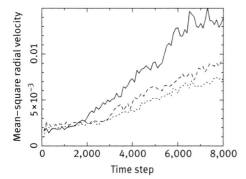

Figure 3.20: Evolution of mean-square radial velocity for $N = 100{,}000$ (solid), $N = 200{,}000$ (dashed), and $N = 300{,}000$ (dotted) runs. About 500 particles are used in deriving the mean-square radial velocity in the first case, 1,000 particles in the second case, and 1,500 particles in the third case, all have the same mean radius of $r = 13.5$ (adapted from Zhang 1998).

are no longer close to one another as was the case for the total mass evolution, but rather diverge from one another as the time step increases. Furthermore, the trend of variation of the heating curves, as we increase the number of particles, does not tend to zero as would be expected if the heating is entirely due to the particle Poisson noise, but rather to a finite slope or a finite heating rate. We expect that this converged heating rate is equal to that given by the theoretical heating rate $D^{1d}_{(\text{theoretical})}$, which is the systematic heating due to the spiral pattern (Section 2.6.2).

We now try to verify this speculation. During the time period between step 3000 and step 6000, the spiral pattern in all the runs are found to have similar amplitudes. The measured heating rates, however, give $D^{1d}(N = 100{,}000) = 2.3 \cdot 10^{-6}$, $D^{1d}(N = 200{,}000) = 1.5 \cdot 10^{-6}$, and $D^{1d}(N = 300{,}000) = 1.3 \cdot 10^{-6}$. Using the theoretical value of $D^{1d}_{(\text{theoretical})} = 0.8 \cdot 10^{-6}$ gives the amount of heating which is produced by the spiral pattern itself, we arrived at that the amount of *excess heating* due the particle Poisson noise, $\Delta^{1d} \equiv = D^{1d} - D^{1d}_{\text{theoretical}}$, is $\Delta D^{1d}(N = 100{,}000) = 1.5 \cdot 10^{-6}$, $\Delta D^{1d}(N = 200{,}000) = 0.7 \cdot 10^{-6}$, and $\Delta D^{1d}(N = 300{,}000) = 0.5 \cdot 10^{-6}$, respectively. We see right away that the amount of excess heating in the three cases satisfy

$$\frac{\Delta D^{1d}(N_1 = 100{,}000)}{\Delta D^{1d}(N_2 = 200{,}000)} \approx \frac{N_2}{N_1} \approx 2, \qquad (3.14)$$

and

$$\frac{\Delta D^{1d}(N_2 = 200{,}000)}{\Delta D^{1d}(N_3 = 300{,}000)} \approx \frac{N_3}{N_2} \approx 1.5, \qquad (3.15)$$

which is just the expected behavior of the Poisson random noise[‡]!

Therefore, we arrive at the conclusion that besides incorporating the intrinsic heating due to collective dissipation in the large-scale spiral pattern itself, the N-body

---

[‡] Note that since D is the rate of increase of mean-square velocity, it has a $1/N$ dependence on the number of particles instead of a $1/\sqrt{N}$ dependence.

simulation introduces *excess heating* due to the small number of particles used. It is this excess heating that scales with the particle number N as the behavior of Poisson noise. The collective heating effect is independent of the particle number N.

3.6.3 Lifetime of a Spiral Pattern Inferred from N-Body Simulations

For physical spiral galaxies, the number of stars present is on the order of 10^{11}, therefore the scaling relation we had in the last subsection would give $\Delta D^{1d}(N = 10^{11}) = 7 \times 10^{-13}$, in the normalized units of the simulation. Therefore, $\Delta D^{1d}(N = 10^{11})/D^{1d}_{\text{(theoretical)}} = 8 \times 10^{-7}$, which shows that in physical galaxies, heating due to the random fluctuations of individual stars is negligible compared to that due to a quasi-stationary spiral structure. Recall that the $m = 2$ mode in all of our N-body simulations has survived throughout the simulation runs. This indicates that the (nonlinear) spiral mode would be prominent if not because of the (exaggerated) large epicycle radius of the particles towards the end of the simulation runs. We can therefore estimate the lifetime of the nonlinear spiral patterns in physical galaxies from the lifetime of our N-body spiral patterns, through a proper rescaling of the excess heating effect due to the small number of particles. For example, the spiral shock pattern (especially the part inside corotation) for the run using 200,000 particles was found to last about 10 revolutions (6000 time steps) with the exaggerated heating rate of $D^{1d} = 1.5 \times 10^6$. Therefore with the realistic heating rate $D^{1d}(\text{theoretical}) = 0.8 \times 10^6$, the spiral pattern in physical galaxies could last (but may gradually tighten up in pitch angle, as well as develop nested nuclear patterns) for at least 20 revolution periods if the secular mass redistribution effect is not too drastic[†]. This is usually comparable to 2/3 of the Hubble time, which is also close to the time scale whereby the dominant disk galaxy-type changes from Sc to S0 in groups and clusters. The secular evolution induced by the galactic spiral structure causes both the central concentration of stellar mass, as well as the increase in velocity dispersion of stars. In a quasi-equilibrium state the two should satisfy the virial relation.

The result of our N-body simulations therefore indicates that the spiral patterns in physical galaxies *could in principle* survive on the order of a Hubble time, with the exact lifetime of an individual spiral galaxy depending upon its total mass, as well as on its environment, both of which influencing the type of spiral patterns present and therefore the effective evolution speed of the basic state, which depends on the pitch angle and the equilibrium amplitude of the spiral mode. The variation of the evolution speeds for galaxies with different initial and environmental conditions in

---

[†] The actual heating rate in physical galaxies is in fact even smaller, when we realize that in a realistic disk the amount of random energy input due to the spiral pattern is distributed in all three spatial dimensions, instead of in the two dimensions as in our 2D simulations. The net heating rate in physical galaxies can also be lowered through gas cooling.

fact is responsible for us being able to observe a whole spectrum of galaxies of differing Hubble types in the nearby universe, as well as the correlation of the average Hubble types with the environment. Furthermore, at the advanced stage of a disk galaxy's secular evolution, the type of spiral modes present is usually tightly wound and has small equilibrium amplitude, so the speed for further evolution is reduced, which means that the spiral pattern could survive even longer than if the pattern is a vigorous one throughout its lifetime.

We thus see that the result of our current work finally vindicates the quasi-stationary-spiral-structure hypothesis of Lindblad (1963) and Lin & Shu (1964, 1966) made half a century ago, though the nature of these quasi-steady waves is now shown to be unstable modes in a dynamical equilibrium state, rather than the neutral waves as were first conceived.

3.7 Role of Gas[†]

It was the dust lanes at the inner edges of grand design spirals that first impressed many about the potential importance of gaseous shock waves of galactic scale on the dynamics of galactic density waves (Roberts 1969). Many had placed hopes for the damping of the violently unstable stellar density waves on the dissipative gaseous density waves (Kalnajs 1972; Roberts & Shu 1972; Bertin et al. 1989a, 1989b). In these early studies, as a result of the dissipation of gas under the applied stellar potential, an azimuthal phase shift between the gaseous mass response and the forcing stellar potential was observed. This occurs largely because the studies were not done self-consistently (i. e., the gas was modeled as passively responding to the forcing stellar potential). When the self-gravity of the gas was considered together with the stars, the phase shift between the gaseous and stellar potential was found to become insignificant (Lubow, Balbus, & Cowie 1986; Balbus 1988).

In the earlier discussions of this chapter, we have seen that an N-body quasi-stationary spiral mode can form without a separate dissipative gas component. However, one still wonders what role the interstellar medium, if present together with the stars, plays in a physical galaxy. We show in the following that in such a case stars and gas play more or less parallel roles in limiting the wave amplitude growth. In particular, we argue that even gas clouds contribute to the wave damping process mostly through participating in the density wave collisionless shock that the stars are supporting.

We first present the result of a two-component (star plus gas) N-body simulation in a two-dimensional disk geometry, with the gas component going through inelastic collisions when approaching each other at a short distance (Zhang 1998). The details of the N-body gas collisional cooling treatment can be found in Thomasson (1989).

[†] Portions of this section used material previously published in Zhang (1998), reproduced with modifications from The Astrophysical Journal @ AAS. Reproduced with permission.

We have chosen to use a gas cloud collisional cross section of half of the mesh length unit L and a coefficient of restitution of 0.5 (i. e., the approaching gas clouds, when their separation is smaller than a prescribed value, lose about 1/2 of their kinetic energy after the "collision" process). We use 25,000 particles to represent stars and another 25,000 particles to represent gas clouds, with the two components together make up the active disk mass. The star and gas "particles" are chosen to have the same mass. The computation mesh as well as the basic state used are also identical to that used in the stellar simulations presented earlier in Section 3.4 demonstrating the long-lasting mode.

In Figure 3.21, we plot the stellar and gaseous morphologies at three different time steps. We observe that the gas clouds are condensed into much narrower spiral arms than stars and show filamentary structures. A "twin-peaks-and-bar" morphology near the inner Lindblad resonances, similar to that observed for galaxy NGC 3627 (= M66, Zhang, Wright, & Alexander 1993; see also Section 4.11), is seen to develop at time step 2400. This inner nested modal pattern is also reflected in the inner oscillations on the phase-shift plot we will be presenting next.

In Figure 3.22, we plot the calculated phase shifts of the stellar and gaseous spiral densities with respect to a common spiral potential, at time step 1600. We observe

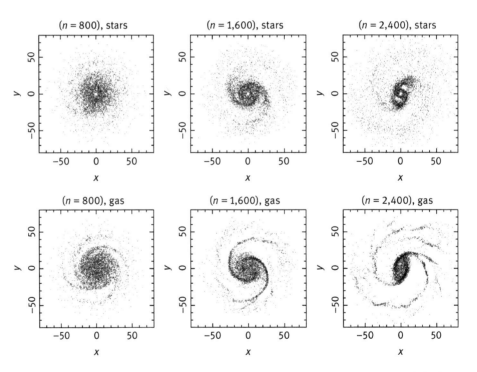

Figure 3.21: Star and gas morphologies at three different time steps. The simulation and the basic state parameters are described in Section 3.2 of the text. About 5000 randomly chosen particles are plotted in each frame (adapted from Zhang 1998).

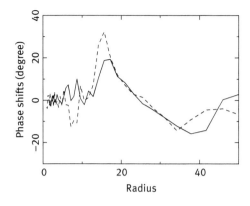

Figure 3.22: Potential and density phase shift for the stellar component (solid) and the gas component (dash), respectively, for the spiral mode of this section at time step 1600 (adapted from Zhang 1998).

that the two phase shifts have similar profiles, though the amplitude of the phase shift for gas is slightly larger than for stars. However, the relative phase shift between the stellar and gaseous densities is obviously much smaller than the phase shift of each component with respect to their common potential. Since at the quasi-steady state the potential and density phase shift of each component indicates the contribution of each component to the dissipative damping of the potential wave, the result in Figure 3.22 shows that stars and gas play parallel roles in limiting the wave amplitude growth, and the stellar wave is not damped solely at the expense of gas. A plot of surface density evolution of the two components (not shown here) reinforces this conclusion.

How then was the idea of using gaseous waves to damp stellar waves ever took existence, and persisted as a major wave damping mechanism for more than two decades? Here we give a brief historical account. One of the earliest versions of the spiral density wave theory treats the tightly wrapped waves in stellar disks using the lowest order WKBJ approximation (Lin & Shu 1964, 1966), as we had mentioned before. In this approximation, the potential and density spiral patterns are everywhere proportional to each other, so it is in essence a local theory, and no self-consistent potential/density phase shift is allowed in this theory. Roberts (1969) calculated the nonlinear gaseous response under the *enforced* (non-self-consistent) WKBJ spiral potential, and found that shocks usually form. Kalnajs (1972) correctly pointed out that such an induced gaseous shock is phase shifted from the forcing spiral potential as a result of the action of the shock dissipation, and therefore should tend to damp the forcing spiral potential of the stars. Kalnajs' work was supplemented by that of Roberts and Shu (1972), and this mechanism has since become one of the major candidates for the damping of growing spiral wave modes.

However, what was not realized at the time was the fact that these WKBJ shock waves can *never* be made self-consistent within the framework of WKBJ analysis. The presence of the phase shift between the potential and density spirals of a global spiral shock solution contradicts the WKBJ form of the Poisson equation, which requires the potential and density to be exactly in phase. Unfortunately, this subtle point of non-self-consistency in the WKBJ shock solutions was not appreciated by the majority of the workers in the field, and the impression that a gaseous spiral wave serves to damp the stellar wave has become wide spread, even though all the self-consistent two-component N-body simulations conducted in the interim had shown no evidence of a significant star-gas phase shift (see, e. g., Carlberg & Freedman 1985).

Our current work shows that only *open* spiral waves can self-consistently admit a spiral shock since only open waves can self-consistently admit a potential/density phase shift, which is an unavoidable element in a global viscous spiral shock solution. Furthermore, the amount of shock dissipation at the quasi-steady state of the wave is always given by the global phase-shift closure relation, whether a separate dissipative gas component is present or not. This characteristic of global-determines-local, or hierarchical control, is typical of the coherent structures formed in nonequilibrium systems.

In physical spiral galaxies, most of the mass of the interstellar medium are in the form of molecular clouds, especially at the spiral arm region. A good fraction of these clouds are self-gravitating, with the mean free path of cloud–cloud collisions on the order of a few hundred parsec. In Section 2.2.2, we showed that the size of star-gas two-fluid instability at the spiral arms, which is the effective particle mean free path in the spiral collisionless shock, is on the order of 1 kpc, which is not much larger than the gas-cloud mean free path in the absence of the shock. Therefore, it appears that a large fraction of the gas-clouds' contribution to the dissipative damping of the wave is through participating in the spiral gravitational shock, and the scattering of these clouds in the collisionless shock is at least as important as the direct collisions of clouds due to their random velocities. In fact, a good fraction of the cloud–cloud collisions themselves happen as a result of the gas' response to the spiral-arm local gravitational instability, as we will show later in Section 4.5.4 for the observational studies of the Carina molecular cloud complex.

Within the confines of the 1 kpc-sized spiral gravitational instability, a series of secondary shocks are expected to develop in the gas component, as a result of turbulence cascade. This could underlie the observed cloud-size/velocity-dispersion relation for the molecular clouds of our Galaxy (Larson 1981), as we will show in Section 4.5. Radiative cooling is only initiated when the spiral shock energy input is cascaded down to hydrodynamic shock scales.

Therefore it is most convenient to think of gas just as another gravitational mass component, slightly more dissipative than stars (see Figure 3.22), but with only a quantitative rather than a qualitative difference when participating in the collective interactions with the density wave potential field. Note that to properly model the collective dissipation effect the gas must be modeled as dissipative particles which lose energy during close encounters/collisions at the density wave crest (as modeled in Zhang (1998) based on the algorithm described in Thomasson 1989), rather than using a smooth-particle-hydrodynamics (SPH) approach. In the former the gravitational viscosity of the gas (as was that of the stars) self-consistently emerges as a result of the global collective interaction of the wave and the basic state matter through the mediation of spiral collisionless shock, whereas in the latter approach the viscosity is put in "by hand" by the simulator, thus is not modeled self-consistently. Incidentally, modeling interstellar clouds as collisional particles was also the preferred simulation approach suggested in Binney and Tremaine (2008, pp. 520, 521).

Through the presence of the potential-density phase shift, allowed by both the Poisson integral and the equations of motion for a self-organized unstable density wave mode, the star-gas two-fluid mode is thus self-damped: the growth tendency of the global resonance cavity due to over-reflection at corotation and feedback at galactic center is offset by the dissipation tendency as is revealed and supported by the potential-density phase shift, and mediated by the collisionless shock that happens partly as a result of the phase shift (i. e., owing to this phase shift, matter rams unexpectedly into the potential field of the density wave at a supersonic velocity, during spiral arm crossing orbital motion). The phase shift required by dissipation is provided by the Poisson integral naturally due to the skewness of the pattern and the nonlocal nature of the Poisson integral. Stars and gas on the basic state thus both can have dissipative interaction with their common density wave potential.

3.8 Implication on Orbits as "Building Blocks"[†]

In this section we comment on the past passive orbit approach for constructing galaxy models, including the differences in behavior of these models from true self-consistent N-body models, as well as the dynamical origin of these differences.

In the past efforts on building galaxy models (e. g., the extensive work by Contopoulos and collaborators in the 1980s), the single-particle orbits solved in passive (i. e., with a rigid potential forcing) analyses were commonly used as the "building blocks" of these models. From the analysis in the previous chapter, we see that while this approach is acceptable for describing the behavior of disk galaxies in the linear regime of their density wave patterns, such models are expected to fail once the collisionless shock behavior sets in at the nonlinear regime, and especially at the quasi-steady state of the wave modes. The passive approach is essentially a variant of the mean field theories for treating many-body problems, and mean field theories are known to simplify the many-body interactions into effectively a one-body (in response to the forcing mean field potential) problem, thus are not able to deal with the correlated interactions among particles that are responsible for the new physics emerging at nonequilibrium phase transition point.

Stellar trajectories in a self-sustained density wave pattern can show drastically different behavior compared to the periodic or quasi-periodic orbits calculated in an

---

[†] Portions of this section used material previously published in Zhang & Buta (2007), reproduced with modifications from The Astronomical Journal @ AAS. Reproduced with permission.

applied density wave potential. While both types of orbits become chaotic at large wave amplitudes, the former also exhibit secular mean radius change, which is not displayed in the latter. Such a difference in the behavior is due to the constraint of the potential-density phase shift applied by the self-consistent spiral pattern on the individual stars' trajectories, which causes the spiral wave to steepen into large-scale collisionless spiral shocks that further induce secular migration and secular heating of the stellar orbits. We now present a quantitative example to illustrate the difference in behavior between a passive orbit calculated in a rigidly rotating, applied spiral/bar potential, and a stellar trajectory obtained from a self-consistent N-body calculation.

In Figure 3.23, we plot the nonlinear orbit solution, together with the corresponding surface of section, for spiral forcing of strength 10, 20, and 30%, respectively, under an axisymmetric mean potential similar to that in the solar neighborhood. The equations used for calculating the stellar orbit (with galactic coordinate r, ϕ) are the standard nonlinear coupled second-order ordinary differential equations in the corotating frame of the spiral (Binney & Tremaine 2008 eqs (3.135a), (3.135b)), i. e.

$$\ddot{r} - r\dot{\phi}^2 = -\frac{\partial \Phi}{\partial r} + 2r\dot{\phi}\Omega_p + \Omega_p^2 r, \tag{3.16}$$

$$r\ddot{\phi} + 2\dot{r}\dot{\phi} = -\frac{1}{r}\frac{\partial \Phi}{\partial \phi} - 2\dot{r}\Omega_p, \tag{3.17}$$

and with the forcing potential being of the form

$$\Phi = v_c^2 \ln r + A \cdot \cos\left(2\phi + \frac{2\ln r}{\tan i}\right), \tag{3.18}$$

where v_c is the circular speed, Ω_p is the pattern speed, and i is the pitch angle of the spiral. In all the calculations performed for Figure 3.23, we have used $r = 8.5$ kpc, $v_c = 220$ km s^{-1}, $\Omega_p = 13.5$ km s^{-1} kpc^{-1}, and $i = 20°$. The fractional forcing strength f is defined as $f \equiv 2A/(\tan i \cdot v_c^2)$.

We observe from Figure 3.23 that for all the orbits calculated there is no secular change in the mean orbital radius, which is a result consistent with the constancy of the Jacobi integral in a rigidly rotating spiral potential (Binney & Tremaine 2008, p. 179). Furthermore, we notice that orbits are only regular at small spiral forcing amplitude. For forcing amplitudes which exceed 20%, the orbits gradually become chaotic, and their traces on the surface of section begin to cover an entire two-dimensional area.

It is obvious that a coherent spiral density distribution cannot be obtained by the straightforward superposition of orbits of the type given by Figure 3.23(e). However, 20–30% is a realistic forcing level in the observed spiral and barred galaxies. Therefore at least in the case of (but not limited to) large amplitude spirals, orbits can no longer be considered as the "building blocks" of the global spiral pattern.

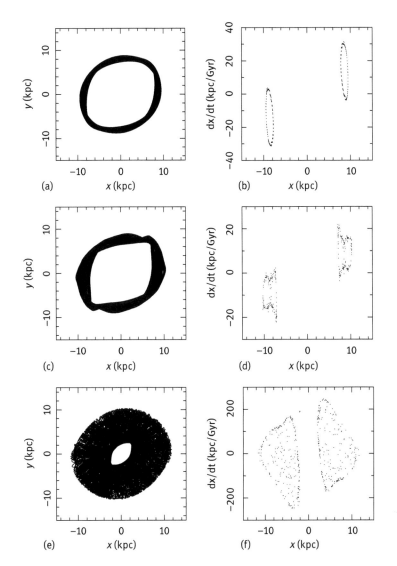

Figure 3.23: Orbit solution, plotted in the pattern-stationary frame ([a],[c],[e]), and surface of section ([b],[d],[f]) for spiral forcing strengths of 10% (a and b), 20% (c and d), and 30% (e and f), respectively. Parameters common to these three cases are: the pitch angle of the spiral $i = 20°$, $\Omega_p = 13.5$ km s^{-1} kpc^{-1}, $r(t = 0) = 8.5$ kpc, $v_c = 220$ km s^{-1}, and total integration time $t = 80$ Gyr. The test particle was initially on a circular orbit (adapted from Zhang & Buta 2007).

Certain collective effects have to be present in order to "collapse the chaos" inherent in the stellar trajectories, to arrive at a coherent spiral density response. Contopoulos and collaborators had in later years also included more of the role of chaotic orbits in constructing galaxy models (Contopoulos & Voglis 1996 and references therein; Voglis, Stavropoulos, & Kalapotharakos 2006), though the full collective effects are still missing in these passive calculations because of the lack of mutual correlations among the chaotic orbits used[†].

N-body simulations of spontaneously formed spiral modes revealed that typical particle trajectories in such finite-amplitude, self-sustained spiral patterns are again chaotic in the corotating frame of the pattern; however, a new phenomenon not present in the passive orbit calculations is now also evident. A typical trajectory of a particle in the simulation of Zhang (1996) in the corotating frame of a quasi-stationary spiral pattern, for a radial location inside corotation, is plotted in Figure 3.24. This simulation run obtained spontaneously formed spiral mode of equilibrium amplitude around 15%. It is clear that this particle trajectory covers a two-dimensional area in the reference frame corotating with the density wave pattern, signaling the presence of chaos (without chaos, for a 15% forcing the orbit should look like the tubes in

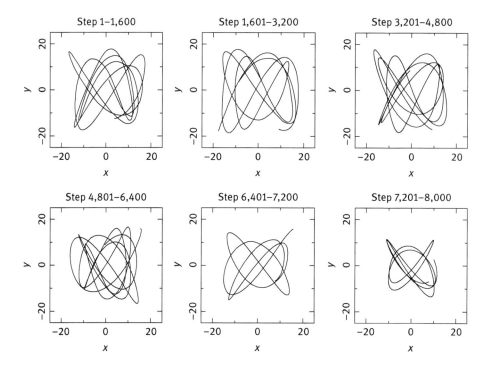

Figure 3.24: Time evolution of the trajectory of a typical star inside corotation, for a spontaneously formed N-body spiral mode simulated in Zhang (1996). The trajectory sections are plotted with respect to the pattern-stationary (corotating) frame. The forcing amplitude is around 15% (adapted from Zhang & Buta 2007).

[†] The chaos present in the passive orbit response is in general of a different nature than the chaotic behavior observed in collective systems, that is, the former lacks the hidden intricate correlations among the N-particles' trajectories in collective systems, yet displays coherent patterns as well as correlation only with respect to the mean forcing potential.

Figure 3.23, between the top two frames in the left column). The invasion of chaos is thus found to occur at a much smaller spiral forcing amplitude for the spontaneously formed N-body spirals than for the non-self-consistent case, a result consistent with the theoretical finding that collective relaxation induced by large-scale global instability patterns generally dominates the relaxation behavior of stellar trajectories in galaxies (Pfenniger 1986; Weinberg 1993; Zhang 1996).

Another new phenomenon evident from Figure 3.24 is that there is now a secular decay of the mean orbital radius for stars inside corotation. A corresponding secular increase of the mean orbital radius for stars outside corotation is also found (Zhang 1996). Both these secular changes in mean trajectory sizes are absent from the non-self-consistent forced orbit calculations.

The presence of stochasticity in an individual star's orbit is an integral element in a self-organized and self-sustained spiral pattern. The stochasticity derives from the exponential sensitivity of the stellar trajectory to the small perturbations in initial/boundary conditions, with the sensitivity itself a result of the presence of global instabilities which creates an effective "collisional relaxation" condition. The large-scale coherence of a spontaneously formed spiral pattern is nonetheless not sensitive to the presence of the chaotic behavior in an individual star's trajectory. This is because the spontaneous emergence, and the subsequent maintenance and secular evolution of these coherent patterns, is a sequence of events which corresponds to a large-phase space volume in all the possible initial conditions that a proto-galactic-disk could choose from. The large increase in coarse-grained entropy when a disk develops a spiral mode means that the pattern itself is an *attractor* to the evolution of these particles' collective motions.

Therefore in spiral disks we witness three levels of organization: deterministic in the governing equations, chaotic in the individual star's trajectory, and deterministic again in the large-scale patterns formed in such a many-body system. Collective dissipation is what made possible the existence of large-scale coherence in a sea of underlying chaotic orbits. Each time at the crossing of the spiral or bar crest, the orbital parameters of the individual stellar trajectories are reorganized to support the wave pattern, and they are completely decorrelated from their parameters in the previous cycle of arm-crossing due to the presence of gravitational instability at the arms. The presence of global resonances (such as ILRs and CRs) in such self-consistent patterns will no longer act on individual orbits, but rather on the collective motions of all the particles constituting the density wave[†]. In a sense, the collection of stars behaves rather like molecules in a fluid, and the collisionless shock in the spiral arms and bars

---

[†] The fact that an individual star's resonance condition is totally destroyed in a disk galaxy containing a global instability can be understood as due to the decorrelation of the phases of the trajectory each time it experiences a gravitational-instability-induced scattering/collision event at the spiral arm crest. Like the example of a child's swing, both the exact frequency condition and the exact phase condition need to be satisfied for the swing to gain in amplitude. In the case of individual trajectories

create conditions of particle interaction akin to true collisions in a true fluid, albeit here not in the sense of two-body collision, but rather the scatterings of particles in the collective instability at the density wave crest.

The past work on galaxy model constructions was mostly done under an applied potential, and not the kind of self-consistent potential in the many-particle systems. Even for several supposedly self-consistent calculations using N-body simulations, a closer examination of the ways some of the final analyses were conducted revealed that it was only the forced/passive orbital behavior that had been uncovered in these analyses. For example, Athanassoula (2003) studied the resonance exchanges between bars and active halos. Even though she had started out with a self-consistent three-dimensional N-body simulation, when the time comes to calculate the resonant exchanges between the bar-stars and halo-stars, she had to halt the N-body simulation and revert to using a *rigid, uniformly rotating*, and *enforced* bar potential which best approximates the bar potential she had obtained at the end of the N-body simulation, so as to obtain the sought-after orbital resonance exchange with the halo. It is not surprising that such a practice should be needed: for resonance conditions exist for individual orbits only under an *applied* and *rigidly rotating* potential, and not for a self-consistent and (microscopically) time-fluctuating potential from the actual N-body simulations where collective effects completely wipe out the individual orbit's resonance behavior. Likewise, other investigators who had conducted periodic- and quasi-periodic-orbit-searches in self-consistent N-body models had also used either *the N-body potential at one instant of the simulation* (Pfenniger & Friedli 1991) or a *time-averaged potential* (Sparke & Sellwood 1987) to search for periodic orbit families, which effectively are equivalent to using a forced, rigidly rotating and non-self-consistent potential over the orbital period under study. It is no wonder that the secular mean trajectory changes we have found in the true, fully self-consistent N-body simulations were never found in those orbit searches which used the N-body potential at a fixed instant to approximate the potential over the entire orbital period.

The formation of density wave collisionless shocks effectively introduces singularity behavior in the passive orbit solutions, even at galactic radial locations away from wave-particle resonances. This allows new phenomena previously thought impossible in the passive treatments to emerge, including density wave patterns such as the "super fast bars" discovered in Zhang & Buta (2007), and Buta & Zhang (2009), where the length of the density wave bar pattern extends beyond their corotation radii. We will address this phenomenon more fully in Section 4.2.6.

The true collective behavior is encoded in, and is enabled by, the spatial and temporal correlations of the N-particles' positions and velocities. While the phase correlation of a single star's trajectory with the mean forcing potential has been

---

in galaxies, it is the phase coherence condition that is being destroyed by the density wave-induced gravitational instability.

destroyed by the presence of collective instabilities[†], the multi-faceted correlations among the entire collection of N-particles are being established by the very collision process itself to drive the emergence, maintenance and long-term evolution of the collective instability. For galaxies possessing a large-scale density wave mode, the actual particles' trajectories supporting the pattern acquire global coherence during the spontaneous growth process of the mode, despite their chaotic appearance when each is viewed individually. In some sense, the apparently "ordered" forms of the periodic orbits for galactic systems are static and inert, and as a result they lack the life force to produce the correlated collective effects, whereas the apparently more chaotic-looking trajectories in the N-body systems in fact contain innate orders derived from the self-organization process, which facilitate the further maintenance and evolution of the parent galactic system. The equilibrium in physical galaxies is a dynamical equilibrium, maintained as a balance of the growth and dissipation tendencies of the wave mode, rather than the static equilibrium described by the passive orbit analysis and model-building approaches.

3.9 Second-Generation Tests

The first-generation N-body tests presented in the last few sections, which were conducted during the 1990s, established both the collisionless shock nature of the spiral density wave modes and the pattern of mass inflow inside corotation and outflow outside corotation, as well as quantitatively confirmed the analytically derived secular evolution rate closure relation. However, this confirmation is for a much smaller wave amplitude in the simulated galaxies compared to the wave amplitudes typical of observed galaxies. This smallness in simulated density wave amplitude pertains not only to the author's simulations, but to nearly all of the N-body galactic simulations conducted during the past few decades. In the recent years, the author has made another attempt at examining the common procedures in N-body simulations of galaxies, and uncovered the cause of the small-wave-amplitude problem in simulated galaxy disks (Zhang 2016). In what follows, we describe these more recent simulation results.

[†] Note that this behavior appears to have violated the conclusion of the celebrated KAM theorem (after Kolmogorov, Arnold, and Moser), which prescribes the conditions for the stability of periodic and quasi-periodic orbits under small perturbations. We point out that the KAM theorem was also derived for orbit structures under an applied potential, and not for self-organized collective instabilities. Or in other words, for self-organized collective processes, the perturbations cannot be regarded as small already from the outset of the growth of the instability.

3.9.1 The Critical Roles of the Softening Parameter

Observational studies conducted by Zhang & Buta (2007, 2012, 2015) have indicated that large radial mass radial flow rates are possible in observed galaxies with strong density wave patterns. Yet, past numerical simulations had failed to account for such high mass flow rates. Zhang (2016) showed that the main culprit for the discrepancy between the observationally inferred mass accretion rate and the N-body generated rates is the improper treatment of "gravitational softening." A particle softening constant is an artificial parameter inserted by numerical simulators into the formula for gravitational potential to control the magnitude of relaxation in simulations with smaller numbers of particles compared to a real galaxy[†]. Excess softening reduces the collective effects underlying the significant secular evolution inferred for physical galaxies. Lesser softening, coupled with an increase in the number of particles and increased grid resolution, allows the N-body simulations to reveal significant morphological transformation of galaxies over a Hubble time.

As we have shown in the previous sections of this chapter, the simulated small mass flow rates were associated with the small amplitudes of the spiral patterns formed in these simulations. Zhang (2003) has speculated that the small simulated wave amplitudes may be caused partly by the 2D nature of the simulations, and that active bulge, halo, as well as active thick disk may help to boost up the wave amplitude. Recent simulations conducted by the author showed that partially active bulge, halo, and thick disk only boosted up the wave amplitude and mass flow rate moderately while keeping other simulation parameters the same.

As eventually realized after extensive testing, the main reason for the low mass accretion rate in N-body simulations of disk galaxies lies in the choice of the so-called softening parameter. The softening parameter a_{soft} is generally defined as follows: The gravitational potential in an N-body system is calculated as

$$\Phi(r) = \frac{-Gm}{\sqrt{(r^2 + a_{\text{soft}}^2)}}, \quad (3.19)$$

where Φ is the gravitational potential, G is Newton's constant of gravitation, and m is the mass of the field particle contributing to the potential, and a_{soft} is an artificial parameter introduced into Newtonion potential/force equation to circumvent the fact that N-body simulations usually employ significant less number of particles than the physical systems they try to emulate. A finite softening parameter reduces the amount of unrealistic close encounters in N-body simulations which would be more often for a small N system, and thus reduces the level of artificial relaxation in these small N

[†] The particles in the N-body simulations of disk galaxies are thus composite particles, each usually possessing $10^4 - 10^6 M_\odot$ of mass, depending on the total number N of particles used to represent disk mass.

systems. The hope is that a collisionless configuration would then result from a well-matched N and a_soft. In particle-mesh-based N-body approaches (which the N-body polar grid code is a member of), the grid size serves as an additional softening parameter that acts in consort with a_soft in controlling relaxation (see Sellwood 1987 and the references therein).

However, with the realization that galaxies which possess large-scale density wave modes are governed by collective dissipation processes, one must take into account the fact that it is impossible to completely avoid collision-like behavior (Pfenniger 1986; Romeo 1990; Weinberg 1993), no matter how large the particle numbers are used: after all, the collective effects in these systems *depend* on the near-collision or scattering of particles in the global instabilities to set up long-range correlations to achieve self-organization and to induce secular evolution. In some sense, these galactic systems are in a *forced relaxation configuration*, with the forcing accomplished by the density wave collisionless shock, which forces the *effective* instability parameter Q in the spiral arms to be less than one, which further leads to interparticle interactions and correlations. Thus, in hindsight, it is obvious that excess softening reduces the very interparticle interaction that is the backbone support of collective effects.

In essence, overly softened gravity is *modified gravity*, and thus is modified Newtonian interaction (not to be confused with the modified Newtonian dynamics proposed to account for the flat rotation curves). The system one simulates with softened gravity is no longer exactly the same system governed by Newton's gravitational law, but rather a more sluggishly-interacting system. This might not be a serious concern if one's interest is in obtaining a modal morphology that mimics the observed galaxy morphology (as shown in the work of Donner & Thomasson (1994), which is the forerunner of long-lasting N-body spiral modes), but as we will show here, it is detrimental to determining the realistic secular mass flow rates that are relevant to physical galaxies. Pioneering studies of the effect of softening in N-body simulations by Romeo (1994a, 1997, 1998) had also cautioned against the use of large gravity softening in disk galaxy simulations.

3.9.2 The Impact of Softening on N-Body Simulated Mass Flow Rates

In what follows in this subsection, we use a set of N-body simulations to demonstrate the powerful effect of gravity softening on the ability of the N-body codes to properly model the secular mass redistribution process in disk galaxies, even for those basic state choices that allow unstable global modes to form, which is a prerequisite condition for obtaining the coordinated large-scale mass flow pattern of inflow inside corotation, and outflow outside corotation, needed to achieve secular morphological evolution along the Hubble sequence.

We chose a set of basic state parameters similar to that first explored in Donner and Thomasson (1994), and subsequently used for the study of collective effects and secular evolution in Zhang (1996, 1998, 1999). The grid size is 220 by 256 in the radial and azimuthal direction, respectively. The mass in the active disk, the inert halo, and inert bulge are 0.4, 0.5, and 0.1, respectively, in the normalized unit, the same as that used in Zhang (1998). The scaling of the time unit is such that 1256 time steps represent one rotation period at $r = 20$. Both the particle numbers and the time resolution used are higher than that in Zhang (1998). These grid and time-step resolutions are used for the rest of the simulations presented in this section as well, only the particle numbers and the softening parameters are changed and will be noted in each case.

In Figure 3.25 we show the evolution of enclosed mass within the central $r = 3.5$ region (in the scale of this disk which has corotation radius around 20–25 in normalized unit, this corresponds to the central bulge region). The three curves, from bottom to top, correspond to softening parameter value of 1.5 (as used in Donner & Thomasson 1994; and Zhang 1998), 0.75, and 0.25 in the unit of the smallest grid size of 1. As we can see, the choice of softening of 1.5 (as was commonly used by N-body practitioners up until now) produced a barely noticeable mass inflow (to really see the small amount of mass inflow for this choice of softening, one is referred to Figure 10 of Zhang 1998, reproduced as Figure 3.19 in this chapter, which zoomed in on the vertical scale, even though that particular figure was for the enclosed mass within central $r = 13.5$, rather than the central $r = 3.5$ as for the current figure). For this particular choice of softening, which was regarded by the N-body simulation community as being adequate in suppressing unwanted relaxation effects as well as in matching the grid softening, the central mass growth of only 6.5% was observed over about 25 pattern rotation periods, which is obviously quite inadequate in transforming the Hubble type of a galaxy. Gradually reducing softening is seen to systematically lift up the mass inflow rate. For the range of softening tried by the current author (1.5–0.1), the trend of increasing mass inflow with reduced softening continues. It is possible that realistic galaxies, which have zero particle-softening in the form of gravitational law but do have much larger numbers of particles, as well as the finite thickness of the

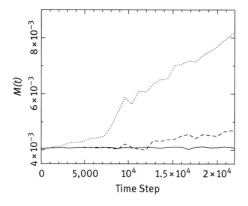

Figure 3.25: Evolution of enclosed disk mass within central $r = 3.5$ radius of a set of 2D N-body simulations with different softening parameters. Solid: $a_{soft} = 1.5$, Dashed: $a_{soft} = 0.75$, Dotted: $a_{soft} = 0.25$. 1 million particles are used to represent the active disk (adapted from Zhang 2016).

disk (another effective softening parameter but its role to inhibiting collective effects is not nearly as pronounced as the particle-softening) to keep relaxation in check, the mass flow rates much exceed those we have seen in these recent simulations are possible – as was indeed observed in the mass flow rates calculated for the sample galaxies presented in Zhang & Buta (2007, 2012, 2015).

The price one pays for reduced softening in small-N simulations is the increase in relaxation rate. To compensate, one needs to use a correspondingly increased number of particles, for otherwise the collision-induced heating will swamp the density wave activity and reduce the wave amplitude. How much the particle number needs to be increased for a given choice of softening can in fact be found empirically (see below), and the empirical result turns out to agree with the analytical expectation of an N^2 dependence of relaxation effect for 2D N-body simulations (Thomasson, Donner, & Elmegreen 1991).

In what follows, we show a set of simulations with changing particle numbers while holding the softening parameter $a_{soft} = 0.1$. In Figure 3.26, we present the enclosed mass within the central $r = 3.5$ radius, with the different runs having a constant $a_{soft} = 0.1$ but varying the number of particle used. For the smaller particle number runs, the rapid mass inflow is seen to saturate (i.e., the slope of the line changes from a finite value to a near-zero value) at an earlier time step, which corresponds to the time when spiral activity is damped by heating due to the insufficient number of particles for this extremely small softening, especially in the outer disk region where the surface density is low and the particle numbers are small. For large particle runs, the rapid mass inflow is seen to remain at a constant rate all the way until the end of the run, which corresponds to 25 rotations.

Furthermore, we note that the mass inflow rate observed here (the constant slope part of the figures) corresponds to about 60% increase in enclosed mass within $r = 3.5$ (the bulge region) over 10 rotation periods or roughly 1/5 of a Hubble time. This level of mass accretion is more than sufficient to transform the Hubble type of a galaxy by several stages in a Hubble time. Plus, the mass inflow rate conceivably could be even higher for even smaller choices of a_{soft}. The natural limit obviously is set only when

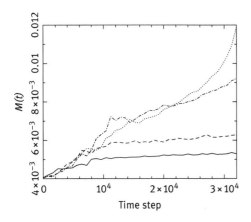

Figure 3.26: Enclosed mass within central $r = 3.5$ of the N-body disk for runs with different numbers of particles and $a_{soft} = 0.1$. Solid: 1 million particles. Dashed: 10 million particles. Dotted: 20 million particles. Dash-dotted: 40 million particles (adapted from Zhang 2016).

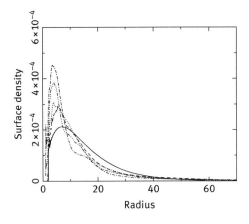

Figure 3.27: Basic state surface densities at the beginning (solid), and the end of four N-body runs with different number of particles. Dashed: end surface density for 1 million particles. Dotted: end surface density for 10 million particles. Dash-and-single-dotted: end surface density for 20 million particles. Dash-and-triple-dotted: end surface density for 40 million particles. $a_{\text{soft}} = 0.1$ (adapted from Zhang 2016).

one has the actually physical galaxy in its 3D representation with the full interparticle interactions taken into account – that is, we know what the actual density wave amplitudes in galaxies are from the observation of physical galaxies. Note that having active halo, bulge, and thick disk components will further increase the mass flow rate from a pure 2D simulation, as the author has confirmed in separate N-body tests.

In Figure 3.27, the disk surface density at the beginning of the run, and at the end of the four runs with different numbers of particles (as used in Figure 3.26, but with different line style due to the need to represent the "before" surface density), is presented. It can be seen that the large inflow of the large-particle-number runs indeed builds up a more substantial bulge. However, the 20 million particle run in fact produced more substantial bulge-building than the 40 million particle run. This is consistent with the trend observed in Figure 3.26, where it is seen that the 40 million particle run has somewhat reduced mass inflow rate near the end of the run. This may be related to the emergence of different kinds of local instabilities in the different particle runs.

3.9.3 Four Runs with Differing Softening Parameters

The initial tests of the last subsection show that by significantly decreasing the amount of particle softening, and simultaneously increasing both the number of simulation particles used and the degree of grid resolution, we can achieve increased radial mass flow rates to levels comparable to that needed for a significant change of galaxy morphological types over a Hubble time, as indicated in observational studies (Zhang & Buta 2007, 2015).

A natural question arrives as to whether this observed increase in radial mass flow rate is a true physical effect or else is a numerical artifact due to noise brought about by the small softening and the corresponding increased collisional relaxation. To answer this question, in this subsection we present detailed correlations of density wave modal characteristics and the observed mass flow behavior in the N-body simulations.

Softening will be shown to change the equilibrium amplitude of the modes formed, but it hardly changes the *level of agreement* between the mass flow rate *predicted* using the modal parameters found in the simulations, and the corresponding mass flow rate *measured* in the same simulations (i.e., smaller equilibrium wave amplitudes are found to correspond to smaller measured mass flow rates, given precisely by the theoretical predictions using the corresponding wave amplitudes).

In what follows, we will first establish the modal nature of the density wave patterns formed from originally featureless (axisymmetric) basic state of the galactic disk. After presenting the modal characteristics, we will go on to demonstrate the presence of the potential-density phase shift distribution, which leads to the secular torque interaction between the basic state mass distribution and the wave-mode potential field. This torque interaction leads to the angular momentum exchange between the basic state and the density wave, which further leads to the radial mass flow behavior. We demonstrate that both the qualitative and quantitative mass flow behavior observed in these simulations are consistent solely with the mode/basic-state interaction picture, and are inconsistent with either the transient-wave- or noise-induced mass flow behavior.

Four sets of 2D *N*-body simulations are performed utilizing the same polar grid (220 radial grid cells, 256 azimuthal grid cells), the same number of particles (20 million), the same duration of simulation run (32768 steps, corresponding to roughly 25 galactic rotation periods at radius 20), and the same time resolution of 1256 time steps per galaxy rotation period time at $r = 20$, but with the particle-softening parameter having the values of $a_{soft} = 1.5, 0.75, 0.25, 0.1$, respectively, in the Plummer softening scheme (eq. 3.19).

All the runs have the same basic state specification as that of Zhang (1998) paper, which was also the basic state used in the previous subsection. This basic state specification allows the setup of a galactic resonant cavity between the inner bulge region and the corotation radius in the mid-disk, and thus the formation of unstable density wave mode. The resulting corotation radius ranges from $r_{co} \approx 28$ (for $a_{soft} = 1.5$) to $r_{co} \approx 23$ (for $a_{soft} = 0.1$) at the initial emergence stage of the mode, and the corotation radius tends to decrease in the late stages as the secular mass inflow leads to a more centrally concentrated mass distribution.

3.9.3.1 Morphological Evolution

In Figures 3.28–3.31, we present the morphological evolution of the density wave pattern with basic state and numerical grid specifications given previously, and with particle-softening parameter $a_{soft} = 1.5, 0.75, 0.25$ and 0.1, respectively. The initial disk mass assignment is an axisymmetric modified exponential disk, and the rotation curve is nearly constant at a value $v_c \approx 0.1$ in the normalized unit. The initial velocity dispersion assignment corresponds to the instability parameter $Q_T \approx 1$ across the disk.

A coherent spiral-bar pattern spontaneously emerges from the originally featureless disk in each case. The morphology of the $m = 2$ (two-armed) dominant mode is

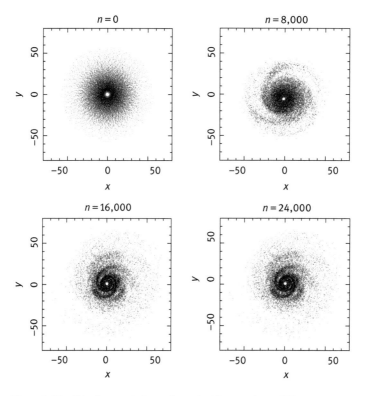

Figure 3.28: N-body morphology of a spiral/bar mode at different time steps. The rotation period at $r = 20$ is about 1256 steps. The averaged pattern speed is 0.0035 radian/step. The softening parameter is $a_{soft} = 1.5$ (adapted from Zhang 2016).

shown in Figure 3.36, during its initial emergence phase. The mode in fact emerges at progressively earlier time steps as softening decreases; this will be shown more clearly later during detailed analyses. This is mainly due to the larger seed of noise present for modal amplification in the smaller softening cases.

The patterns in various cases contain other contaminants besides the $m = 2$ mode, for example, the $a_{soft} = 1.5$ case has some $m = 3$ component, and the two small-softening cases ($a_{soft} = 0.25$, $a_{soft} = 0.1$) are seen to have local instability clumps formed in the outer region of the simulation disk. The small softening cases also tend to have higher heating effect which submerges the nonlinear pattern within noise at the later stages of the simulation, though the underlying $m = 2$ mode can be shown to be persistently present in every case, for the entire duration of the simulation run (corresponding to 25 rotation periods at radius 20).

In the small softening case of $a_{soft} = 0.1$, the pattern is seen to have evolved from a spiral-like morphology in the early stage of the simulation to a bar-like morphology in the later stage, partly due to the rapid mass inflow in this small softening case. The increased central concentration of mass then favors a bar mode.

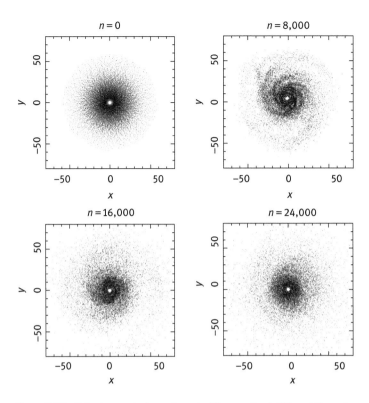

Figure 3.29: *N*-body morphology of a spiral/bar mode at different time steps. The rotation period at $r = 20$ is about 1256 steps. The averaged pattern speed is about 0.0035 radian/step. The softening parameter is $a_{\text{soft}} = 0.75$ (adapted from Zhang 2016).

3.9.3.2 Phase Evolution

In Figure 3.32, we plot the time evolution of the azimuthal phase of the $m = 2$ (two-armed) perturbation-potential component from the above four runs, at a series of radii across the simulation disk.

The first thing we notice from this plot is the coherent, approximately linear, evolution of the phase at the various radial locations, for all softening choices, indicating that the patterns involved are rotating at nearly constant angular speed at each radius (though the pattern speed may change across the different radii, indicating the winding up of the pattern, especially for inner galaxy disk, and for large softening cases). This shows that the patterns that spontaneously emerged are likely to be unstable *modes* of the disk rather than random transient waves, with the latter not expected to have a coherent phase evolution over the lifetime of a galaxy.

Furthermore, we observe that while the phase evolution curves for the large softening runs diverge at the different radial locations, indicating a change of pattern speed for the different portions of the spiral arms (or the wrapping up of the

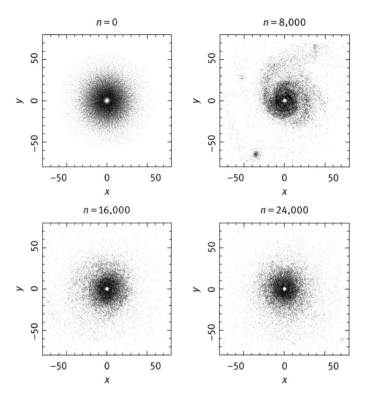

Figure 3.30: N-body morphology of a spiral/bar mode at different time steps. The rotation period at $r = 20$ is about 1256 steps. The averaged pattern speed is about 0.0035 radian/step. The softening parameter is $a_{soft} = 0.25$ (adapted from Zhang 2016).

pattern with time), the run with the smallest softening length ($a_{soft} = 0.1$ case) shows an extremely constant pattern speed between $r = 10$ and $r = 30$ (i.e., between the inner Lindblad resonance and slightly beyond corotation radius). As a matter of fact, two of the six curves nearly overlap (that is why there appears to be only five curves).

Note also that in all cases the more rapid phase advance between $r = 5$ and $r = 10$ locations is likely to be due to a nuclear pattern with a separate pattern speed since the outer main pattern has its inner Lindblad resonance at around $r = 10$.

3.9.3.3 Pitch Angles of the Patterns

In Figure 3.33 we plot the $m = 2$ potential phase versus radius at different time steps for the four runs in order to study the variation of pitch angle of the spiral-bar pattern versus radius.

For most plots, it can be seen that the slopes of the $m = 2$ phase versus radius starting from $r = 10$ (which is close to the inner Lindblad resonance location of the dominant mode) to $r = 40$ (which is close to the outer Lindblad resonance or the outer

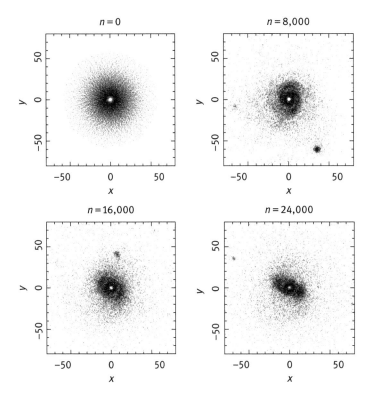

Figure 3.31: *N*-body morphology of a spiral/bar mode at different time steps. The rotation period at $r = 20$ is about 1256 steps. The averaged pattern speed is about 0.0035 radian/step. The softening parameter is $a_{soft} = 0.1$ (adapted from Zhang 2016).

limit of the mode) are within a narrow range, especially if we exclude the step $n = 8000$ (solid) lines, which may be contaminated by the noisy features arose during the mode emergence phase. This near constancy of the slopes of the phase versus r curves indicates the presence of logarithmic spirals, or else skewed bars, with nearly constant pitch angle. Some of the regional bumps on the curves could be due to contamination from local instability clumps or noise.

Of particular interest is the plot for $a_{soft} = 0.1$, which, as we had previously shown, harbors a pronounced bar pattern in the later stage of the run. The evidence of the spiral-to-bar morphological change can be seen here from the phase-versus-radius plot. At the earlier stages of the run (step 8000 and earlier) there is a more steep gradient for the phase versus radius curves, indicating the presence of a (skewed) spiral pattern, whereas for the later stage of the run the phase versus radius curves flatten out, reflecting the reduced skewness of the bar. However, a comparison with the $a_{soft} = 0.1$ phase-versus-time plot (Figure 3.32) shows that between $r = 10$ to $r = 30$ the phase-versus-time curves have nearly constant slope throughout the simulation

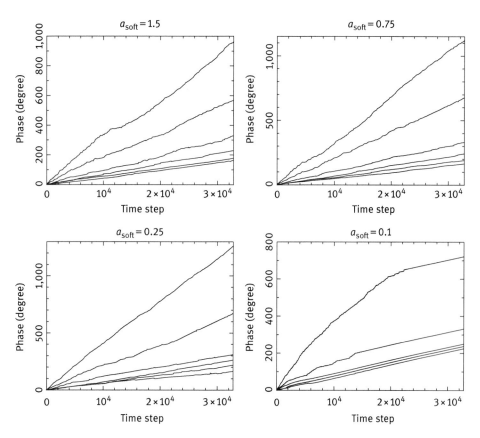

Figure 3.32: Phase evolution of the $m = 2$ potential at radial locations (from top to bottom) $r = 5$, 10, 15, 20, 25, 30, respectively, using four different softening parameters (adapted from Zhang 2016).

run, indicating a constant pattern speed throughout both the spiral and bar phases for this evolving mode. This constant pattern speed during the modal morphological evolution is also supported by the time-segmented power spectrum analysis of Figure 3.35, which we will present next. These results thus corroborating one another to show that the spiral and bar modes are from the same modal family (i.e., the spiral evolves smoothly to a bar, rather than spiral disappearing and a bar of a different pattern speed later appearing). Therefore, even with the modal shape evolving from spiral to bar, we are still led to the quasi-steady modal picture, rather than a transient and recurrent pattern picture[†].

[†] In essentially all the numerical simulations of spontaneously formed density waves in disk galaxies, the unstable bar modes were obtained from initially going through a spiral phase (see, e.g., Sparke & Sellwood 1987). This is because the formation of density wave modes requires the removal of angular momentum from across the inner disk (within corotation) and the deposit of angular momentum to

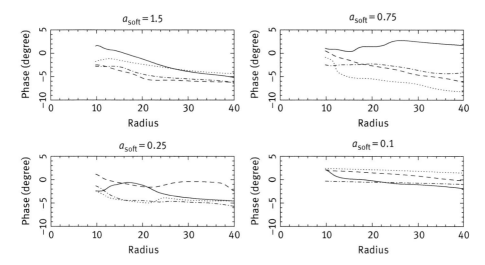

Figure 3.33: Phase versus radius of the $m = 2$ potential, plotted only for $r > 10$, at time steps 8000 (solid), 16000 (dashed), 24000 (dotted), 32000 (dash-dotted), for runs with four different softening length choices (adapted from Zhang 2016).

3.9.3.4 Power Spectra

In Figure 3.34 we present the contour plots of the power spectra of the $m = 2$ potential perturbations, from the above four runs with differing softening parameters during selected time intervals, and over the galactic radial range where the dominant mode is expected to have significant amplitude. The solid line in each figure indicates the galactic rotation curve in angular speed, and the dashed curves delineate the range where modal amplification within the galactic resonant cavity is most likely to occur (Bertin et al. 1989a, 1989b).

For the small softening length $a = 0.1$ run, despite the transition from spiral-shaped to bar-shaped morphology, the pattern speed of the mode is seen to remain

---

the outer disk region outside corotation, and this removal and deposition of angular momentum in turn depends on the presence of potential-density phase shift of the pattern involved (Zhang 1998), and the phase shift value is smaller for bars than for spirals (since bars generally are less skewed than spirals). Therefore a bar mode employs a spiral modal shape during its youthful developmental stage to remove angular momentum needed for its growth, until it reaches the adequate amplitude and modal shape to become its mature self. This process incidentally lends support to many of the so-called "super-fast bars" found in the Zhang & Buta (2007) and Buta & Zhang (2009) samples, which have the bar corotation radius located intermediate in the bar rather than at the bar end as dictated by conventional wisdom from passive orbit analysis. If all bars are formed through an initial spiral phase, and if the corotation radius during spiral phase is located midway in the spiral arms, then we should not be surprised to find a fraction of bars to have this property as well, since the transition stage between spirals and bars can appear as either an open spiral or a skewed bar.

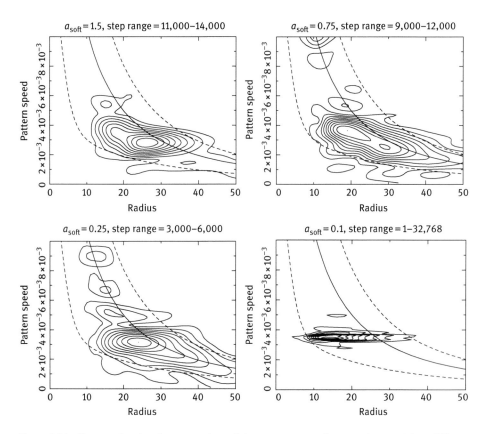

Figure 3.34: Contour plot showing comparison of the power spectra from $m = 2$ potentials at different radii during indicated time intervals, for the four different softening parameters and with 20 million particles. The solid line indicates galactic rotation speed Ω, the two dashed lines are $\Omega \pm \kappa/2$, where κ is the epicycle frequency (adapted from Zhang 2016).

constant. This can be seen especially clearly if we break up the time duration of the power spectrum calculation into two segments, as shown in Figure 3.35. It can be seen here that the main power spectrum speak stayed at nearly the same vertical scale (indicating the same pattern speed) for both time segments. The conclusion of the constancy of the pattern speed throughout the spiral-to-bar morphological evolution during the $a_{\text{soft}} = 0.1$ run is further confirmed when we reexamine the $m = 2$ phase-versus-time plot (Figure 3.32). Note that for the larger softening runs ($a_{\text{soft}} = 1.5, 0.75, 0.25$) we have only plotted during the time interval where the first dominant mode emerges and reaches quasi-steady state. This mode will further evolve and change pattern speed during the remainder of the simulation run[†]. Note also that

---

[†] This is the reason we have only plotted the mode for limited time range so as to avoid pattern-speed smearing. The end of the above plotting range does not indicate the end of the modal activity.

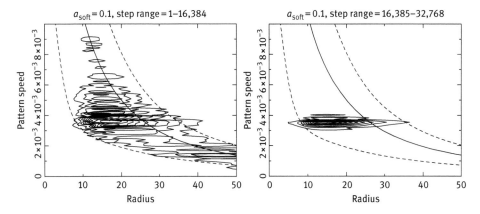

Figure 3.35: Power spectra from $m = 2$ potential at different radii for the 20 million particle and $a_{soft} = 0.1$ run. The solid line indicates galactic rotation speed Ω, the two dashed lines are $\Omega \pm \kappa/2$ (adapted from Zhang 2016).

the pattern speed for the $a_{soft} = 0.75$ case varies slightly with radius (i. e., the power spectrum has a nonzero slope), indicating the tendency for the winding up of the pattern with time for this softening run.

Cross-comparison with the phase-versus-time/phase-versus-radius plots presented previously shows that changes in pattern speed and spiral winding angle appear *continuously* both in time as well as in space throughout the simulation run, rather than appear as random fluctuations that a transient spiral pattern would produce.

This shows that with the choice of small particle softening (and in physical galaxies, the *zero* particle softening in the Newtonian force law), and with the thickness of the disk here partly represented by grid softening, allow self-organization and global self consistency of the galaxy density wave mode to be accomplished and maintained during both the spiral and bar evolution phases.

3.9.3.5 Potential-Density Phase Shift and Radial Mass Flow

Having established the modal characteristics for the simulated density wave patterns, we now look into the issue of radial mass flow in the parent disk galaxies containing these patterns.

In Figure 3.36, we present the contour plots of the morphologies of the $m = 2$ modal density and potential (with the potential negated so as to ease the comparison with density) for the four runs at a progressively decreasing time step for each case, corresponding to the initial emergence period of the mode for the particular softening-parameter choice. We can see that a spiral-like morphology is present in all cases (this morphology will later evolve into a more bar-like morphology for the $a_{soft} = 0.1$ run). Furthermore, we see that the potential patterns display a more open appearance

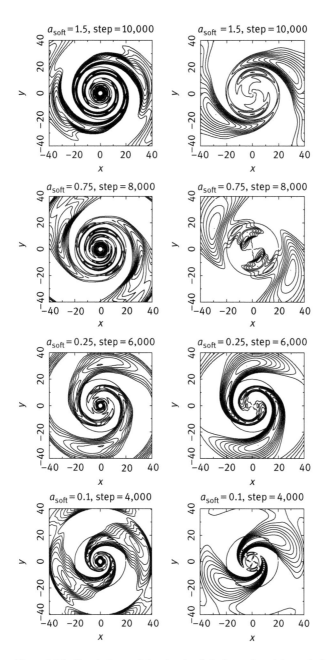

Figure 3.36: Morphology of the density (left frames) and (negative) potential (right frames) of the $m = 2$ modal component for the four runs with different softening at the respective time steps (10000, 8000, 6000, 4000) as indicated on the figure. The averaged pattern speed is about 0.0035 radian/step. Each frame is individually scaled to have 10 contours. Only positive contours are plotted. The circle in each case corresponds to the approximate location of the corotation resonance for that particular choice of the softening (adapted from Zhang 2016).

than their corresponding density patterns, which are more tightly wound. *This difference in the winding of the density and potential patterns reveals the radial distribution of the azimuthal potential-density phase shift*, which we had plotted schematically in Figure 2.2, and which we will analyze quantitatively below.

The existence of the quasi-steady azimuthal phase shift distribution, with potential lagging density within the corotation radius, and potential leading density outside corotation, means that there is a secular torque applied by the perturbation potential on the basic state mass distribution, thus a resulting secular angular momentum exchange between the wave and the basic state at the quasi-steady state of the wave mode. This in turn leads to the inflow of basic state mass inside corotation and the outflow of matter outside corotation.

In Figures 3.37–3.40, we plot the results of the radial dependence of the azimuthal potential-density phase shift, as well as the corresponding instantaneous mass flow rate, calculated using the perturbation density Σ_1 and perturbation potential $\mathscr{V}_1$ at each radius according to the expressions previously (eqs 2.27, 2.124), at a progressively decreasing time step for each of the four softening choices of 1.5, 0.75, 0.25, and 0.1, respectively. The time step for plotting is chosen near the peak of the first emerged density wave mode in the respective run.

Note that if the density wave modal surface-density and kinematic distributions are truly globally self-consistent and quasi-stationary, and if the galaxy admits only one dominant mode, the phase shift (as well as mass flow rate) distribution will experience positive-to-negative zero-crossing exactly at one corotation radius. The radial distribution of the phase shift would be a positive hump followed by a negative hump, with the transition between the two humps occurring at the unique corotation radius. In our simulations, we observe secondary oscillations in the phase-shift curve in the central region of the disk as well as in the outer region. These have various

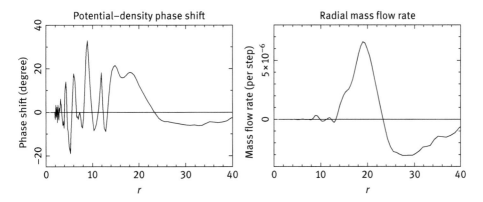

Figure 3.37: Potential-density phase shift (left frame) and mass flow rate (right frame) versus r at time step 12800 for $a_{\text{soft}} = 1.5$ run. These two figures have identical zero crossings though different amplitude distributions (adapted from Zhang 2016).

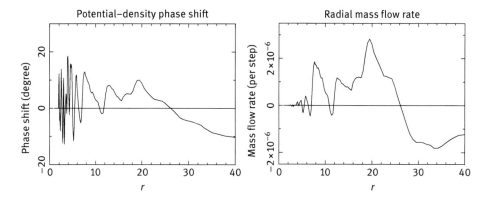

Figure 3.38: Potential-density phase shift (left frame) and mass flow rate (right frame) versus r at time step 10400 for $a_{\text{soft}} = 0.75$ run. These two figures have identical zero crossings though different amplitude distributions (adapted from Zhang 2016).

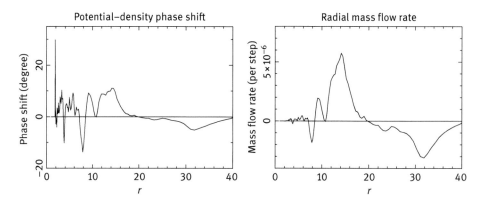

Figure 3.39: Potential-density phase shift (left frame) and mass flow rate (right frame) versus r at time step 5,600 for $a_{\text{soft}} = 0.25$ run. These two figures have identical zero crossings though different amplitude distributions (adapted from Zhang 2016).

causes: (1) noise in the perturbation density and potential used, especially near the region where (and when) the perturbed variables themselves are small; (2) outer disk truncation effect in a finite-domain simulation (Zhang 1998); (3) nonsteady nature of the modes formed; and (4) the formation of nested patterns inside the primary pattern.

From the derivation of expressions for mass flow rates (eq. 2.124) and for phase shift (eq. 2.27), we see that the mass flow pattern *has exactly the same zero crossings* as the phase-shift pattern, yet different radial amplitude profile, due to the fact that for the mass flow rate expression the local surface density, the fractional density wave amplitude, and the galactic rotation curve also play a role, in addition to the phase-shift or normalized torque-integral contribution. Thus the appearance of large-value, multiply-oscillating phase-shift features near the galactic central (or outer) region

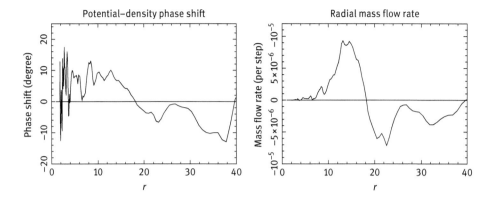

Figure 3.40: Potential-density phase shift (left frame) and mass flow rate (right frame) versus r at time step 8000 for $a_{\text{soft}} = 0.1$ run. These two figures have identical zero crossings though different amplitude distributions (adapted from Zhang 2016).

does not necessarily signal their importance if the wave amplitude there is small, and the phase-shift oscillations are mainly due to noise. This point is made self-evident from comparison of the left and right frames in Figures 3.37–3.40.

3.9.3.6 Comparison of Analytical and Numerical Accretion Rates

The evidence we have presented so far supports the view that the spiral and bar patterns obtained in these N-body simulations were intrinsic modes of the basic state of the disk, and they emerge spontaneously out of originally featureless (axisymmetric) disks beyond the threshold for nonaxisymmetric instabilities. These instabilities display long-range correlation and are self-organized. That natural laws implicitly allow such organized patterns to spontaneously appear when proper initial-boundary conditions are met is part of the reason the universe displays the observed complexity and hierarchical organization, starting from the very simple set of fundamental laws and initial conditions of the Big Bang.

In what follows, we present another set of analyses of N-body results, specifically to check whether the radial mass flow rates obtained in these simulations (responsible for bulge building as well as for the spreading out of the outer disk) agree with analytical prediction using the parameters of the global density wave modes in the same simulations. An agreement between the two lends support both to the analytical formalism, as well as to the relevance of using N-body simulations to characterize the self-organized density wave modes.

In Figure 3.41 we plot the time evolution of enclosed mass at two different radii (one inside and one outside corotation), after subtracting their respective values at the beginning of the simulation for each time step represented, for the previous simulation run using softening length $a_{\text{soft}} = 1.5$. We observe a general trend of mass inflow inside corotation resonance (CR) radius, and mass outflow outside CR. Note

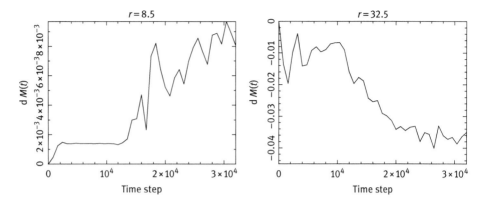

Figure 3.41: Time evolution of the enclosed mass for a typical radius inside corotation, and a typical radius outside corotation, relative to the initial time step $n = 0$ mass at the same respective radius, for the softening length $a_{soft} = 1.5$ run (adapted from Zhang 2016).

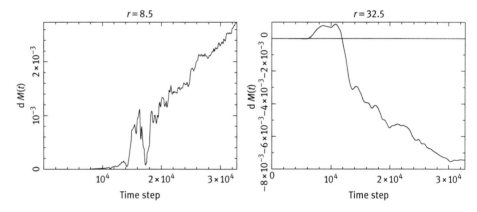

Figure 3.42: Predicted integrated mass evolution according to eq. (2.124) within the two different radii, for softening length $a_{soft} = 1.5$ run (adapted from Zhang 2016).

that the CR radius for this large softening choice is between 25 and 30, with larger value toward the earlier part of the simulation run. We therefore choose to plot the outside-CR mass flow curve at radius 32.5. For progressive smaller softening choices, the CR radius reduces accordingly. So for later plots (a_{soft} = 0.75, 0.25, 0.1) we will choose to plot the outside-CR mass flow curves at radius 27.5.

In Figure 3.42, we plot the *expected* mass increase within the above two radii by *integrating in time* of the instantaneous mass inflow rate predicted by eq. (2.124), using the density wave modal characteristics (i.e., the perturbation potential and density distributions) as well as galaxy parameters obtained in the same simulation run.

Note that the time resolutions for data points in Figures 3.41 and 3.42, as well as for subsequent similar pairs of figures, are slightly different. Figure 3.42 and similar figures below are obtained from instantaneous mass flow rates calculated at every step of the simulation, and these are only recorded at selected radii. Figure 3.41 and similar figures, on the other hand, are obtained using the whole disk surface mass maps written out at selected time intervals (Δt = 800 time steps in this case), so the time resolution is coarser. Nonetheless the gross features of these two kinds of plots can be compared without losing crucial information.

We observe from the comparison of Figures 3.41 and 3.42 that at the location inside CR, r = 8.5, the observed mass increase rate is about three times as large as the analytically predicted rate, though the trends of the mass increase are similar between the two. This large quantitative discrepancy is particular to the a_{soft} = 1.5 run, likely related to the prohibition of the growth of the dominant m = 2 intrinsic instability by the large softening. With the suppression of the growth rate of the dominant m = 2 mode, some three-armed as well as other noisy patterns are seen to emerge. These non-modal patterns can produce radial mass flow yet may not enter into the analytical mass flow prediction.

The mass flow behavior outside corotation likewise shows disagreement between the two plots: the actually measured mass increase in Figure 3.41 shows a consistent outflow pattern as befitting its location outside corotation, whereas the expected mass flow calculated using the torque equation showed initially a mass inflow, and only afterward changed to outflow. This shows that the nonlinear modal organization does not exactly reproduce the quiescent linear modal picture of the smooth launching of the mass flow activity. Also, once again the magnitude of the actually measured outflow rate is significantly higher than predicted for this case.

We had performed another simulation run with a_{soft} = 1.5, but with *the potential calculation keeping only the even azimuthal harmonics* during the Fourier transform process to obtain the grid potential, to reflect the fact that observed spiral and bar patterns show predominant bi-symmetric patterns. The results for the observed and predicted mass accumulations inside and outside CR are shown in Figures 3.43 and 3.44. Here the predicted and measured mass flow rates within CR are seen to be in better agreement, apparently due to the fact that the enforcement of even-harmonics-only coerced a much more effective growth of the collective instability which has bi-symmetry. It is now clear that as long as the global instability has accomplished its growth potential implicit in the mode/basic-state correspondence, through either the tinkering of symmetry (as shown here) or through the reduction of softening (as shown below), the agreement between the theoretical prediction and the actual measurement in N-body simulations will tend to be good.

The even harmonics enforcement is most effective in encouraging larger wave amplitude for the a_{soft} = 1.5 choice. For the other smaller softening choices we will present next (a_{soft} = 0.75 – 0.1), the even harmonics enforcement in fact hampers the wave amplitude growth in the *later stages* of the calculation because there is the natural tendency in the small a_{soft} runs to form local instability clumps in the outer disk, and through the enforcement of bi-symmetry these would-be local clumps were seen

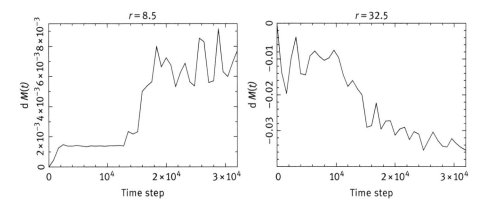

Figure 3.43: Time evolution of the enclosed mass for a typical location inside corotation, and a typical location outside corotation, relative to the initial $n = 0$ mass at the same respective radius, for the softening length $a_{soft} = 1.5$ run, using only the even harmonics in potential calculation throughout the N-body simulation (adapted from Zhang 2016).

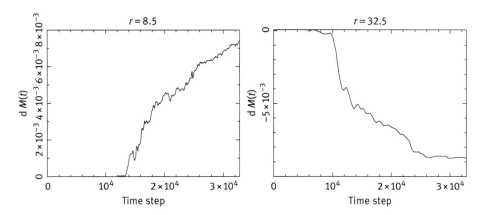

Figure 3.44: Predicted integrated mass evolution according to eq. (2.124) within the two different radii, for softening length $a_{soft} = 1.5$ run, using only the even harmonics in potential calculation throughout the N-body simulation (adapted from Zhang 2016).

to morph into diffused noise which tend to heat the outer disk and inhibit larger wave amplitude to be maintained. Still, tests show that even for $a_{soft} = 0.75$ the enforcement of even harmonics encouraged a more vigorous modal growth during the first 1/3 of the simulation run, and as a result the mass flow rate for that duration of time has better agreement between the predicted and measured values.

In Figure 3.45 we plot the actually measured enclosed mass evolution compared to its value at the beginning of the simulation, for $a_{soft} = 0.75$ run, with the left frame shows the typical mass flow trend inside CR, and the right frame is typical of the behavior outside CR. In Figure 3.46, we plot the expected accumulation of mass within the two radii by integrating the instantaneous mass inflow rate according to eq. (2.124).

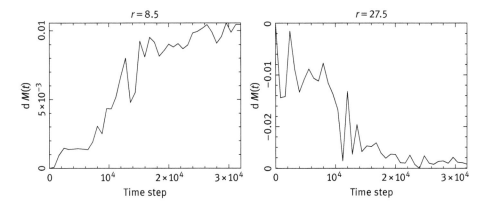

Figure 3.45: Time evolution of the enclosed mass for a typical location inside corotation, and a typical location outside corotation, relative to the initial $n = 0$ mass at the same respective radius, for the softening length $a_{\text{soft}} = 0.75$ run (adapted from Zhang 2016).

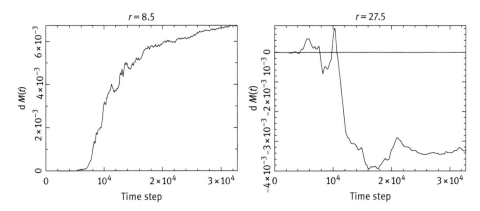

Figure 3.46: Predicted integrated mass evolution according to eq. (2.124) within the two different radii, for softening length $a_{\text{soft}} = 0.75$ run. The second frame (for $r = 27.5$) used only even harmonics in the mass flow calculation (adapted from Zhang 2016).

We see that at $r = 8.5$, the trend of mass accumulation with time between these two figures is extremely similar. Furthermore, the value of analytically inferred rates is only about 30–40% smaller than the actual observed rates. This is a much better agreement than the case we had shown before for the $a_{\text{soft}} = 1.5$ run using all harmonics. This shows that decreasing softening does make a significant difference in our ability to correctly model the collective effects. As we have commented before, enforcing bi-symmetry in the potential calculation will lead to further improved agreement between the predicted and measured mass flow rates, for the initial 1/3 of the simulation duration, before the heating effect sets in.

In the $a_{soft} = 0.75$ run, the predicted mass accretion for $r = 27.5$ has some large fluctuations due to the contamination of multi armed patterns. We therefore chose to use only *even harmonics in the accretion calculation for the $r = 27.5$ location (even though the N-body simulation itself was run with the full harmonics set)*. This procedure in the analysis helps to filter out effect of the transient features and leads to better agreement between the measured and predicted mass accumulation rate at $r = 27.5$, though some residual fluctuations still exist. Once again the magnitude of the outflow is smaller for the predicted than the measured for this run, as for $a_{soft} = 1.5$.

In Figure 3.47 we plot the actually measured enclosed mass evolution at the two different radii compared to its value at the beginning of the simulation, for the

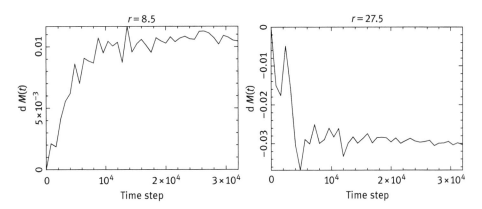

Figure 3.47: Time evolution of the enclosed mass for a typical location inside corotation, and a typical location outside corotation, relative to the initial $n = 0$ mass at the same respective radius, for the softening length $a_{soft} = 0.25$ run (adapted from Zhang 2016).

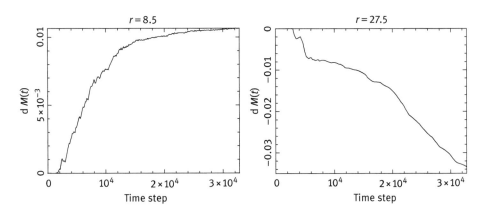

Figure 3.48: Predicted integrated mass evolution according to eq. (2.124) within two different radii, for softening length $a_{soft} = 0.25$ run. The second frame (for $r = 27.5$) used only even harmonics for mass flow calculation (adapted from Zhang 2016).

$a_{soft} = 0.25$ run. In Figure 3.48, we plot the predicted accumulation of mass within the two radii by integrating the theoretical instantaneous mass inflow rate eq. (2.124).

For this run, for the location $r = 8.5$ the measured enclosed mass evolution curve has similar profile as that predicted. Furthermore, we see that the numerical values of analytically inferred and actually measured mass accumulation rates are *nearly identical*. This shows that reducing the softening length to $a_{soft} = 0.25$ allows the interparticle interactions to fully represent the collective effect implicit in the global modal characteristics.

The predicted mass accumulation curve outside CR ($r = 27.5$) was more noisy than the inner region behavior due to the formation of several local instability clumps at the outer region of the computation domain. So we have once again *used only the even harmonics for the mass accretion calculation for the $r = 27.5$ location* (not for the N-body simulation itself, which was performed using full harmonics), which filtered out the contamination of the local transients. Here we witness that the numerical value for the predicted and measured outflow rates is in better agreement than for the previous two larger-softening cases.

In Figure 3.49 we plot the actually measured enclosed mass evolution at the two different radii, compared to their values at the beginning of the simulation, for the softening choice of $a_{soft} = 0.1$, as the most extreme case of softening parameter tests.

Note that the inflection of the curves near time step 5000 for the $a_{soft} = 0.1$ run, inside CR, is linked to the launch of spiral to bar modal shape transition. The spiral phase is seen to lead to higher rate of mass inflow compared to the bar phase due to the more significant skewness and thus the resulting larger potential-density phase shift of the spiral pattern as compared to the bar pattern. This higher mass flow rate for the spirals as compared to bars was also confirmed in the physical galaxy samples analyzed in Zhang and Buta (2007, 2015).

In Figure 3.50, we plot the expected mass accumulation of mass within the two different radii for the $a_{soft} = 0.1$ run by integrating the instantaneous mass inflow rate equation 2.124. We see that the observed and predicted mass flow curves within the $r = 8.5$ region are quite comparable, with the predicted rates smaller than the actual rates by about 30%. Here the difference is obviously not due to the suppression of collective effects because the softening length used is even smaller than in the $a_{soft} = 0.25$ case. The likely cause for this residual discrepancy may be attributed to the viscosity-induced accretion effect (or dynamical friction) not accounted for by the density wave torque calculation.

The mass flow curves outside CR for $a_{soft} = 0.1$ have similar trends, though the predicted rate once again is smaller in magnitude.

Re-examine all of the outside-CR curves in the previous figures, we see that they share the same feature of a more pronounced drop in (measured) enclosed mass at the beginning of the run, signaling a rapid mass outflow episode as the self-organization of the mode begins to take place, that was not reflected in the corresponding predicted mass outflow (this large drop in measured enclosed mass was seen to be the main

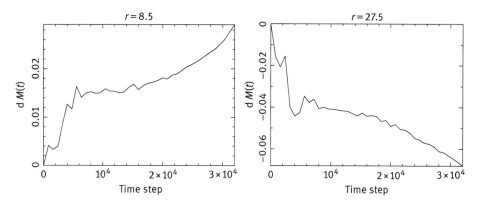

Figure 3.49: Time evolution of the enclosed mass for a typical location inside corotation, and a typical location outside corotation, relative to the initial $n = 0$ mass at the same respective radius, for the softening length $a_{\text{soft}} = 0.1$ run (adapted from Zhang 2016).

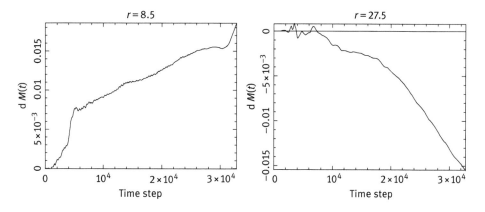

Figure 3.50: Predicted integrated mass evolution according to eq. (2.124) within the two different radii, for softening length $a_{\text{soft}} = 0.1$ run (adapted from Zhang 2016).

reason of the difference in magnitudes of the measured and predicted outflows in most cases because after this initial drop the slopes of the outflows are seen to be quite comparable between the measured and predicted). This common discrepancy may be partly a result of the fact that the outer disk has very low surface density, and the density wave is poorly represented there, so the self-organization process may depart more significantly from the smooth modal-emergence picture.

We also note from the above runs that the net amount of mass *increase* within $r = 8.5$ does not have as substantial a difference between the different softening runs (apart from the extreme case of $a_{\text{soft}} = 0.1$), compared to that given in Figures 3.25 and 3.26 for $r = 3.5$. This is partly accounted for by the shape of the modified exponential surface density for the choice of basic state: as the secular evolution proceeds the entire curve squeezes inwardly, as opposed to a surface density increase *at every radius* as a result of mass inflow (see below the surface density evolution in Figure 3.52).

We choose this larger radius $r = 8.5$ to confirm the analytical mass flow equation because near the very central region the presence of nested resonances complicates the analysis. Even $r = 8.5$ itself is close to the inner Lindblad resonance of the outer mode[†]. A location further out in radius turned out to exhibit even earlier saturation of the accretion behavior because the disk surface density is depleted as a result of the accretion process (see Figure 3.52), thus becoming unable to support the full amount of mass inflow required by the density wave mode.

That $r = 8.5$ location can still be a good representative of the accretion flow due to the outer mode is seen also in the $m = 2$ morphology plot we presented earlier (Figure 3.36). There we see that the dominant mode does penetrate to within the central $r = 8$ region (the nuclear nested mode is well within the central $r = 5$ region), so the mass inflow predictions calculated for $r = 8.5$ are still those for the outer dominant mode. This dilemma of the simulation can be avoided in physical galaxies through the vertical accretion of cold gas onto the galaxy disk, followed by its inward channeling to replenish the depleted surface density of the disk, and to support continued mass accretion throughout the lifetime of a galaxy.

In Figure 3.51, we plot the evolution of the fractional wave amplitude (defined as the *geometric mean* of the density and potential wave fractional amplitudes) during the simulation run, for the four softening choices and at the radius $r = 8.5$. Note the progressive delayed emergence of the wave mode, as well as the decrease of equilibrium amplitude as the softening parameter is increased. Note also that for

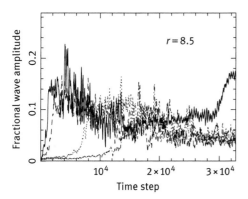

Figure 3.51: Evolution of the fractional wave amplitude for the four softening choices, at radius $r = 8.5$. Solid: $a_{soft} = 0.1$; dashed: $a_{soft} = 0.25$; dotted: $a_{soft} = 0.75$; dash-dotted: $a_{soft} = 1.5$ (adapted from Zhang 2016).

[†] The resonance itself is shielded from being active to the wave by the rapid increase in velocity dispersion in this region, which forms the so-called "Q-barrier" that allows the inwardly propagating (short trailing) density wave train to be reflected to become outward propagating (long trailing) wave train toward corotation region to complete the feedback loop and to form growing modes (Mark 1976).

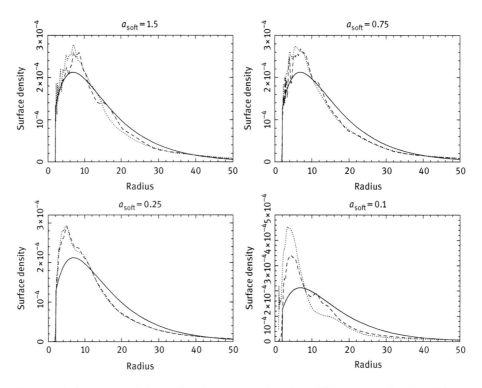

Figure 3.52: Comparison of the surface density evolution of the N-body runs using four different softening parameters, and all with 20 million particles. The different curves in each figure are for time steps of 0 (solid), 16384 (dashed), and 32768 (dotted), respectively. Note that the last frame (for a_{soft} = 0.1) has a different vertical scale than the other three frames (adapted from Zhang 2016).

a_{soft} = 0.1 case the amplitude is reinvigorated toward the end of the simulation run due to the strengthening of the central bar as a result of previous mass inflow – this fact can also be discerned from Figures 3.49 and 3.50 as the steepening of the enclosed mass slope in the last quarter of the run, as compared to the corresponding figures for a_{soft} = 0.75 (Figures 3.45 and 3.46), and for a_{soft} = 0.25 (Figures 3.47, 3.48), where the enclosed mass flattens out toward the end of the simulation, indicating the slowing down of accretion behavior (which is matched by their respective decreased wave amplitudes as shown in Figure 3.51). Of course, the mass flow rate is determined not only by the fractional wave amplitude, but also by the potential-density phase shift, as well as by other wave and basic state parameters (eq. (2.112)), all of them working in concert to produce the final radial mass flow rate.

In Figure 3.52, we plot the surface density evolution of the above four runs. This plot corroborates what we had found before from the plots of mass inflow within the central r = 3.5 and r = 8.5 radius, as well as outside the corotation radius. The a_{soft} = 0.1 choice is seen to be the most effective in producing enhanced mass flow rates due to its larger wave amplitude and most persistent density wave activity.

3.9.4 Grid Noise Associated with the Use of Small Particle Softening Parameter

In this subsection we address the issue of whether the quality of the simulation results presented in this work is negatively affected by the use of small particle-softening parameter a_{soft}. We will show that the *effective* disk thickness in these 2D simulations is maintained mainly by the finite size of the grid, whereas the effective interparticle interaction which allows the operation of collective instability is enhanced by the choice of small particle softening parameter a_{soft}. The combined effects of the two, coupled with sufficient number of simulation particles and finer grid, allow the *macroscopic* features of the modes to be faithfully reproduced despite of the increased noise in individual particle's orbit.

3.9.4.1 The Differing Roles of Grid and Particle Softening

In the simulations described in the last subsection, while the particle softening parameter a_{soft} had been gradually reduced to significantly lower value than commonly used, the corresponding grid softening had not been reduced to a similar extent (the newer grid is only about a factor of 4 finer in linear resolution than used in Zhang 1998, whereas the particle softening is reduced by a factor of 15 for the extreme choice of $a_{\text{soft}} = 0.1$). It appears from our explorations that particle softening is the main inhibitor to obtaining higher mass flow rates in N-body disks containing collective modes, the grid softening has only marginal effect in this regard.

The finite amount of grid softening, on the other hand, helps to maintain the desirable finite thickness effect in these 2D simulations. The finite *effective* disk thickness achieved in our new set of simulations manifests both in the N^2 dependence of the noise behavior (Figure 3.26) since the relaxation behavior in a razor-thin disk would otherwise have showed no particle-number dependence (Rybicki 1971); as well as in the longevity of the spiral-bar patterns formed since a razor-thin disk is expected to have its noise relaxation time scale on the order of a mean orbital period (Rybicki 1971), whereas the patterns in our simulations using sufficient number of particles were shown to last more than 25 orbital periods.

In fact, Rybicki (1971) already pointed out the role of finite grid size on mimicking a finite-thickness disk. He derived that for a numerical disk which has grid size of h, the relaxation time t_R can be expressed as

$$t_R = \frac{\lambda^3 N h t_M}{2R} \qquad (3.20)$$

where λ is the ratio of random velocity σ to circular velocity V, that is, $\sigma = \lambda V, \lambda \leq 1$, and t_M is the orbital crossing time, that is, $t_M = R/V$ where R is the disk radius.

The above expression of Rybicki's can be shown to be nearly equivalent to another expression derived in White (1988)

$$t_R = \frac{\sigma^3 h}{5G^2 \mu m}, \tag{3.21}$$

where μ is the surface density of the disk, G is the gravitational constant, and m is the mass of the particle. White's expression can be shown to be equivalent to Rybicki's, after replacing the factor of 5 in the denominator of White's expression by a factor of 2π in Rybicki's (or π if the crossing time is defined using the diameter instead of the radius of the system). Rybicki (1971) stated that the factor h is either the particle softening length a_{soft} or the grid size h, whichever is greater. Romeo (1994b) and Donner and Thomasson (1994) argued that the effective softening length is rather given by an expression combining both particle and grid softening, that is, $h = \sqrt{s^2 + a_{soft}^2}$, where s is the local grid size. After this substitution, this expression is essentially the same as that used earlier, eq. (3.10).

The above quoted analyses show that relaxation rate is controlled equally by the finite grid size as well as by the particle softening length. The finite grid size can simulate a finite-thickness disk (in terms of inhibiting the unwanted fast relaxation) just as effectively as the particle softening, yet finite grid size does not seem to significantly diminish the desired collective dissipation effect as the choice of large particle softening length does. This may partly be due to the fact that grid softening is implemented in an anisotropic fashion (i. e., the mass and force assignments through the CIC scheme retain to some degree the angular inhomogeneities due to the correlated particle distributions, not only for the particle positions, but also for the correlation between the positions and kinematics of particles), whereas the particle softening is an isotropic scheme that uniformly evens out the fluctuations in the force, thus has a more detrimental effect on suppressing the correlated collective interactions.

Figure 3.53 gives the relaxation time for our previous 20 million particle, $a_{soft} = 0.1$ simulation calculated using the White (1988) expression above, with h given by the root-mean-square of the particle and grid softening (times $\sqrt{2}$). The straight line in the figure shows the maximum time step duration (32768) used in the actual simulation presented here. It can be seen that over most of the radial range the relaxation time is much longer than the duration of the simulation, so two-body relaxation effect should be minimum, as was also borne out by the consistently vigorous spiral/bar activity shown earlier.

3.9.4.2 Grid Noise in Particle Mesh Simulations

By choosing to use grid sizes larger than the particle softening parameter (due to our choice of the smaller particle softening), we do pay a price: that of increased grid noise. This manifests as fluctuations of the interpolated forces on individual particles

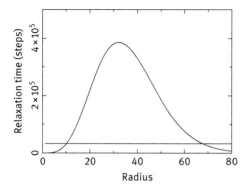

Figure 3.53: Relaxation time in the 20 million particle simulation using $a_{soft} = 0.1$. The straight line in the lower part of the figure indicates the maximum calculation duration (32768 steps) used in the simulations.

as we traverse across grid boundaries, which can also be attributed to an aliasing effect in the Fourier transform domain (Hockney & Eastwood 1988, Section 5-6-3 and Section 5-6-4).

Grid noise is more pronounced for the smallest softening choices ($a_{soft} = 0.1$) than for larger softening, and it is also more pronounced in the outer disk region than in the inner disk region, for the polar grid used here, since the grid spacings in the outer disk are exponentially larger than in the inner disk (therefore, for *all* the past polar grid simulations presented in the literature, the grid noise issue was already present, since even when the particle softening parameter was chosen to be of the order of the grid size in the inner disk, for the outer disk region the grid size is much larger than the particle softening parameter, due to the exponential growth in grid size in polar grid design).

In what follows the results of several Monte Carlo calculations are presented to characterize the magnitude of grid noise for the simulation parameters we have used in this work. A source particle is placed at a select radial location in the disk (here we test two radius values, roughly corresponding to the inner Lindblad resonance and the corotation resonance of the mode, which give approximately the best-case and the worst-case error bounds for a modal resonant cavity between the inner disk and the corotation radius), and 100 test particles are placed at each of a range of distances from the source particles. Analytical results (from the Newtonian equation with softening) are compared with that obtained in the N-body (interpolated) grid force calculation scheme, averaged for the 100 test particles for each distance, at a particular choice of a_{soft}. In Figures 3.54 and 3.55, we plot the noise performance of the Plummer softening scheme for the choices of softening parameters and grid sizes used in this study. Note that even though the larger softening case ($a_{soft} = 1.5$) has its CR closer to 30, this large softening case has relatively small grid noise so the extent of the noise performance in that case (plotted only up to radius of 20 but can be easily extrapolated up to 30) is no worse than the worse-case scenario (for $a_{soft} = 0.1$) represented in these figures. The trend of increasing grid noise with decreasing softening, with increasing galactic radius (and thus grid size), and with overall grid coarseness, are evident in these plots.

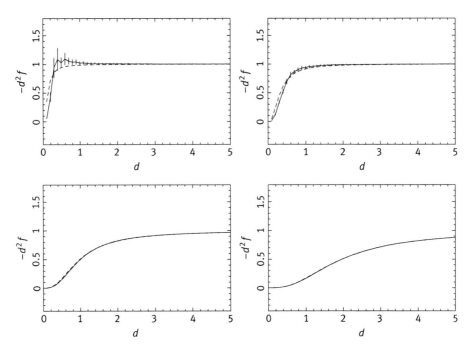

Figure 3.54: Grid noise estimates for N-body particle-mesh simulations using Plummer sphere softening with $a_{\text{soft}} = 0.1$ (top left), $a_{\text{soft}} = 0.25$ (top right), $a_{\text{soft}} = 0.75$ (bottom left), and $a_{\text{soft}} = 1.5$ (bottom right), respectively. All the calculations start with a point-mass source particle at $r = 10$, and with 100 test particles at an average distance d (varying from near zero to about five in the normalized unit of the disk) from the source particle randomly placed within the computational grid of 220 exponentially spaced radial sections and 256 equally spaced azimuthal sections. Solid lines are expected N-body performance with $1\,\sigma$ rms fluctuations, and dashed lines are the corresponding theoretical softened forces (adapted from Zhang 2016).

3.9.4.3 Effects of Grid Size and Grid Noise to Mass Flow Rates in N-Body Simulations

In this section we demonstrate further that the increased mass flow rates in less-softened N-body runs are due to the collective instabilities brought about by the decreased softening, and not due to increased grid noise brought about by the same softening reduction.

The first hint for this conclusion can already be found in the previous simulation result shown in Figure 3.26. It was seen there that when an insufficient number of particles are used for a given a_{soft} (which generally *increases* the few-particle relaxation effect, or increases the noise), the mass inflow tapers off at an earlier time step. Whereas for very large particle numbers (which generally *decreases* the few-particle relaxation effect, or decreases the noise), both the spiral/bar activity and the mass inflow continues until the end of the run. This positive correlation between the number of simulation particles and the longevity of efficient mass inflow shows that noise

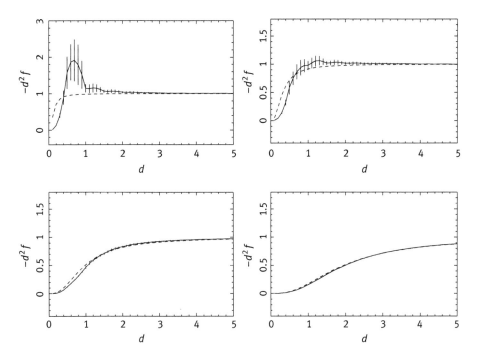

Figure 3.55: Grid noise estimates for N-body particle-mesh simulations using Plummer-sphere softening with $a_{soft} = 0.1$ (top left), $a_{soft} = 0.25$ (top right), $a_{soft} = 0.75$ (bottom left), and $a_{soft} = 1.5$ (bottom right), respectively. All the calculations start with a point-mass source particle at $r = 20$, and with 100 test particles at an average distance d (varying from near zero to about five in the normalized unit of the disk) from the source particle randomly placed within the computational grid of 220 exponentially spaced radial sections and 256 equally spaced azimuthal sections. Solid lines are expected N-body performance with 1 σ rms fluctuations, and dashed lines are the corresponding theoretical softened forces (adapted from Zhang 2016).

is unlikely to be the *cause* of the increased mass flow rate. Instead, the survivability of the density wave modes appears to be the key to the continued mass inflow.

In Figures 3.56 and 3.57, we plot the same mass flow evolution as previously simulated in Figures 3.45 and 3.46 for the softening parameter choice of $a_{soft} = 0.75$, this time reducing the grid linear resolution by one half, and time resolution by one half correspondingly (i.e., 16384 steps here cover the same time duration in terms of pattern rotation periods as 32768 steps before). The corresponding grid noise estimates are given in Figure 3.58.

Decreasing grid resolution by one half should significantly *increase* the grid noise, from the result we presented above. Yet the results here show comparable mass flow rates compared to that obtained from the higher grid resolution ones (the slight *decrease* in mass flow rate is due to the contribution of grid-softening effect to mass flow, which is apparently not as prominent as particle softening; yet even this decrease is in the opposite direction to what one would expect if grid noise is the main

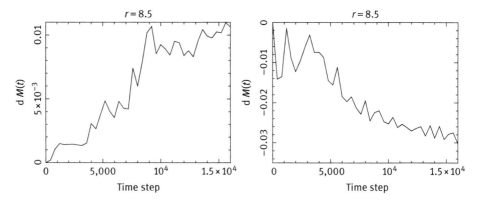

Figure 3.56: Time evolution of the enclosed mass for a typical location inside corotation, and a typical location outside corotation, relative to the initial $n = 0$ mass at the same respective radius, for the softening length $a_{\text{soft}} = 0.75$ run, with a coarser mesh (110 radial sections, 128 azimuthal sections) and twice as coarse a time resolution as before (adapted from Zhang 2016).

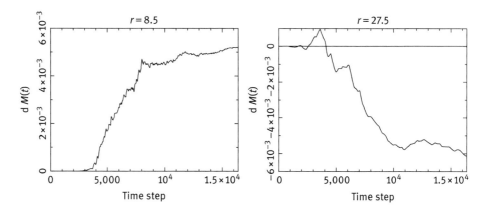

Figure 3.57: Predicted integrated mass evolution according to eq. (2.124) within the two different radii, for softening length $a_{\text{soft}} = 0.75$ run, but with a smaller grid (110 radial sections, 128 azimuthal sections) and twice as coarse a time resolution as before. The second frame (for $r = 27.5$) used only even harmonics in the mass flow calculation (adapted from Zhang 2016).

contributor to the *increase* in mass flow rate in simulated galaxies), especially for the first one-half of the duration of the simulation run, when the heating due to noise has not clamped the amplitude of the pattern for this coarser-grid run. This shows that we are indeed revealing the role of increased collective effects brought about by decreased particle softening, and the increased mass flow rate is not due to increased grid noise (if anything, increased grid noise, and the resulting increased heating of disk particles, slightly hampered the mass flow in the later half of the simulation period, as evidenced by the flattening of the mass-increase curves in both Figures 3.56 and 3.57).

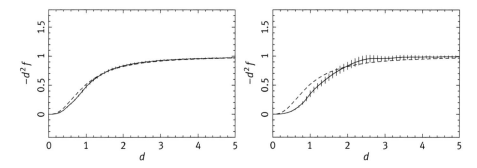

Figure 3.58: Grid noise estimates for N-body particle-mesh simulations using Plummer-sphere softening with $a_{\text{soft}} = 0.75$. All the calculations start with a point-mass source particle at $r = 10$ (left) or $r = 20$ (right), and with 100 test particles at an average distance d (varying from near zero to about five in the normalized unit of the disk) from the source particle, randomly placed with the computational grid of 110 exponentially spaced radial sections and 128 equally spaced azimuthal sections. Solid lines are N-body performance with 1σ rms fluctuations, and dashed lines are the corresponding theoretical softened forces (adapted from Zhang 2016).

Therefore, we conclude that as long as our interests are in the comparison of macroscopic characteristics of the mode (equilibrium amplitude, pitch angle, potential-density phase shift and torque) in the simulations with that in theoretical predictions, as well as with that in observed galaxies, the grid noise present in simulations with smaller softening is not a debilitating inconvenience.

3.9.5 Accuracy and Implications of the Secular Radial Mass Flow Rates Obtained in N-Body Simulations

In the last subsection, we have presented detailed analyses of the effect of the choice of softening parameter on the accuracy of particle-mesh N-body simulations, particularly for the polar-grid configuration. We see there that grid noise can be of concern for small-softening, large-grid-size simulations (see also Efstathiou et al. 1985, Figure 1). For simulation parameter combinations where a_{soft} is small, the grid used is coarse, and for outer disk region where the cell size is large, grid noise has a significant impact on the accuracy of individual star's orbit – if the accuracy of individual orbit is what we are after in the simulation effort.

However, as we have shown in this work, when simulating global density wave modes in galactic disks, grid noise does not appear to have a detrimental effect on the macroscopic properties of the modes, as well as on the *correlation* of modal parameters and their associated basic state evolution rates.

This positive outcome becomes easier to understand if we recall that in a disk system possessing collective instabilities, individual simulated orbits had long been known to diverge exponentially from true orbits in the presence of random noise, especially for simulations dealing with collective instabilities (see, e. g., Miller 1971, Pfenniger 1986, Romeo 1990, Weinberg 1993). This exponential divergence between the simulated individual orbits and their true counterparts in globally unstable systems is present *even when large softening length is used*.

On the other hand, the collective behaviors of the modes being modeled were found not to be impacted by the microscopic inaccuracies of the simulated individual orbits – otherwise none of the previously published N-body simulations of galaxies containing unstable density wave patterns can be trusted! This fidelity of simulated global pattern in the face of drastic (exponential) individual orbit inaccuracy comes about because self-organized global structures such as spiral and bar modes are "dissipative structures" in the sense of Prigogine and coworkers (Prigogine 1980), or else "strange attractors" in the sense used in nonlinear dynamical systems (Wiggins 2003 and the references therein). These systems automatically seek their attractor/modal solutions even if perturbed from corresponding nonequilibrium steady states. This property is the well-known "asymptotic stability" of the dissipative nonequilibrium quasi-steady state. In the case of galaxies possessing density wave modes, the effective collision/scattering processes at the crest of density wave patterns reestablish the correlations needed for the self-organization and collective dissipation processes, even if microscopically these orbits are not exactly what they would be for an infinite-precision calculation. From this perspective it also becomes clear why a smaller-softening choice which allows the proper establishment of correlations among particles in a simulated galaxy disk is more important than the accurate modeling of a large-softening disk which lacks the proper correlations among particles: the near-collision (or the scattering) condition allowed by small softening (or efficient interparticle interaction) is what establishes the correlations among particles to allow the collective instability to successfully operate.

As we have seen, macroscopic properties of the evolution of the global density wave patterns, as well as the evolution of the basic state, are not impacted by the inaccuracies in the behavior of the individual particles' orbits. Furthermore, if the noise in individual stellar orbits had mattered, even in physical galaxies we should not have been able to observe the beautifully organized grand-design density wave structures since in these galaxies the stellar orbits are perturbed by chance encounters with the Giant Molecular Cloud Complexes, the tidal companion galaxies, as well as the formation of disk open and globular star clusters. The stability of the global pattern characteristics against random fluctuation is indeed what allowed these patterns to emerge as a predominant form of organization in disk galaxies, and to have the modal characteristics correlate with that of the basic state[†], rather than with the chance elements of internal and external random perturbations.

[†] We recall that one of the criteria of the Hubble classification scheme is the empirically observed correlation of spiral pitch angle with the degree of prominence of the bulge. This correlation has been reproduced in the modal picture of the galactic density waves (Bertin et al. 1989a, 1989b).

One question that may arise from examining the results of this work is that since the different softening-parameter runs gave different mass flow rates, what magnitude of the mass flow should we associate with the secular morphological evolution of physical galaxies?

We note that the aim of these simulations is not to derive an *absolute* mass flow rate and claim that it corresponds to that in *all* physical galaxies, but rather to demonstrate the *correlation* between the N-body simulated mass flow rate and the rate predicted by the volume-torque/mass-flow-rate equation (i. e., eq. (2.124)). A confirmation of this correlation shows that the analytical equation predicts the correct mass flow rates for *all* ranges of the wave and basic state parameters, thus the rate equation can be used in physical galaxies with the actually measured parameters for the wave and the basic state.

In the simulation results presented in this section, we had indeed shown that once the collective effects are allowed to be fully represented either by the use of sufficiently small a_{soft}, or else by the enforcement of special bi-symmetry condition, *all* of the measured mass flow rates in N-body simulations at a wide range of softening lengths have the correct correspondence with the respective analytical predictions, especially for the radial range inside corotation, which is our main interest in the secular evolution studies. Therefore, these pattern morphologies and mass flow rates can *all* find correspondence with physical galaxies of varying Hubble types. In fact, a little contemplation will show that the larger softening choices correspond to the thicker inner disk scale heights and smaller secular mass flow rates of the earlier Hubble-type galaxies, whereas the smaller softening choices correspond to the thinner disk and the larger secular mass flow rates of later Hubble-type galaxies (Zhang & Buta 2007, 2015).

4 Astrophysical Implications of the Dynamical Theory

The framework provided by the dynamical theory allows a greater synthesis on a diverse range of astrophysical phenomena pertaining to the observed features of galaxies. In the following sections, we will present further astrophysical inferences from the theory, with the content already discussed in the previous chapters summarized as needed. Most of the results presented in this chapter had previously appeared in Zhang (1996, 1998, 1999, 2002, 2003, 2004, 2008, 2016); Zhang & Buta (2007, 2012, 2015); Buta & Zhang (2009); as well as Zhang, Wright, & Alexander (1993) and Zhang et al. (2001).

4.1 Motivations and General Outline

The central thread in the presentations that will follow is once again the radial distribution of the azimuthal phase shift between the potential and density waveforms for self-organized and self-sustained density wave modes in disk galaxies. This central role is due to the fact that a secular angular momentum exchange process between the wave mode and the basic state disk matter, which serves as the driver of the secular morphological evolution of galaxies, will necessarily manifest as a characteristic potential-density phase shift (PDPS) distribution. This distribution is both the symptom and the dynamical agent for the secular evolution process.

As a result of the presence of the PDPS for an unstable density wave mode, the disk material (including both stars and gas) inside corotation spirals inward, and the material outside corotation drifts outward, and this leads to the secular evolution of the disk surface density. The time scale of the radial mass accretion process for a galaxy like our own is estimated to be on the order of a Hubble time, which could account for the formation of the quasi-exponential disk surface density profile and the formation of bulges in galaxies.

For stars moving on nearly circular orbits, the ratio of their energy loss to the angular momentum loss is proportional to Ω, the circular speed of the disk. On the other hand, the ratio of energy to angular momentum that can be absorbed by the density wave is proportional to Ω_p, the pattern speed of the wave. Since inside corotation $\Omega > \Omega_p$, the complete transfer of angular momentum from the disk material to the wave means that the energy released by the disk material cannot be completely absorbed by the density wave. The surplus of this released energy contributes to the secular heating of the disk stars as they migrate inward. This is found to quantitatively explain the observed age–velocity dispersion relation of the solar neighborhood stars. A corresponding energy injection into the interstellar medium (ISM) serves as the top-level energy input to fuel the cascade of interstellar turbulence and can quantitatively account for the so-called Larson law or the size–line-width correlation of the molecular clouds in our Galaxy (Larson 1981). Similar exchanges between the wave and the disk matter also occur outside corotation.

The trend of evolution of the basic state of a spiral galaxy due to the spiral-induced collective dissipation process coincides with the trend of the Hubble sequence of galaxy classification from the late to the early types, whereby a galaxy gradually acquires a thicker inner disk and a larger bulge. This change in the basic state property, in turn, results in the change of the kind of spiral/bar modes present, from the more open spiral type to the more tightly wound spiral type, or else from the open spiral to a nearly straight bar, both leading to smaller PDPS (recall that the phase shift is largest for spiral pitch angle close to 45°), which in turn reduces the speed for further morphological evolution. This correlation of the basic state mass distribution and the density wave modal shape present is consistent with the correlation observed in the Hubble classification scheme (Hubble 1936) and is also consistent with predictions from the linear modal calculations (Bertin et al. 1989a, 1989b).

The secular evolution mechanism presented in this work provides a natural framework for understanding the results of many space- and ground-based observations. It also provides solutions to several long-standing problems in the theory and observations of galaxy formation and evolution, among them the cause of the steepening of the galaxy morphology–environment relation with time; the similarity between the Tully–Fisher and Faber–Jackson relations for spiral and elliptical galaxies, respectively; the existence of a fundamental–plane relation for elliptical galaxies; as well as the mass-angular momentum anticorrelation observed along the Hubble sequence.

One issue is that, for galaxies that have been observed in the Hubble Deep and Ultra Deep Fields, many do not seem to conform to the classical Hubble classification. One might thus wonder: what is the rationale for studying the evolution along the Hubble sequence of nearby galaxies if the majority of the high-z galaxies do not even fall onto the Hubble sequence?

We want to say at the outset that what we are trying to explore and establish is a broad trend that as the universe evolves, the galaxy population as a whole progressively evolves from disk-dominated systems to bulge-dominated systems. There is plenty of observational evidence now that the disk fraction was higher at $z = 1$ and has since decreased (Lilly et al. 1998). This is most naturally explained by secular evolution since the number density of galaxies of all types has not evolved much over the same time range (Cohen 2002; Conselice et al. 2016), and the merger fraction since $z = 1$ is low (Conselice et al. 2003; López-Sanjuan et al. 2009). The secular evolution produced by the PDPS is due mainly to the skewness of the density wave pattern and does not require that these patterns fall into the "grand design" category; even some 3D twisted isophotes observed in high-z proto-galaxies may be modal patterns in the making. As long as galaxies possess the innate capability for self-organization and collective dissipation, which naturally produces skewed modal patterns, the mechanism we discussed in this work will be generally applicable.

The recently discovered evidence of the early assembly (since $z = 2$) of the backbone of the Hubble sequence (Lee et al. 2013) in the GOODS-South field galaxy sample observed by HST/ACS and HST/WFPC3 is consistent with the overwhelming evidence from observational studies by several teams, as compiled in Conselice et al. (2016), that in fact up to $z = 2$ the evolution of co-moving total number density of galaxies is hardly discernible. These new evidence, coupled with the previously known facts that roughly two-thirds of all observed galaxies are disks, and three-quarters of bright galaxies are disk galaxies (Freeman & McNamara 2006), all point to that a galaxy evolution study in mainly disk-dominated configurations is of high relevance.

Another natural question to ask is: what would serve as a starting point of secular evolution along the Hubble sequence? An obvious choice would be the very late type disks, that is, Sd/Scs, and low surface brightness galaxies (LSBs), since these are at the tail end of the Hubble sequence. One subtle point here is that the present-day LSB and late-type galaxies could not have been the ancestors of the present-day early-type galaxies, both because of their formation epoch difference and because of their statistical size/mass difference. We want to bring up here an analogy of the galaxy evolution scenario with the Darwinian evolution theory. According to this theory, humans evolved from an ape-like ancestor, but this is not the same as saying modern-day apes will one day evolve into modern-day humans. This difference is the well-known distinction between an evolution-tree (which has historic connotation) and a ladder-of-life (which is used for present-day classification of species). The two phrases have close connections, but should not be confused with each other. For the problem of the secular evolution of galaxies, the same principle applies: we can fully envision that the ancestors of the present-day early-type galaxies were disk systems much larger in size than present-day late-type disks, and they appear much earlier in the evolution history of the universe. Some of these ancient large disks may be the recently discovered rotationally supported high-z disks (Genzel et al. 2006; Ceverino et al. 2012), as well as the large rotationally supported Damped Lyman Alpha (DLA) systems (Wolfe 2001). Many of the fainter outer disks of these high-z systems may lie beyond detection (Hopkins et al. 2009), and thus we need also to take into account the possible selection effects in interpreting high-z results (this aspect is related to the difficulty in detecting the high-z quasar host galaxies). The observed "downsizing" trend of galaxy mass assembly (Nelan et al. 2005; Cimatti, Daddi, & Renzini 2006) also gives support to the idea that the first disk galaxies formed are likely to be larger ones.

There are theoretical reasons why galaxy evolution tends to proceed through a disk phase. In a dissipative self-gravitating system, the natural step of relaxation is first to settle onto a disk with the axis of rotation corresponding to the axis of maximum momentum of inertia (Zeldovich 1970). Such a system subsequently continues its entropy-increasing evolution toward the direction of increasing central concentration together with an extended outer envelope, and it gets rid of the excess angular momentum during this secular evolution process through the formation of large-scale collective instabilities (i. e., density wave modes).

Besides considering purely gravitational processes for galaxy formation, it is essential to point out that we need also to take into account the evidence that early galaxy formation may involve a significant role played by primordial turbulence and shocks (Ozernoy 1974a, 1974b; Bershadskii & Sreenivasan 2003; Zhang 2008 and the references therein). This more violent disk formation scenario is different from the quiescent formation process of the LSB galaxies in the nearby universe, with these recent formation processes mostly due to gravitational collapse alone.

Despite the heterogeneity of the initial conditions, which allow some galaxies to bypass certain stages of the evolution along the Hubble sequence, or else allow the galaxies to fast-evolve through the initial stages to build up a substantial bulge relatively early in time, the general trend of galaxy formation and evolution still appears to be through a disk-dominated configuration (van Dokkum et al. 2011; Lee et al. 2013). This is only reasonable because the large-scale gravitational instabilities in the form of density wave modes, which form spontaneously in disk galaxies, can accelerate the speed of entropy evolution by many orders of magnitude compared to that for a smooth distribution, and nature, as it appears, has always chosen her dominant evolution configuration that maximizes the entropy production rate.

It is for these reasons that the study of the evolution of nearby galaxies along the Hubble sequence can be used as a template for what could have happened in galaxies in the more distant universe, even knowing that for individual galaxies this trend may not be all smooth and identical, that is, the initial conditions may not all have been that of a thin disk, and the smooth disk-dominated evolution could be interrupted by a merger or satellite accretion. The same phase-shift-induced collective dissipation process is likely to operate in configurations that have a vertical extent as well: that is, the halo and bulge regions of galaxies may possess skewness, which leads to the same kind of PDPS as in thin disks, and thus will have accelerated evolution according to the same basic dynamical process here shown for flat systems.

4.2 PDPS Method for CR Determination[†]

We begin our observational study by applying our theoretical results on the determination of the corotation radii in physical galaxies containing density wave modes, using the so-called Potential-Density Phase Shift (PDPS) method. This apparent kinematic application reveals the deep dynamic roots for galaxy morphology.

4.2.1 Dynamical Basis and Practical Considerations for the PDPS Method

The determination of the corotation resonance (CR) radii or pattern speeds of the galactic density wave patterns was a difficult problem, and few general methods existed for model-independent determination especially of the multiple pattern

---

[†] Portions of this section used material previously published in Zhang & Buta (2007) and Buta & Zhang (2009), reproduced with modifications from The Astronomical Journal @ AAS and The Astrophysical Supplement Series @ AAS. Reproduced with permission.

speeds of the nested wave patterns in galaxies (see Zhang & Buta 2007 for a compiled list of the past proposed methods for pattern speeds and corotation radii determination).

The new dynamical mechanism presented in the current work serves as the basis for a novel and general approach to locating the corotation radii of density waves in disk galaxies: the PDPS method (Zhang & Buta 2007, 2015; Buta & Zhang 2009). The technique employs the use of near-infrared (NIR) and the shorter-wavelength bands of the mid-infrared (MIR) images from ground- and space-based surveys to infer densities and gravitational potentials. These regions of the spectrum contain radiation contributions mainly from old giant and supergiant stars (Frogel et al. 1996), and are considered much better tracers of the stellar mass distribution than optical bands. The effects of dust extinction in these spectral regions are also considerably reduced.

As we have shown, one of the predictions of the new dynamical theory is that for a self-sustained spiral or bar mode *the radial distribution of the PDPS should be such that it changes sign at the corotation radius*. The physical basis for this distribution lies in the fact that since the wave rotates slower than the disk stars inside corotation, and thus has negative energy and angular momentum density with respect to the disk stars, a positive sign of the phase shift (defined as when the potential pattern lags the density pattern in azimuth) for a radial location inside CR leads to an angular momentum loss of the disk stars, which further leads to the spontaneous growth of the wave mode in the linear regime, and to the damping and stabilization of the growing wave amplitude toward its quasi-steady-state value in the nonlinear regime. Similarly the negative sign of the phase shift outside CR also leads to the growth and the stabilization of the wave mode in that region since outside CR the wave has positive energy and angular momentum density relative to the basic state, and a dumping of angular momentum by the wave to the disk, consistent with the potential-leading-density sense of the phase shift distribution, also promotes the wave growth and stabilization. Therefore, the shape and sign change at CR of the PDPS distribution is perfectly matched to the need for the mode to spontaneously emerge as a global instability pattern initially, and to subsequently be stabilized through the dissipative interaction with the basic state.

As we have shown in Section 2.3.1, at the quasi-steady state, the rate of angular momentum exchange between a skewed density wave pattern and the basic state of the disk, per unit area, due to the torquing of the wave on the disk, is given by

$$\overline{\frac{dL}{dt}}(r) = \overline{\mathcal{T}}(r) = -\frac{1}{2\pi}\int_0^{2\pi} \Sigma_1(r,\phi)\frac{\partial \mathcal{V}_1(r,\phi)}{\partial \phi}d\phi, \quad (4.1)$$

where Σ_1 represents the perturbation density waveform and $\mathcal{V}_1$ the perturbation potential waveform (those with the circularly symmetric $m = 0$ component already subtracted). For two sinusoidal waveforms, we have also shown in Section 2.6.1 that the angular momentum exchange rate can alternatively be written as

$$\frac{dL}{dt}(r) = (m/2) A_\Sigma(r) A_{\mathscr{V}}(r) \sin[m\phi_0(r)], \qquad (4.2)$$

where

$$A_{\mathscr{V}} = \frac{\sqrt{\int_0^{2\pi} \mathscr{V}_1^2 d\phi}}{\sqrt{2\pi}} \qquad (4.3)$$

and

$$A_\Sigma = \frac{\sqrt{\int_0^{2\pi} \Sigma_1^2 d\phi}}{\sqrt{2\pi}} \qquad (4.4)$$

are the amplitudes of the potential and density waves, respectively, m is the number of spiral arms, and ϕ_0 is the phase shift between these two waveforms, defined as being positive if the potential lags density in the azimuthal direction in the sense of the galactic rotation. For most galaxies the sense of galactic rotation can be determined by the assumption that the density wave pattern is trailing, although exceptions do exist which will need to be studied on a case-by-case basis.

Using the above two equations, the radial distribution of an equivalent azimuthal phase shift $\phi_0(r, \phi)$ between the (generally nonlinear and nonsinusoidal) potential and density patterns for spirals or bars in a disk galaxy can be calculated from

$$\phi_0(r) \equiv \frac{1}{m} \sin^{-1}\left(\frac{1}{m} \frac{\int_0^{2\pi} \Sigma_1 \frac{\partial \mathscr{V}_1}{\partial \phi} d\phi}{\sqrt{\int_0^{2\pi} \mathscr{V}_1^2 d\phi} \sqrt{\int_0^{2\pi} \Sigma_1^2 d\phi}} \right), \qquad (4.5)$$

where the potential $\mathscr{V}$ is calculated from the Poisson integral of the density distribution, with the density distribution itself calculated from the NIR image of a galaxy for an assumed mass-to-light (M/L) ratio distribution. The equivalent phase shift is the amount of phase shift which would be present between two sinusoidal waveforms if each is endowed with the same energy as the corresponding nonlinear waveform and which would lead to the same value for the torque integral as would the nonlinear waveforms. The phase shift is in general nonzero as long as the density pattern is skewed, that is, in the case of spirals, twisted or offset bars, or even some twisted three-dimensional mass distributions as observed in many high-redshift proto galaxies.

Note that even though these waveforms are given in their perturbational form (with subscript 1 in the above equations), in carrying out the actual calculation for the numerator we can simply use the nonperturbed variable-equivalent in the numerator since the effect of the circularly symmetric component in $\mathscr{V}$ and Σ will (coupled with differentiation) integrate out to zero. This reduces the sensitivity to uncertainties in the M/L ratio of any axisymmetric mass component (or in fact, any nonskewed component, including triaxial luminous or dark halo component), including the axisymmetric bulge, halo, and disk components. The normalization factor in the denominator, however, will be affected by the axisymmetric components, so these should be subtracted out before doing the normalization. However, any uncertainty in the normalization has no effect on the locations of the positive-to-negative *zero crossings* of the phase shift versus radius curve, which are what we use in the corotation determination.

We assume the phase shift to be positive when the potential pattern lags the density pattern in the direction of galactic rotation. Since the face-on view of the spiral or twisted bar in disk galaxies comes in two flavors, that is, either the S-sensed pattern or the Z-sensed pattern, these in general indicate a counterclockwise or a clockwise rotation direction, respectively, if the pattern is assumed to be trailing (the exception to this assumption does exist, such as for NGC 4622 analyzed in Zhang & Buta 2007). In the above definitions, therefore, the sense of the phase shift is correct only for the S-sensed trailing spiral or bar (which winds and rotates in the same counterclockwise direction as used in the azimuthal angle definition). For a Z-sensed trailing pattern, the calculated phase shift curve should be negated to obtain the correct sense of the phase shift in the direction of galactic rotation. Care must also be taken so that the occasional leading pattern such as NGC 4622 is weeded out and treated specially (i. e., by negating the sense of phase shift definition from the above).

After obtaining the phase shift versus galactic radius plot, the positions of the successive corotation radii can be read off as the positive-to-negative (P/N) zero crossings of the phase shift plot. These mark the locations where the direction of angular momentum flow between the disk matter and the density wave changes sign at the quasi-steady state of the wave. The negative-to-positive (N/P) crossings of the phase shift, on the other hand, generally mark the transition locations of nested modal structures, that is, each such negative-to-positive crossing is likely to be where the pattern speeds of the density waves change abruptly. This is because the density wave should have negative angular momentum inside corotation and positive angular momentum outside corotation. The negative-to-positive crossings mark the boundaries where the wave deposits angular momentum onto the disk (for the inner mode outside its corotation) and the wave takes away angular momentum from the disk (for the outer mode inside its own corotation), at the quasi-steady state of the wave mode. So the phase shift distribution provides not only a way for the objective determination of density wave corotation radii, but also for the objective determination of the radial extent of the individual modes in the nested-modes cases, that is, what are the effective radii where adjacent modes decouple: something which methods such as Tremaine & Weinberg (1984) could not determine in a model-independent way. This assertion of course is based on the assumption that the radial penetration of the different sets of modes is minimal, which should be expected if the morphology of the modes are quasi-steady, since the interpenetration of modes with different pattern speeds will result in rapidly changing overall density wave morphology.

NIR (as well as lower-frequency MIR) images can be used to measure the phase shifts because, as previously noted, such images trace the stellar mass distribution better than do optical images, and for this reason are also the best to use for calculating gravitational potentials (e. g, Quillen, Frogel, & González 1994). For most of the calculations given below, the M/L ratio has been assumed to be a constant independent of radius unless otherwise noted. The exact value of this constant is not important since it cancels out between the numerator and the denominator in the phase shift definition equation. The validity of this assumption is further justified in Section 4.2.2. Following Quillen et al., we calculate the 2D potential at the plane of the galaxy from

$$\Phi(x, y, z) = -G \int \Sigma(x', y') g(x - x', y - y') dx' dy' \quad (4.6)$$

where the 2D Green's function $g(r) = g\left(\sqrt{x^2 + y^2}\right)$ is given by

$$g(r) = \int_{-\infty}^{\infty} \frac{\rho(z) dz}{\sqrt{r^2 + z^2}}, \quad (4.7)$$

where $\rho_z(z)$ is the normalized z distribution of the volume density of matter, assumed to be independent of galactic radii r. In practice, we found that the several different forms of the assumed $\rho_z(z)$ distribution given in Quillen et al. (1994) do not give significantly different phase shift results, especially for the zero crossings, even though they do change the absolute values of the calculated potential by about 20%. This insensitivity to the exact z profile of the density distribution is related to the fact that phase shift is determined mostly by the global distribution of the pattern pitch angle and radial density variation, and is insensitive to how "puffed up" the pattern is in the z-direction.

From the phase shift CR determination we had carried out for more than 150 nearby spiral and barred galaxies in the OSUBGS sample (Zhang & Buta 2007; Buta & Zhang 2009), in cases of overlap we found good agreement with the past results of CR determination using Tremaine–Weinberg (TW) method (inferred from the pattern speed of the density wave), especially in cases where there is only a single major CR resonance in a given galaxy, and the galaxy is also of earlier Hubble types. This is apparently due to the fact that the TW method is not ideally suited for determining the pattern speeds of multiple nested CRs, whereas the phase shift method gives the most secure determination for the CRs of density wave modes that have reached quasi-steady state, which occur most often in galaxies of earlier Hubble types. Subsequent studies by Haan et al. (2009) and Martínez-García et al. (2009, 2011) have also found good agreement between the PDPS methods and other independent approaches of CR determination in cases of overlap.

4.2.2 First Application of the Method: NGC 1530

In this subsection, we use the example of a nearby barred spiral galaxy to illustrate the working of the PDPS method. NGC 1530 is a high luminosity SB(rs)b spiral with one of the strongest known bars. We use a K_s-band (2.15 µm) image of this galaxy (Block et al. 2004) to illustrate the effect of M/L ratio on the phase shift and corotation determination, as well as on determining the mass flow rate as a function of radius.

The M/L ratio variations were determined using the color-dependent formulae of Bell and de Jong (2001). The details of this derivation for NGC 1530 are given in Zhang and Buta (2007). The main effect of the M/L correction was to reduce the surface density of the spiral compared to the bar, thereby weakening the effect of the spiral. The azimuthally averaged M/L ratio for NGC 1530 is given in Figure 4.1 (left), along with the breakdown of the M/L along the bar and 90° from the bar in Figure 4.1 (right). This procedure gave the surface mass density map in Figure 4.2 (right), which is in units of $M_\odot$ pc^{-2}.

Figure 4.2, left frame, shows the calculated phase shift versus radius for NGC 1530 using either a constant M/L ratio (dashed curve) or a position-dependent M/L ratio (solid curve). We observe that the effect of varying the M/L ratio changed mostly the scalings of the phase shift results, and the zero crossings changed by only about 1%. This is to be expected from the arguments given before, which showed that the radial M/L ratio change for the axisymmetric components does not impact the values of the zero crossings of the phase shift, and much of the M/L variation in physical galaxies is due to the M/L change of the axisymmetric components only, since the population of stars supporting the density wave is relatively homogeneous in composition. The right frame of Figure 4.2 displays the converted surface density map of the galaxy image in log scale, with two superimposed corotation circles determined using the phase shift method (*arrows* on the left frame). The radial scalings used for NGC 1530 in the above calculations are based on a distance of 36.6 Mpc from Tully (1988).

For reference, we also present in Figure 4.3 an image of the calculated potential distribution used in the phase shift calculations for this galaxy. From this figure one

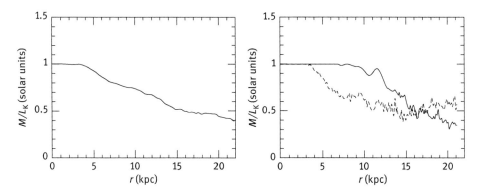

Figure 4.1: *Left*: Radial dependence of the azimuthally averaged M/L ratio for NGC 1530. *Right*: Interpolated M/L profile along the bar (solid curve) and 90° from the bar (dashed curve) (adapted from Zhang & Buta 2007).

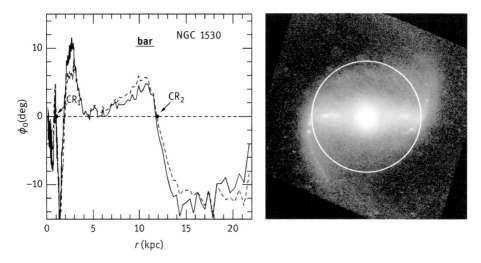

Figure 4.2: *Left*: Phase shift versus galactic radius calculated for the SBb galaxy NGC 1530 from a K_s-band image. The solid curve was calculated using a position-dependent M/L map, and the dashed curve was calculated using the surface density map assuming constant M/L ratio, Two CR circles are indicated. The short horizontal line indicates the range for the end of the bar as determined based on the two definitions given before. *Right*: Deprojected K_s-band image (Block et al. 2004) of NGC 1530 in log scale, with superposed corotation circles, rotated such that the bar is horizontal. This image has been corrected for stellar M/L ratio variations following Bell and de Jong (2001). The radius scale is based on a distance of 36.6 Mpc from Tully (1988) (adapted from Zhang & Buta 2007).

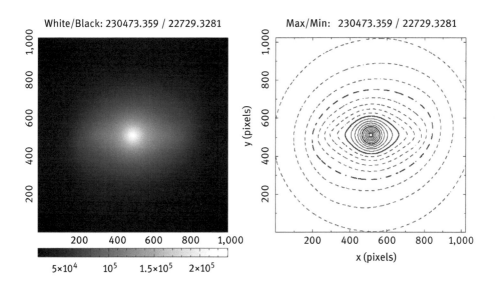

Figure 4.3: Calculated potential distribution for NGC 1530 in gray scale (left) and in contours (right). The map unit is $(km/s)^2$ (adapted from Zhang & Buta 2007).

can discern the slow isophotal twist, with a level of skewness that is more gradual than the density pattern skewness observed in Figure 4.2 (right), consistent with the sense of phase-shift-sign-change inside and outside corotation (which would mean that a potential pattern is in general *straighter* than the density pattern).

The innermost CR for NGC 1530 is associated with a small central pattern (its CR radius is unmarked on the phase shift plot), and the overall extent of this central feature appears to be delineated by the negative-to-positive crossing at about 1.1 kpc in radius on the phase shift plot. Immediately following the central pattern is the intermediate oval pattern between 1 and 2 kpc, with its own CR at about 1.3 kpc (this was marked as CR_1 since the innermost corotation radius is not very well resolved by the current calculation). The main CR encircles the ends of the bar at a radius about 13 kpc, with a pair of bar-driven spiral arms emanating from the end of the bar outside of the main CR.

Figure 4.4 uses published kinematic information to calculate and plot several kinematic diagnostic curves for this galaxy, which we use to compare with the nested resonance features determined using the phase shift method. The left frame shows the rotation curve compiled from the data in Regan et al. (1996), based on the atomic and ionized gas kinematics, while the right frame shows the calculated angular speed Ω, $\Omega - \kappa/2$, and the two pattern speeds determined by the phase shift method. We see from the frequency curves that the location for the inner-inner-Lindblad resonance (IILR) of the outer pattern (signified by the *inner* intersection of Ω_p of the outer pattern, i. e., the lower dotted curve, with the $\Omega - \kappa/2$ curve, or the dashed curve, at about 1.3 kpc radius) is simultaneously the corotation radius of the inner nested secondary pattern (signified by the intersection of the Ω_p of the inner pattern, i. e., the higher dotted curve, with the Ω curve, or the solid line). This kind of nested resonance coupling has been found in other observed galaxies (Zhang, Wright, & Alexander 1993, see further the discussions in Section 4.11; Erwin & Sparke 1999; Laine et al. 2002), as well as in N-body simulations (Schwarz 1984; Zhang 1998, see also Section 3.7; Rautiainen &

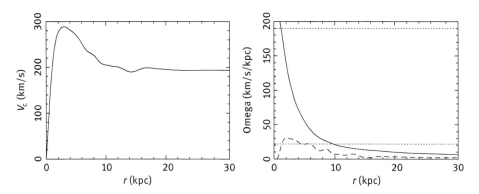

Figure 4.4: *Left:* The rotation curve of NGC 1530 (Regan et al. 1996). *Right:* Plot showing Ω (solid curve), Ω_p (dotted curves) and $\Omega - \kappa/2$ (dashed curve) (adapted from Zhang & Buta 2007).

Salo 1999), and various theoretical mechanisms had been proposed for its interpretation (Contopoulos & Mertzanides 1977; Tagger et al. 1987; Friedli & Martinet 1993). The coincidence of nested resonances with the PDPS predictions in the case of NGC 1530 further confirms the accuracy of the phase shift method in determining the locations of the multiple CRs.

Also from Figure 4.4, the *outer* intersection of the lower Ω_p curve and the $\Omega - \kappa/2$ curve, which signifies the outer-inner-Lindblad resonance (OILR) at radius about 5 kpc, coincides with a depression spot on the phase shift curve of Figure 4.2, left, indicating a region of low rate of angular momentum exchange between the bar density wave and the disk matter. The same radius corresponds roughly to a region of low surface density as well on the image of NGC 1530 in Figure 4.2, right, where the central oval and the straight bar section seemed to be disjoint. This may be evidence for the process of ongoing modal decoupling which will eventually form a new set of nested modes. We will comment further on the process of mode-decoupling in Section 4.2.6.

4.2.3 Phase Shift in a Pure Spiral Galaxy: NGC 5247

In Figure 4.5 we show the classic case of an ordinary grand design spiral pattern NGC 5247, and the corotation circles determined through the phase shift method (Zhang & Buta 2007). The outer crossing, which indicates the location of the main CR circle, is in the middle of the outer spiral arms. The outer corotation radius we determined

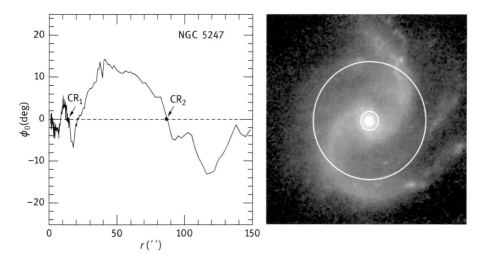

Figure 4.5: *Left*: Phase shift versus radius for the ordinary spiral NGC 5247, showing two major positive-to-negative crossings at $r = 87''$ and at $r = 14''$. *Right*: Deprojected near-infrared H-band (1.65 μm) image of NGC 5247 in log scale, with the corotations determined by the phase shift method superimposed as circles (adapted from Zhang & Buta 2007).

appears to be close to where the inner dust lanes truncate on the two spiral arms (this effect is more obvious from a B-band image of this galaxy), which is usually a good and reliable indication of the corotation radius.

Contopoulos & Grosbol (1986, 1988) had previously determined the corotation radius for this galaxy under the assumption that the 4:1 resonance (where $\Omega - \kappa/4 = \Omega_p$, with κ the epicycle frequency and Ω the angular speed of the disk matter, and Ω_p the pattern speed of the density wave) lies near the end of the strong spiral pattern. Their analysis placed this resonance at 12 kpc, which corresponds to 116″ assuming a distance of 21.3 Mpc. The corresponding location of the corotation radius from their analysis is 23 kpc, or 222″, well outside the NIR image of the spiral pattern. We note that Contopoulos and Grosbol's original analysis was based on matching the surface density obtained from passive orbital responses calculated under a forced potential to that of the given galaxy morphology. They had also assumed that the sharp truncation of the spiral pattern occurs near the 4:1 resonance as the starting point of their model construction process. As we have discussed before, the passive orbit approach is not expected to be able to predict the detailed kinematic features in galaxies. The discernible migration of dust lanes from the inner edge of the pattern to the outer edge on an optical image across the CR location determined by the PDPS method also confirms that this is the more likely CR position.

4.2.4 Multiple Nested Resonances: NGC 4321

In this subsection, we show another example of the phase shift calculation for galaxy NGC 4321 (M100), which has multiple CRs. The PDPS method can determine all the resonance locations simultaneously down to the limit set by the resolution of the image.

In Figure 4.6 we plot a comparison of corotation radii calculated by the phase shift method (circles) with the corotation bounds derived with the TW method (hatched regions, (Hernandez et al. 2005)) for a late-type barred galaxy M100/NGC 4321. The phase shift results are calculated using a Spitzer SINGS (The Spitzer Infrared Nearby Galaxies Survey) 3.6 μm Legacy image (Kennicutt et al. 2003). The phase shift method finds four well-resolved corotation radii for this galaxy, whereas Hernandez et al. (2005) previously derived three corotation radii (hashed regions in this figure) using a 2D velocity field and restricting their analyses to the different regions assumed for different modes.

The CR radius for the outer spiral pattern determined by both the Hernandez et al. analysis (outer hatched region, 140″ – 160″) and by the phase shift analysis (outer red circle, 138″) falls in the middle of the outer faint spiral arms, close to where the inner, stronger spiral pattern appears to truncate. For this CR the phase shift method and the TW method give results which seem to be in good agreement.

The innermost corotation determined by the phase shift method (13″), which surrounds the nuclear bar-ring pattern, was not detected by the Hernandez et al.

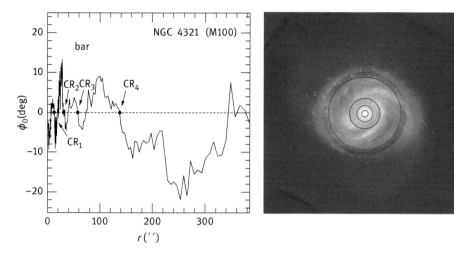

Figure 4.6: *Left*: Phase shift plot for M100 (NGC 4321) based on a deprojected 3.6 μm SINGS image. *Right*: Comparison of the CR radii (four circles) derived from the phase shift method, with the CR bounds determined by the TW method (three hatched regions) from Hernandez et al. (2005) for M100. The box size on the right is 12.'8 by 12.'8. These CR result comparisons are superposed on the 3.6 μm Spitzer SINGS image of M100 (Kennicutt et al. 2003) (adapted from Zhang & Buta 2007). For an unobstructed view of this galaxy, see Figure 4.10, top-right frame.

analysis, possibly because of the lack of spatial resolution for the data set they used. For the intermediate corotations, the phase shift method found two (31″ and 59″), which sandwich the first corotation circle determined by Hernandez et al. (32″ – 61″), shown as the innermost hatched region. It is open to question whether these two closely spaced intermediate crossings indicate two sets of quasi-steady modes with two different pattern speeds, or else are caused by noise or the non-quasi-steady nature of the pattern. However, the middle hatched region (80″ – 110″) as determined by Hernandez et al. has no corresponding feature in the phase shift calculation. This middle hatched region was suggested by Hernandez et al. to be the corotation of the bar of M100.

To deduce the bar radius, Zhang & Buta (2007) used the SINGS image to perform a bar-spiral separation using the methods described by Buta et al. (2005), the results of which are shown with our four CR circles superposed in the top two panels of Figure 4.7.

The lower two frames of Figure 4.7 present the original image (without bar-spiral separation) zoomed-in by a linear factor of 2 and 4, respectively, compared to the top two frames. From these two frames we can see more clearly that the innermost corotation circle (CR_1) encloses the strong secondary bar. Between the next two CRs (CR_2 and CR_3) there appear to be faint spiral structures. The correspondence between the CR circles and morphological patterns highlights the possibility that all four CRs could be real, although we do not exclude the possibility that some of the structures in the intermediate regions may be transient.

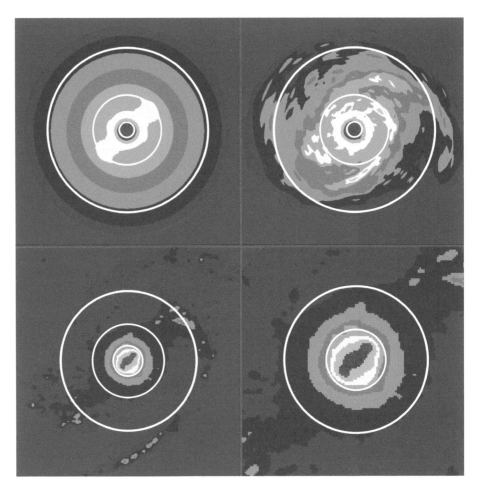

Figure 4.7: *Top left:* Bar-separated SINGS image of the bright inner region of NGC 4321 superimposed with the four corotation circles determined using the phase shift method. The box size of the image is $6'.4$ by $6'.4$. *Top right:* Spiral-separated SINGS image of NGC 4321 of the same region as at left superimposed with the four corotation circles determined using the phase shift method. The box size of this image is also $6'.4$ by $6'.4$. *Bottom left:* SINGS image (without bar-spiral separation) of NGC 4321 with a factor of 2 linear zoom compared to the top panels (box size $3'.2$ by $3'.2$), superimposed with the central three corotation circles determined using the phase shift method. *Bottom right:* SINGS image (without bar-spiral separation) of NGC 4321 with a factor of four linear zoom compared to the top panels (box size $1'.6$ by $1'.6$), superimposed with the central two corotation circles determined using the phase shift method (adapted from Zhang & Buta 2007).

Zhang & Buta (2007) had tentatively associated CR_3 as the CR for the bar pattern. Our later assessment, given in Zhang & Buta (2015), revised that choice, and we now tend toward the conclusion that CR_2 which encloses the oval is also the CR for the two sections of the straight bar emanating from it (or at least a significant fraction of the bar – the ends of the bar does thicken to two dumbbell-shaped blobs on Figure 4.7, top left frame, indicating orbit crowding or two pattern speeds intersecting, so it is possible the inner bar is interacting with an outer bar-driven spiral). This later conclusion makes the straight bar of the "super-fast" type to be discussed further in Section 4.2.6. CR_3 on the other hand is associated with the inner spiral pattern, which shows most prominently on the bar-spiral separated image of Figure 4.7, top-right frame (as well as bottom-left frame), whereas CR_4 corresponds to the outer faint spiral pattern, also most easily discernible from Figure 4.7, top-right frame.

Oey et al. (2003) used an HII region spatial isochrone method to determine the corotation radii for the outer region of M100. They found for the outermost corotation a similar value (154″) as Hernandez et al. using the TW method (140″ – 160″), as well as ours (138″) using the phase shift method, but they did not find in their data the intermediate corotation location found by Hernandez et al. and others (80″ – 110″), which is also absent in our results. The HII region isochrones are expected to be fairly reliable for determining the corotation crossings since the density-wave-triggered star-formation behavior has been well-confirmed from many previous studies.

4.2.5 Phase Shift in an Interacting Galaxy: M51

M51 (NGC 5194) is one of the most well-known interacting galaxies. In Figure 4.8 we present here the CRs predicted from the PDPS method. By focusing on the area that just excludes the small companion NGC 5195, which is likely to lie outside of the M51 galactic plane and thus have minor influence on the internal dynamics of M51 at the epoch of observation (a conjecture which is confirmed by our analysis), the phase shift analysis gives two major CR radii (positive-to-negative, or P/N crossings

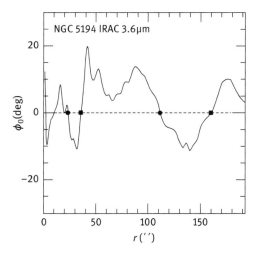
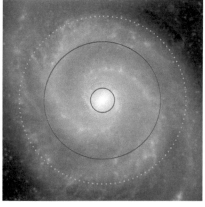

Figure 4.8: *Left*: Calculated phase shift versus galaxy radii for NGC 5194 (M51). Two corotation radii are indicated, as well as the mode decoupling points. *Right*: Deprojected mid-infrared 3.6 μm SINGS image of NGC 5194 (M51), with the CRs determined by the phase shift method superimposed as solid red circles, and mode decoupling radii as dashed green circles (adapted from Zhang & Buta 2012).

on the phase shift plot, represented by red circles on the overlay image) followed by two negative-to-positive (N/P) crossing radii, represented by the green circles on the image; the latter are believed to be where the inner mode decouples from the outer mode. These radii match very well the galaxy morphological features (i. e., the inner CR circle lies near the end of the bar, and the first N/P crossing circle is where the two modes are seen to decouple). For the outer mode, the CR circle seems to just bisect the regions where the star-formation clumps are either concentrated on the inner edge of the arm, or on the outer edge of the arm – a strong indication that this second CR is located right near where the pattern speed of the wave and the angular speed of the stars match each other. As a matter of fact, if we place the overlay image for M51 from Figure 4.8, next to Figure 3.7 which is the image of an *N*-body-simulated spiral mode, we will notice a great deal of similarity in the modal features both within and outside the corotation radius (which is represented by the second red circle in Figure 4.8, and is at $r = 30$ in Figure 3.7) for the outer mode: i.e., the shocked appearances of the spiral arms inside corotation, with density maxima at the inner edges of the arms; and the thickening of the spiral arms outside corotation, with the density maxima moving to the outer edges of the arms, with respect to the direction of galactic rotation, assuming the pattern is trailing.

Note that we are not using the simulation to fit the observed morphology of a spiral galaxy: The *N*-body simulation of Zhang (1996), which produced Figure 3.7, was conducted more than a decade before the observational studies of Zhang & Buta (2012) which produced Figure 4.8. This makes the agreement between simulated and observed modal features and their correspondence with the CR predictions all the more impressive. This supports the hypothesis that the spiral patterns in this galaxy are intrinsic modes rather than tidal transients, and that tidal perturbation serves to enhance the prominence of the intrinsic mode, but does not alter its modal shape.

4.2.6 Regarding the So-Called "Super-Fast Bars"

The phrase "super-fast bars" are used to denote the type of galactic bar patterns whose lengths extend significantly beyond their CR radii. The existence of super-fast bars is in contradiction to the earlier conclusion reached in Contopoulos (1980) that bars must termination at their own CRs. This conclusion of Contopoulos', however, was reached through passive orbit analysis, which could be invalidated when collective effects of a globally self-consistent analysis are taken into account.

In Zhang & Buta (2007), we have shown a case of a single isolated bar in the galaxy NGC 4665, where the bar ends extend about 10–20% beyond the CR radius, and argued that in this case it is reasonable to expect the bar to be longer than the CR radius because the SWING amplified, over-reflected waves from the inner disk must penetrate CR into the outer disk as transmitted wave, in order to have the overall angular momentum budget balance. In examples shown in the appendix of Zhang & Buta (2015), some of the super-fast section of the bars are straight segments emanating from the location of an inner oval, much like the case for NGC 1530 if the modal decoupling process in that galaxy would eventually go to completion. The ends of these straight bars coincide not with CR but with the next N/P phase shift crossing after the CR. This morphological configuration is reasonable since modal growth requires that for a complete self-sustained mode there must be a positive phase shift packet followed by the negative phase shift packet, with the two packets joining at CR. Therefore we see that for the kind of morphology of twin-straight-bars joining the central oval, the mode has little choice but to have the N/P phase shift crossing be at the end of the bar, if the central oval itself ends at its own CR.

The evidence that bars can extend beyond its corotation had in fact already showed up in the earlier self-consistent N-body simulations. Sparke & Sellwood (1987) had produced one of the earliest self-sustained 2D N-body bar modes, which settled onto quasi-steady state after first emerging as a spiral pattern. During much of the simulation run after the mode had reached nonlinear saturation amplitude, the extent of the bar in fact is longer than the corotation radius at the corresponding time step. For example, from their Figure 3(a), we can see that the inner bar mode has contours extending to about 1.7 in radius at time step 40, whereas the corotation radius for the same time step is of size 1. In a later stage (step 103) of the simulation (their Figure 3b) the corotation grows to 1.3 (due to the slowing down of the bar pattern), and the bar on the other hand had reduced in length so it is indeed enclosed within the corotation radius at this late stage of the simulation. Note that even though the morphology of the bar and spiral pattern in the Sparke & Sellwood (1987) simulation is evolving continuously, this evolution is the evolution of the quasi-steady modes as a result of the evolution of the underlying basic state distribution due to the wave-induced angular momentum transport and exchange. The wave modes themselves had achieved quasi-steady state throughout most of the simulation run, as reflected in the well-defined pattern speeds that delineate precisely the extents of the inner and outer modes (see their Figure 3). That the Sparke & Sellwood (1987) simulation had the longest bar length (compared to the size of the modal corotation radius) at the intermediate stage of the simulation is also consistent with our finding that the smallest $r(CR)/r(bar)$ ratios are observed for intermediate Hubble-type galaxies.

The fact that bars and spirals can extend beyond their own corotation radii is partly due to the fact that when self-consistent patterns and collective effects are taken into account, many of the conclusions reached through passive orbit calculations are no longer valid. As a matter of fact, there will in general not be a clear dividing line between bars and spirals (apart from the fact that bars tend to occur in the inner regions of galaxies and spirals the outer regions): most realistic bars in physical galaxies (and this includes the more than 150 galaxies from the OSUBGS sample examined by Zhang & Buta 2007; Buta & Zhang 2009) all possess a certain degree of skewness, which enables the presence of the phase shift, and the associated collective dissipation effect which leads to the secular evolution of the basic state. In cases of certain bar-driven spirals it might even be hard to tell where the bar ends and the spiral starts, especially during the intermediate stage of the evolution when the spiral is being transformed into a bar-driven spiral. For other galaxies the scenario of bar-extending-beyond corotation is achieved through the decoupling of the inner and outer nested modes, as what appears to be happening for NGC 1530.

Another piece of evidence supporting the claim that an N/P crossing is the location where the modal pattern speed changes discontinuously is that there is often a pronounced ring-like structure at the radius of the N/P phase shift crossing, obviously caused by the snow-plough effect of the interaction of the inner and outer modes (since these two sets of modes have different pattern speeds). Similar configuration of a central oval joined by straight super-fast bars is also observed in NGC 3351, NGC 1073, NGC 5643, and in the central region of NGC 4321 (which we had shown earlier in this chapter). As a matter of fact, since this configuration requires the central oval connecting to the straight bars, it is always found in the central configuration of the nested modes.

It is no coincidence that the super-fast bars, which often appear in early-type galaxies, have a more rounded nuclear pattern and very straight two segment of bars connecting to it: both of these patterns are not much skewed and they correspond to very small phase shifts, and slow secular evolution rates as befit the early-type galaxies. The fact that bar-driven spirals mostly have skewed nuclear bar patterns followed by trailing spiral segments that taper into narrow tails, whereas super-fast bars most often have rounded nuclear patterns followed by very straight bar segments which broaden into dumbbell-shaped pile-up of material at the mode-decoupling radii, shows that super-fast bars are real and are of completely different modal category than bar-driven spirals.

4.2.7 Physical Basis Underlying the Validity and Accuracy of the Phase Shift Method

The validity of the phase shift method for corotation determination is rooted in the global self-consistency requirement of the wave mode (i. e., both the Poisson integral and the equations of motion need to be satisfied at the same time). Therefore, the appearance that the corotation can be determined from the Poisson integral alone, without knowledge of the kinematic state of the disk matter, is only an illusion. In actuality the global self-consistency between the Poisson integral and the equations of motion has been enforced by nature itself when the galactic resonant cavity filters and selects the set of modes it can support. The phase shift resulting from the Poisson integral is not only determined by the spiral pitch angle, but also by the radial density variations of the nonaxisymmetric density wave features. For a self-sustained global spiral mode, the radial density variation of the modal perturbation density as well as the pitch angle variation together would be such that the Poisson integral will lead to the positive-to-negative zero crossing of the phase shift curve to be located exactly at the corotation radius of the mode, since that's where the peak of the bell-shaped angular momentum flux is, and the radial gradient of the angular momentum flux is proportional to the sine of the phase shift (Chapter 2). Since the bell-shaped angular momentum flux distribution is what allowed the spontaneous growth of the mode in the linear regime, only modes with density distributions which result in the correct phase shift distributions are naturally admitted by the galactic resonant cavity.

The self-consistency requirement of a spontaneously formed density wave mode explains not only why we can use the Poisson integral alone to determine the corotation radius, but also why we can use the continuity equation alone to determine the pattern speed as in the TW approach: each is only seeing one side of the coin, but since both sides join at a unique boundary, we can determine the circumference of the coin from the measurement conducted on either side.

The phase shift distribution calculated through the Poisson integral would be the same whether the pattern is rotating or not as long as we used the same surface density distribution. The phase shift method is thus inherently not a rotation detection method. Rather, it is a diagnostic tool for characterizing the features of self-organized modes in galactic disks. We can imagine, for the sake of the argument, that a mode had initially reached quasi-steady state with a given pattern speed. If, suddenly, we apply a forced potential of exactly the same shape as the original modal potential, but with a somewhat faster pattern speed than the original natural pattern speed, the pattern density will tend to follow this applied potential's forcing with only minor changes in shape. The phase shift method would not be able to predict that the rotation speed of the pattern has changed in this case, or equivalently the corotation radius has changed. But once the forcing is removed, the pattern will quickly recover to its original pattern speed. The reason is that the faster pattern speed had moved the corotation radius from its original, natural position, which previously allowed the sign of the angular momentum deposition by the wave everywhere across the disk to match the sign of the angular momentum density of the wave mode itself. This matching condition is violated as in the forced case, and once we remove the forcing, the faster mode will be damped out because of the wrong sign and wrong amounts of angular momentum being deposited by the wave mode over much of its extent. The original mode will reappear since it is naturally amplified in the galactic resonant cavity determined by the basic state. Or to put it in another way, the spiral shock and the associated collective dissipation tends to destroy or damp the wave at every instant of the time. It is only through the continued amplification tendency of the original unstable mode in the galactic resonant cavity that a quasi-steady state can be maintained. Therefore, we arrive at the conclusion that *the modes that satisfy the self-sustainability condition have a special density distribution that matches its kinematic characteristics, so that the growth tendency balances the dissipation tendency to reach a dynamical equilibrium at the quasi-steady state.*

We thus see that as a general method for corotation determination the passive approach of visually matching the *forced* pattern response with that of the observed galaxy morphology, as practiced by some researchers in the past, is not likely to be reliable since the response pattern tends to look almost identical for a quite wide range of pattern speed of the forcing potential. The limitations of the passive simulation approach become all the more prominent when we realize that realistic galaxies contain nested density wave patterns which all rotate at different pattern speeds.

On the other hand, the phase shift method itself is only expected to be reliable for cases of spontaneously formed density wave modes that have reached quasi-steady state. We have found empirically from analyzing the OSUBGS sample that usually the early-type disk galaxies gave the best fit between the predicted corotations and the resonance features, likely related to the fact that early-type galaxies have had the longest time to evolve and reach quasi-steady state. The latest Hubble types usually have more flocculent patterns, and the phase shift result correspondingly does not show a well-organized pattern, reflecting partly the fact that the wave mode pattern speeds and corotation radii in these galaxies are themselves poorly defined (Buta & Zhang 2009). The intermediate Hubble types have strong and open spiral arms, but the results of the phase shift calculation tend to be more sensitive to the uncertainty in inclination-angle deprojections which can result in inaccurate spiral pitch angle representation. The intrinsic accuracy of the phase shift method should be high if we are dealing with an ideal, spontaneously formed mode that has reached quasi-steady state, and if we have accurate knowledge of the orientation parameters of the galaxy. In practice, however, the mode may not be in quasi-steady state and/or the orientation parameters may be uncertain. Since the errors will likely to be dominated by unknown systematics rather than by random measurement errors, we have not assigned error bars to our results in the past analyses.

We now comment briefly on the pattern speeds and corotation radii determined from previous modal calculations and their relation to the results from applying the PDPS method. In the 1970s and 1980s Lin and collaborators constructed self-consistent spiral modes both through solving the exact fluid set of equations in the linear regime, as well as through the asymptotic approach of higher-order local calculations which obtained piecewise continuous wave trains, and then joined them together to derive the quantum conditions for the pattern speed determination (see, e. g., Lin & Lau 1979 and the references therein; Bertin et al. 1989a, 1989b and the references therein). These two approaches were found to agree with one another as long as one deals with slow-growing modes. In both approaches, the quantum condition for the mode was derived by enforcing the global self-consistency requirement between the Poisson equations and the equations of motion (or the so-called Poisson–Euler set in the fluid approach). In the asymptotic approach, this global self-consistency is acquired in a piecewise, locally self-consistent manner, and by *throwing away* the so-called "out-of-phase" terms (which in fact corresponds to the PDPS we had discussed in this work). The fact that this latter approach leads to results similar to the "exact" numerical solutions which accurately contain all the phase shift information is because the asymptotic approach only solved the weakly growing kind of modes (i. e., either very small pitch angles, as in tightly wound early Hubble-type galaxies, or very large pitch angles, as in very straight bars, again appearing in early Hubble-type galaxies; both of these cases give small phase shifts since the phase shift is largest for pitch angle close to 45° and is zero for pitch angles of 0 and 90°). Therefore, the asymptotic approach worked precisely for situations where the phase shift is small, so the resulting small "out-of-phase" terms (which reveal the effect of PDPS) can be thrown away. For the fast-growing "violently unstable" modes which have a more open spiral morphology (which resemble intermediate Hubble-type galaxies), the exact numerical solutions of linear global modes do implicitly contain the phase shift distribution that changes sign at corotation, and it is precisely this phase shift distribution in the open spiral modes which leads to the large growth rate of the mode in the linear regime, and to the rapid azimuthal steepening of the sinusoidal wave profile into narrow shock fronts in the nonlinear regime.

4.2.8 Implications for the Kinematics and Dynamics of Nearby Galaxies

Spiral and barred galaxies are expected to be systems in near dynamical equilibrium, and thus their kinematic and dynamical states are partly reflected in their morphological appearances: this connection between morphology and kinematics for quasi-equilibrium galaxies is the foundation of the phase shift approach for corotation determination. The success of the phase shift approach as we had demonstrated early on in this section indirectly confirms that a large fraction of the observed disk galaxies in the nearby universe have acquired a quasi-steady state with respect to the formation of spiral/bar modes since only in the quasi-steady state of the wave mode can one expect a correlation between the kinematic characteristics (such as corotation radius) and the pure morphological appearance (i. e., the NIR image). Here by quasi-steady state we mean that the pattern is quasi-stable at least on the local dynamic time scale.

Over the past five decades since the advent of density wave theory, the issue of whether the observed density wave patterns in galaxies are quasi-stationary or else are transient has always been a matter of debate (see, e. g., Lucentini 2002; Binney & Tremaine 2008 and the references therein). Among the N-body simulated galaxies, results also differ between the quasi-steady patterns obtained (Donner & Thomasson 1994; Zhang 1998), and the more transient patterns obtained (Sellwood & Carlberg 1984; Sellwood & Binney 2002). We point out here that at least one of the known causes of the differing results in these N-body simulations is *the differing basic states of the galactic disks being used*. When simulating spiral patterns, Sellwood & Evans (2001), Sellwood & Binney (2002) used a disk model that is over-stable with respect to the formation of large-scale spiral modes. Therefore any spiral perturbations will tend to die out, which is the reason they only obtained transient spiral patterns in these studies. However, growing observational evidence, including our current work, shows that the large-scale spiral and barred patterns in physical galaxies are in fact *gravitationally unstable global modes*, not the transient structures studied in the above work which used overstable basic states. As a matter of fact, even in some of the work conducted by these authors themselves (i. e., Sparke & Sellwood 1987; Debattista & Sellwood

2000), when a basic state model which allowed a genuine global mode to emerge was used, there are invariably collective effects present due to these long-lasting modal patterns.

Sellwood (2011) repeated some simulations done previously in Donner & Thomasson (1994), and Zhang (1996, 1998). He was able to reproduce most of the characteristics of our previous simulations, and this very fact itself shows that these are unstable *modes* insensitive to the differing details of the numerical setup. Sellwood claimed in the same paper that because of the spread of the power spectrum in these repeat simulations, these were transient waves rather than modes. However, a simple plotting of the $m = 2$ mode phase versus time, as we had done in Chapter 3, shows that this could not have been the case since the continuous linear curves of the phase evolution show that the $m = 2$ modal pattern is rotating semi-rigidly throughout the simulation run. This ordered bahavior is not the signature of a set of transient waves, but rather that of a mode or modal set (as Sellwood observed, there sometimes exists nested modes – but these also exist in real galaxies such as NGC 4321. The fact that nested modes exist in simulations does not equate them to transient waves).

The conclusion of quasi-steady co-evolution of the basic state of the disk and the density wave modes in most of the observed galaxies is also supported by another aspect of the current work (Zhang & Buta 2007; Buta & Zhang 2009), which showed that for galaxies that are still at the initial stage of the secular evolution sequence, such as the case of the late-type galaxy NGC 0628 (which we will further analyze in the next section), there is a poorer coherence in both the phase shift plot as well as in the galaxy image, and poorer correspondence between the phase shift positive-to-negative zero crossings with galaxy resonance features. As a galaxy settles down while the secular evolution proceeds, once it reaches intermediate Hubble types, such as for galaxy NGC 4321, there is a very nice correspondence between the phase-shift-plot organization and the galaxy-image resonance organization, signaling that an internal dynamical equilibrium state has been reached. The delicacy of this balance (i. e., sometimes if we use an M/L-corrected *stellar image* alone without adding the gas maps in calculating the phase shift distribution, we would not get as well-organized phase shift curves as when we use the *total mass image*) shows that the equilibrium state is well negotiated between the mass distribution and the kinematic distribution of the density wave pattern, which can only come as a result of the long-term evolution of spontaneously formed density wave modes and is not possible for random transient wave trains. The observed correlation between basic state characteristics (i. e., the bulge-to-disk ratio) and pattern characteristics (i. e., the pattern pitch angle), already known at Hubble's time and is part of the basis of his classification scheme, is also most naturally explained within the modal framework (e. g., Bertin et al. 1989a, 1989b), at least qualitatively – even though the details of the original version of the modal theory of Lin and his collaborators may be challenged, as Sellwood had shown in the first part of his 2011 paper.

To summarize, we note that to be able settle the question of transient versus modal nature of the density wave patterns in disk galaxies, we have to first be clear what we regard as a satisfactory definition of such a mode. We assert that a density wave is a mode if

(i) It is a genuine global instability in the underlying basic state of the disk.
(ii) As a result it spontaneously emerges out of an originally featureless, differentially rotating disk, a process for which the bell-shaped distribution of total angular momentum flux (equivalent to the two-humped distribution of the PDPS) of the mode is chiefly responsible.
(iii) The properties of the emerged density wave patterns are determined solely by the basic state characteristics and not by the accidentals of noise or external perturbation.
(iv) The emergence of these patterns greatly accelerates the entropy evolution of the parent basic state of the disk, as is the requirement for all dissipative structures formed in far-from-equilibrium systems.

Upon examining the evidence in both the simulated and observed disk galaxies possessing grand-design density wave patterns, we have found that these patterns do conform to the definition of density wave modes as stated above. Furthermore, we emphasize that the modes in galaxies are indeed only quasi-steady, and not eternal and unchanging, since the secular mass flow they induce changes the basic state distribution, which in turn affects what kinds of modes are compatible with it. So these modes may change morphology continuously on time scales of a few times of the local dynamical time scale (the exact rate of modal evolution differs depending on the density wave amplitude and pattern opening angle, etc., which will set the mass flow rates). Here we once again bring out the connection with nonequilibrium phase transitions, and the fact that the coherent patterns formed in these transitions are in a state of dynamical rather than static equilibrium.

Many of the more chaotic looking density wave patterns in late type disks, on the other hand, are likely to be in the process of evolving into grand-design ones due to the secular evolution connection among galaxies along the Hubble sequence.

Later, in Chapter 5, we will present an objective approach for judging the quasi-steady modal nature of the observed, as well as simulated, density wave patterns.

4.3 Secular Mass Migration and Bulge Building[†]

At the quasi-steady state of the density wave mode, the PDPS and its associated torque action lead to a secular exchange of energy and angular momentum between the wave and the basic state. The consequence of this exchange process *on the wave mode* is such that it offsets the spontaneous growth tendency of the mode so that the mode can maintain a quasi-steady state despite constantly transporting angular momentum

---

[†] Portions of this section used material previously published in Zhang (1999) and Zhang & Buta (2007), reproduced with modifications from The Astrophysical Journal @ AAS and The Astronomical Journal @ AAS. Reproduced with permission.

outward. One of the consequences *on the basic state stars* of this exchange process, however, is that the stars both inside and outside the corotation radius migrate away from their previous locations inward and outward, respectively, so that with time a more and more centrally concentrated disk surface density distribution is achieved, together with the build-up of an extended outer envelope.

We have obtained in Chapter 2 that the rate of density-wave-induced orbital change for an average star within a nearly flat-rotation-curved galaxy under the influence of a quasi-stationary spiral or bar mode, is given by

$$\frac{dr_*}{dt} = -\frac{1}{2} F^2 v_c \tan i \sin(m\phi_0), \qquad (4.8)$$

where i is the pitch angle of the spiral, m is the number of arms of the spiral pattern, ϕ_0 the potential and density phase shift, F the fractional amplitude of the spiral, and v_c the circular speed of the galaxy at the relevant radius. This equation has been quantitatively verified through the N-body simulations we have presented in the previous chapter.

For a typical two-armed spiral of 20% fractional amplitude and 20° pitch angle, the orbital decay rate is about 0.2 km s^{-1}, or 2 kpc of radial migration in 10^{10} years. Thus the spiral-induced stellar orbital decay inside corotation should have a significant effect on the evolution of the disk morphology over the time span of a Hubble time. In particular, fast evolution is expected for galaxies which contain large amplitude and large pitch-angled spiral patterns because of the effective square-power dependence of mass flow rate on the wave amplitude as well as pitch angle (for pitch angle that is not too large, the phase shift itself turns out to be proportional to the pitch angle).

4.3.1 Formation and Evolution of Galactic Bulges

Due to the spiral-induced collective dissipation process, an average star inside corotation tends to drift toward the central region of a galaxy. During its radial migration, *all three* components of the random velocity of an average star increase in concert with one another, through the very same collective dissipation process (see next section on the secular heating of disk stars[†]). The radial profile of stellar velocity dispersion produced by this heating process has increasing velocity dispersion with decreasing galactic radius. Consequently, when the vertical component of an inwardly accreting

---

[†] Here we are getting a little ahead of ourselves in terms of needing the result of the secular heating process of the next section while discussing bulge-building through the radial mass accretion. This compromise is unavoidable partly due to the very fashion that globally self-consistent processes operate: that is, we are dealing with a system where all physical processes involved are operating simultaneously as an interlocking whole, whereas in book form we can only discuss these processes in a sequential manner.

star's velocity dispersion becomes comparable to its circular velocity, the star will rise out of the galactic plane and become a bulge star.

As an example, we now estimate the rate of mass accretion for the formation of the Bulge of our Galaxy. We assume that the spiral pattern during the past 10^{10} years has an average pitch angle of 20°, and an average value of 20% fractional amplitude around the solar neighborhood. Furthermore, using an average local surface density of $\Sigma \approx 60$ M$_\odot$ pc^{-2} (Bahcall 1984; Kuijken & Gilmore 1989), and also using the orbital delay equation of (4.8), we have that the average rate of radial inflow of mass near the solar neighborhood is

$$\frac{dM}{dt} = 2\pi r \frac{dr_*}{dt} \Sigma \approx 0.6 M_\odot/\text{yr}. \tag{4.9}$$

This means that in a Hubble time the average radial mass accretion across the disk will be able to accumulate an amount of mass toward the center of the Galaxy on the same order as the mass of the Galactic Bulge, which is about 10^{10} M$_\odot$ (Gilmore et al. 1990, p. 224), especially considering the fact that the Bulge was not devoid of matter at the beginning of the Galaxy formation. If indeed the Galactic spiral has four arms instead of two as we has assumed in the above calculation, the radial mass accretion rate will also increase accordingly.

The mean metallicity of the Galactic Bulge stars is found to be twice that of the solar value. Some of the Galactic Bulge stars are more metal-rich than found anywhere else in the Galaxy (Gilmore et al. 1990, p. 53). One of the possible processes to account for this gradient is that the stars which are currently at the bulge region presumably started their radial migration close to the galactic center region, and since the dynamical time scale there is short, the material which made up bulge stars went through more cycles of spiral-or-bar-triggered star formation. This is also consistent with the general knowledge that although early-type spiral galaxies and elliptical galaxies contain a large population of old stars, the central regions of these galaxies also contain some of the most metal-rich stars anywhere in the universe (Barbuy & Grenon 1990).

Some of the past-proposed bulge-building mechanisms involving vertical resonances (e. g., Pfenniger & Friedli 1991). We had noted before that the single-orbit resonance condition is likely to be destroyed by the presence of collective instabilities. In fact, the so-called "buckling instability" attributed to those vertical resonances acts very fast (Pfenniger & Friedli 1991) and would have resulted in two distinct classes of disk galaxies (one before the buckling, with no bulge, and one after the buckling, with bulge), whereas the observed bulges come in all variations of bulge-to-disk ratio, and there is no signature of the above bi-modality which would have resulted if buckling is the underlying mechanism for bulge formation. The buckling behaviors observed in previous 3D N-body simulations are apparently a result of the initial conditions chosen in these simulations, which results in the rapid evolution of the initial configuration to reach a new equilibrium quasi-steady state, not unlike the rapid emergence of the spiral pattern from an originally featureless disk as we had shown in the previous chapter. From the observational side, we know that bulge formation cannot be solely due to the dissolution of a pre-existing stellar bar since the bulge light is found to be added onto the disk light, instead of subtracted from it (Wyse, Gilmore & Franx, 1997).

Andredakis, Peletier and Balcells (1995) found that the gradual variation of the surface density profile of bulges from late to early Hubble types can be fitted very well by the generalize exponential profile of Sersic (1968)

$$\Sigma(r) = \Sigma_0 \exp\{-(r/r_0)^{1/n}\}, \qquad (4.10)$$

whereby as the Hubble type of a galaxy changes from late to early, the fitted index n in the above expression also increases ($n = 1$ is for a pure exponential, and $n = 4$ is for a de Vaucouleurs law).

The current secular evolution theory gives a trend of variation of disk surface density which matches very well that found in the above-mentioned result of Andredakis et al. (1995). As a galaxy evolves through the spiral-induced collective dissipation process, its surface density distribution (i. e., disk plus bulge) will become more and more centrally concentrated, together with the built-up of an extended outer envelope. This trend agrees with that of varying of n-index in the above bulge profile as the Hubble types of galaxies change from late to early.

In Chapter 3, we have presented the simulation results which showed that significant mass inflow in the inner galaxy can result in the significant change of Hubble types of a galaxy within the past Hubble time. In what follows in this section, we will further demonstrate, using the observed galaxy images at the NIR and MIR range, that the density wave patterns in physical galaxies possess the characteristics (amplitude and pitch angle) that allow them to transform the Hubble types of their parent galaxies within their lifetime. Furthermore, gravitational tidal interactions from neighboring galaxies tend to excite large amplitude wave patterns that further accelerate the speed of the secular morphological transformation of galaxies.

4.3.2 Secular Mass Flow Rate Determination Using NIR and MIR Images

The secular morphological transformation mechanisms proposed in the past involve mostly gas accretion under a barred potential (Combes & Sanders 1981; Kormendy & Kennicutt 2004 and the references therein), which were found to be inadequate to transform the Hubble types of galaxies except for the very latest types (Sc, Sd), due to the paucity of gas in most disk galaxies. When the role of stars is also considered, as we have done in this work, a more coherent and theoretically sound basis for the secular evolution paradigm emerges that is powerful enough to account for both the morphological transformation of all galaxy types along the Hubble sequence, and for the differing evolution rates observed for galaxies residing in the field and the cluster environments. Galaxies in group and cluster environments evolve faster compared to isolated field galaxies because of the large-amplitude, open spiral patterns excited during the tidal interactions with neighboring galaxies and with the cluster potential, and in new dynamical mechanism the secular evolution rate is effectively proportional to the wave amplitude squared and pattern pitch angle squared (since for pitch angle not too large the phase shift is proportional to the pitch angle). We will discuss this effect of enhanced secular evolution due to galaxy interactions in different environments later in this chapter.

In Figure 4.9, we give an example of density-wave-driven radial mass flow calculation for the SBb galaxy NGC 1530 (for its morphology, see Figure 4.2, right hand side), using the surface density map in conjunction with the mass flow rate equation (c.f. equation (2.124))

$$\frac{dM(r)}{dt} = \frac{r}{v_c} \int_0^{2\pi} \Sigma_1 \frac{\partial \mathcal{V}_1}{\partial \phi} d\phi \qquad (4.11)$$

which we had derived in Section 2.6.3 before. Mass accretion rate calculations thus require the knowledge of M/L ratios, galaxy rotation velocities, and absolute lengths. From Figure 4.9, we see that the peak accretion rate for this galaxy is over 100 $M_\odot yr^{-1}$ near the end of the main bar. The absolute values of the mass flow rates for this galaxy are revised somewhat from that given in Zhang and Buta (2007) due to new calibrations. The inner accretion rate we obtained for this galaxy (~5–10 $M_\odot yr^{-1}$) is consistent with the gas accretion rate estimated by Regan et al. (1997, see their Figure 2), which is in the range of 2–5 $M_\odot yr^{-1}$ for the inner 4 kpc, since the current result includes both the stellar and gaseous accretions. Also note that this figure indicates both accretion and excretion in the central region due to nested resonance structures, with the exact locations of these resonances expected to evolve over long time as the basic state itself evolves. These accretion and excretion patterns help to establish and maintain the nested resonances, as well as producing continued secular evolution of both the modal set and the basic state, but the overall mass flow trend should still be mostly inward for the inner galaxy, throughout the lifetime of a galaxy: that is, the nested mass flow pattern should have its intrinsic hierarchy to allow the entropy evolution of the parent disk to be realized in the long run. The large excretion rate in the outer

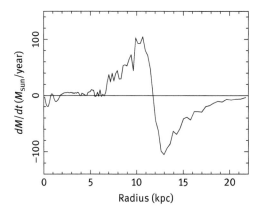

Figure 4.9: Radial mass accretion/excretion rate calculated for NGC 1530 from a K_s-band image (adapted from Zhang & Buta 2007).

disk is expected to contribute to the build-up of an extended outer envelope usually observed for disk galaxies and disky ellipticals.

This particular example of the mass accretion calculation shows that for at least a fraction of the observed disk galaxies such as NGC 1530, which have extremely strong and open density wave patterns, the secular evolution can be very rapid (i. e., a mass accretion rate of 10^{10} solar masses per 10^8 years is possible), compared to a typical galaxy like our own Milky Way, where we have shown that it takes on average about 10^{10} years to accumulate 6×10^9 solar mass through the same mechanism, since the (most likely) Galactic spiral is both less intense and more tightly wound.

Next, we present the result of secular mass flow rate determination for six galaxies of varying Hubble types: NGC 0628, NGC 4321, NGC 5194, NGC 3627, NGC 3351, NGC 4736, first obtained in Zhang & Buta (2015). The parameters of these six galaxies are given in Table 4.1. The total mass maps derived using the methods described in Zhang & Buta (2015) are shown in Figure 4.10.

The stellar mass surface density maps were derived using the SINGS (Kennicutt et al. 2003) 3.6 μm IRAC images, as well as using *SDSS* i-band images, with proper mass-to-light ratio conversion. The gas surface density maps were derived from the corresponding images of VIVA (Chung et al. 2009), BIMA SONG (Helfer et al. 2003), and THINGS (Walter et al. 2008) surveys.

In Figures 4.11 and 4.12, we present the rotation curves (observed, versus disk-total-mass-derived) as well as the total mass flow rates for these galaxies. The sequence of arrangement for the different frames, from left to right, then top to bottom, is chosen to be roughly along the Hubble sequence from the late to the early types, in order to reveal any systematic trends along this sequence. Notice, however, that two of the intermediate types (NGC 5194 and NGC 3627) are strongly interacting galaxies, and thus they might deviate from the quiescent evolution trends.

Table 4.1: Adopted parameters for sample galaxies.[a]

Galaxy	i (degrees)	Φ_n (degrees)	h_R (arcsec)	h_z (arcsec)	Distance (Mpc)
NGC 628	6	25	64 (3.6)	7.1	8.2
NGC 3351	40.6	13.3	44 (3.6)	8.9	10.1
NGC 3627	60	173	66 (3.6)	13.3	10.1
NGC 4321	31.7	153	63 (3.6)	12.6	16.1
NGC 4736	30	116	135 (i)	27.1	5.0
NGC 5194	20±5	170±3	50 (K_s)	10	7.7

[a]Col. 1: galaxy name; col. 2: adopted inclination; col. 3: adopted line of nodes position angle; col. 4: adopted radial scale length, not based on decomposition but from slope of azimuthally-averaged surface brightness profile (filter used in parentheses); col. 5: adopted vertical scale height derived as $h_z = h_R/5$ except for NGC 628 where $h_z = h_R/9$; col. 6: mean redshift-independent distance from NED; typical uncertainty ±1–2Mpc. Sources of adopted orientation parameters can be found in Zhang & Buta (2015).

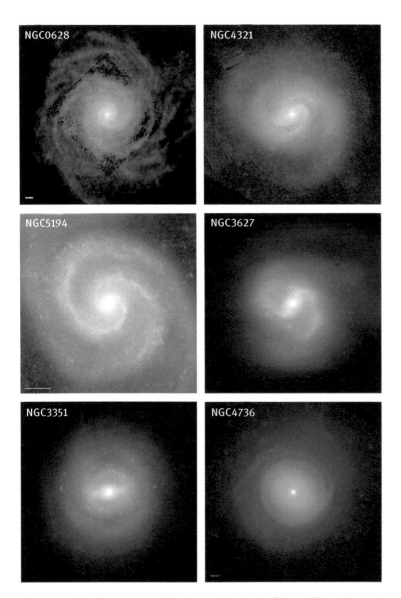

Figure 4.10: Total mass maps of the six sample galaxies (adapted from Zhang & Buta 2015).

From Figure 4.11, we observe that the contribution of the disk baryonic matter (excluding the contribution from Helium and heavy elements) to the total rotation curve increases as the galaxy's Hubble type changes from the late to the early. For early-type galaxies such as NGC 4736, the entire rotation curve may be accounted for by disk baryonic matter (Jalocha, Bratek, & Kutschera 2008). Note however the close match between the observed and disk-inferred rotation curves for NGC 3627, which is unusual because of its intermediate Hubble type, this maybe be a result of the close encounter with NGC 3628 which likely to have striped a large portion of its halo (see Section 4.11).

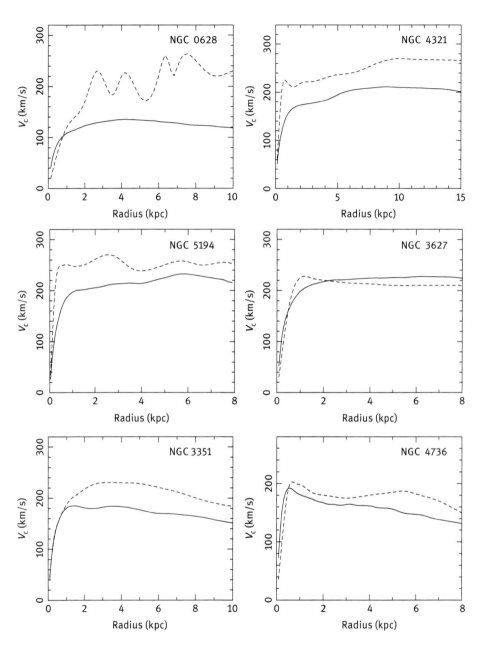

Figure 4.11: Comparison of rotation curves for the six sample galaxies (adapted from Zhang & Buta 2015). *Solid lines:* Disk rotation curves inferred from total disk surface density (stellar plus atomic and molecular hydrogen mass. Note that the corrections to account for helium and metal abundance in the gas mass have not been made). The inferred disk rotation curves were derived using IRAC 3.6 μm data for NGC 0628, using an average of IRAC 3.6 μm and SDSS i-band data for NGC 4321, 3351, 3627, 5194, and using SDSS i-band data for NGC 4736, plus the atomic and molecular gas contribution from the VIVA, THINGS, and BIMA SONG observations. *Dashed lines:* Observed rotation curves. The sources for these observations are given in Zhang & Buta (2015).

In Figure 4.12, we present the mass flow rate for the six sample galaxies calculated according to the rate equation (4.11). We have verified that the calculated mass flow rates are not sensitively dependent on the choice of scale height in the potential calculation (i. e., a change of vertical scale height from 12.6″ to 3.8″ in the case of NGC 4321 changed the scale of mass flow by only 10% or less for the entire radial range considered).

From Figure 4.12, we see that the mass flow rates for the various Hubble types range from a few solar masses per year to about a few tens of solar masses a year, except for the very late type galaxy NGC 628 which is still in the process of coordinating a significant galaxy-wide mass flow pattern. The intermediate-type galaxies appear to have the largest mass flow rates (the one galaxy, NGC 1530, which we have presented before, has mass flow rates on the order of 100 solar mass a year, and is also an intermediate-type galaxy), whereas for the early-type galaxies the mass flow is more concentrated to the central region. Although both inflow and outflow of mass are present at a given instant across a galaxy's radial range, due to the need to support the structure of density wave modes which have alternating positive and negative angular momentum density outside and inside CR, respectively, the general trend is for the mass to concentrate into the inner disk with time, together with the build-up of an extended outer envelope, consistent with the direction of entropy evolution. We have verified this trend already in the N-body simulations presented in Chapter 3.

Note that the radial mass flow rates presented in these curves are actually the lower bounds of the actual radial mass flow rates due to nonaxisymmetric instabilities in physical galaxies since these instabilities are expected to result in skewed distributions in parts of the thick-disk and halo as well, and thus result in the radial redistribution of the halo dark matter along with the luminous disk matter. The nonaxisymmetric instabilities in the spheroidal components are reflected in the twisted isophotes often observed in disky elliptical galaxies. These skewed distributions in the spheroidal components will lead to the same PDPS and collective secular mass redistribution as we discussed for the disk component. Furthermore, accretion flux due to viscous processes as well as due to the emergence and transformation of the density wave modal set will also tend to lead to an increase in the above-estimated radial mass accretion flux.

Even though we have carried out detailed calculations for the mass redistribution process in only seven galaxies in this section (more in fact had been done though not yet published), we have analyzed the phase shift and wave amplitude distributions for more than 150 galaxies in the OSUBGS sample (Zhang & Buta 2007; Buta & Zhang 2009). These galaxies show that the range of phase shift and amplitude values as represented by these seven galaxies we have presented in this section is in fact typical for the nearby bright spiral and barred galaxies. Therefore, it appears that the average mass flow rates in observed galaxies are sufficient for the build-up of Milky Way type bulge in a Hubble time. Furthermore, for some of the strongly interacting galaxies (i.e., M51), or else high-wave-amplitude, isolated galaxies (like NGC 1530), evolution toward disky ellipticals from an initial late to intermediate-type galaxies

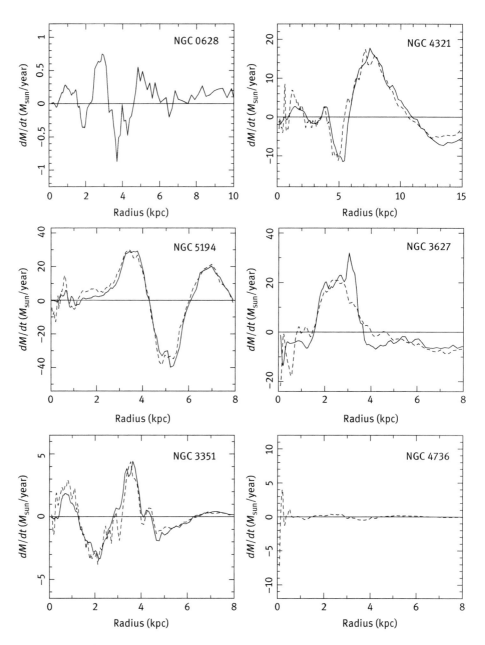

Figure 4.12: Total radial mass flow rate for the six sample galaxies. Positive portion of the curves indicate inflow, and negative portion of the curves indicate outflow. The solid lines have stellar mass derived from IRAC 3.6 μm data, and the dashed lines have stellar mass derived from SDSS i-band data. Atomic and molecular gas mass from VIVA, THINGS, and BIMA SONG were added to obtain the total mass maps which were used to derive these flow rates (adapted from Zhang & Buta 2015).

within a Hubble time is also entirely possible. Zhang & Buta (2015) Appendix B classified major categories of bar-spiral modal patterns, together with their characteristic phase shift distribution patterns. These modal patterns can be regarded to lie in an approximate linear sequence which corresponds to the different stages of secular evolution of the basic state of their parent galactic disks, from the late to the early Hubble types.

4.3.3 Relative Contributions from Gravitational and Advective Torques

In Figure 4.13, we plot the calculated gravitational torque couple for the six sample galaxies. First we discuss the result for NGC 4321. This torque result is very similar in shape to the one calculated for the same galaxy by Gnedin et al. (1995), though the scale factor is more than a factor of 10 smaller than obtained in their paper. Part of the difference can be accounted for by the difference in galaxy images used as well as in the galaxy parameters used between these two studies. We had tried to switch to use an R-band image as in Gnedin et al. and rescaled the galaxy parameters to be in agreement with what they used. However, after these adjustments the resulting scale is still smaller by a factor of ~5 from that in the Gnedin et al. (1995). The same amount of magnitude difference was found by Foyle et al. (2010) as well, when they try to reproduce the Gnedin et al. result. It is possible that there was an internal error in the Gnedin et al. calculation.

As we see from a close inspection of Figure 4.13, for most of the galaxies in the sample, especially those of late to intermediate Hubble types, there is a main peak of the gravitational torque couple, which is centered in the mid disk. Thus the implied mass inflow/outflow should be focused around the mid disk near the peak location of the main bell curve. This location, however, is expected to change with time for a given galaxy as secular evolution proceeds, as we have shown in Chapter 3, since the basic state configuration and the resulting modal configuration will both evolve due to the redistribution of the disk mass. In fact, by examine this small sample of six galaxies, we can already see a trend that the main peak of the C_g curve moves from the outer region of the galaxy (as for NGC 0628) to the mid-disk (as for NGC 4321, 5194 and 3627), and then onward to the central region (NGC 3351 and 4736).

The existence of the multiple humps of the gravitational torque couple (as well as advective and total torque couple since these latter two are found to be of similar shape to the gravitational torque couple for self-sustained modes) in a single galaxy, and by implication multiple nested modes of differing pattern speeds, shows that the secular mass flow process is not a one-way street. During the lifetime of a galaxy a given radial range may experience inflows and outflows of different magnitudes, but there is an overall trend of the gradual central concentrating of matter together with the building up of an extended envelop, consistent with the entropy evolution direction of the self-gravitating systems. Galaxies accomplish this in a well-organized fashion, by first employing a dominant mode across the galaxy's outer disk, and subsequently

180 — 4 Astrophysical Implications of the Dynamical Theory

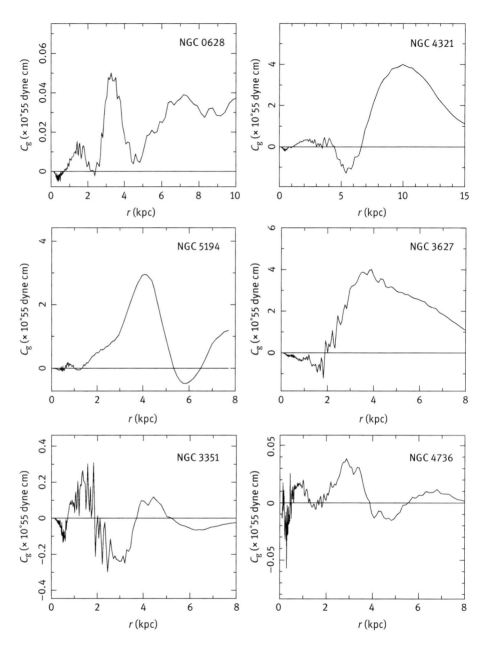

Figure 4.13: Calculated gravitational torque coupling for the six sample galaxies using IRAC 3.6 μm data for NGC 0628, an average of IRAC 3.6 μm and SDSS i-band data for NGC 4321, 3351, 5194, 3627, and SDSS i-band data for NGC 4736, to these the usual gas maps were added (adapted from Zhang & Buta 2015).

with more of the modal activity moving into the central region of a galaxy, and the single dominant mode also bifurcates into multiple nested modes. Furthermore, as our N-body simulation results in the last chapter had shown, nested resonances in the central region may only exert a secondary effect in terms of its own inflow–outflow pattern, compared to the dominant mass flow carried inward by the primary density wave mode and its diffusive mass current.

In Figure 4.14, we show the calculated radial gradient of gravitational torque coupling integral $dC_g(R)/dR$ as compared to the volume torque integral $T_1(R)$. Despite the overall similarity in shapes (though dC_g curves are somewhat smoother than the T_1 curves) and the locations of zero crossings, there is about a factor of 3–4 difference between the $T_1(R)$ and $dC_g(R)/dR$ for most of the galaxies in the sample (for NGC 4736, this ratio in the inner region is as high as 8), indicating that the remainder, which is contributed by the gradient of the advective torque coupling, $dC_a(R)/dR$, is in the same sense but much greater in value than $dC_g(R)/dR$. Note that this difference between the volume torque integral and the gradient of the (surface) gravitational torque couple is only expected in the new theory: in the traditional theory of Lynden-Bell & Kalnajs (1972) these two are supposed to be equal to each other, as we have shown in Chapter 2.

Furthermore, if the Lynden-Bell & Kalnajs (1972) theory is used literally, one should not expect any mass flow rate at all over most of the galactic radii in the quasi-steady state, except at isolated resonance radii – this is because in their theory the total angular momentum flux, $C_a(R) + C_g(R)$, is a constant independent of radius. The existence of a mass flux across the entire galactic disk is thus also contrary to the Lynden-Bell & Kalnajs' original expectations, especially since there is strong evidence that the patterns in these same galaxies were steady modes (i. e., the fact we can use the PDPS to successfully locate resonances in these galaxies shows that these patterns were steady modes).

We had also repeated the calculations for $dC_g(R)/dR$ and $T_1(R)$ using different scale height values. Even though both terms are affected by a particular choice of the scale height, the relative ratios between the two terms appear little affected. This is due to the fact that the potential that is being affected by the scale height choice enters into both the C_g and the $T_1(R)$ results, so their ratio is insensitive to the choice of scale height. So the conclusion of the significant difference between these two terms at the nonlinear regime of the wave mode is robust.

4.3.4 Relative Contributions of Stellar and Gaseous Mass Flows

In the past few decades, secular evolution, bulge growth, and the evolution along the Hubble sequence were studied mainly within the framework of the secular redistribution of the ISM under the influence of a barred potential, and the resulting growth of the so-called "pseudo bulges" (Kormendy & Kennicutt 2004 and the references therein). However, through gas accretion alone one could at most build late-type pseudo bulges, but not intermediate and early-type bulges, since the observed gas inflow rates and nuclear star-formation rates are both insufficient to account for the

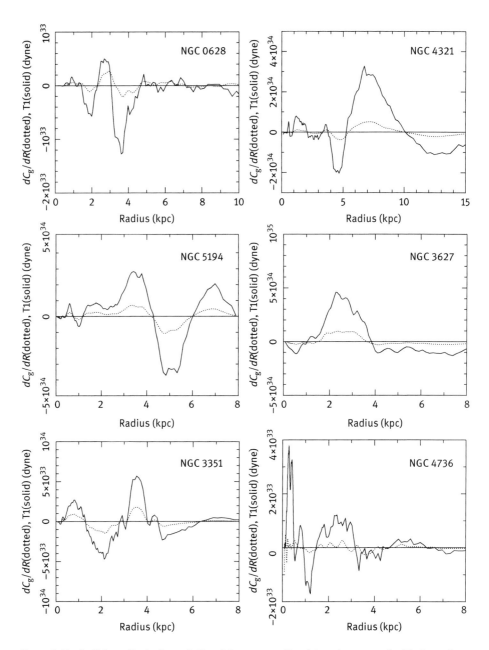

Figure 4.14: Radial gradient of gravitational torque coupling integral compared with the volume torque integral for the six sample galaxies. For NGC 0628, used IRAC 3.6 μm data; for NGC 4321, 3351, 3627, 5194, used the average of IRAC 3.6 μm and SDSS i-band data; for NGC 4736, used SDSS i-band data; to these the total gas maps were added (adapted from Zhang & Buta 2015).

formation of earlier-type bulges through secular inflow of gas and the subsequent conversion of gas to stars during a Hubble time. Yet observationally, early-type galaxies (including disky ellipticals) are known to form a continuum with the intermediate and late-type galaxies in terms of the structural parameters and kinematics (Franx 1993; Jablonka et al. 2002; Cappellari et al. 2013), and thus their formation mechanisms are also expected to be similar.

With the recognition of the role of collective effects in the secular mass redistribution process, the study of secular morphological transformation of galaxies should now give equal emphasis on the role of stellar and gaseous mass redistribution. Due to their different intrinsic characteristics (radial surface density distribution, compressibility, star-formation correlation, and dissipation capability), stars and gas do play somewhat different roles in the secular evolution process. In this subsection we illustrate with our sample galaxies some of the specifics of these roles.

In Figure 4.15 and 4.16, we present the comparison of stellar and gaseous (HI plus H_2) mass flow rates, as well as their respective phase shifts with respect to the total (star plus gas) potential. The phase shift plots tell us the relative efficiency of the stellar and gaseous mass accretion processes, since

$$\overline{\frac{dL}{dt}}(R) = \frac{1}{2}F^2 v_c^2 \tan i \sin(m\phi_0)\Sigma_0, \qquad (4.12)$$

where $F^2 \equiv F_\Sigma F_{\mathcal{V}}$ is the product of the fractional density and potential wave amplitudes, and m is the number of spiral arms (usually taken to be 2). Therefore, using also eqs (2.123), (2.124),

$$\sin(m\phi_0)_i = \frac{1}{\pi v_c R F_i^2} \frac{\frac{dM_i}{dt}}{(\Sigma_0)_i} \qquad (4.13)$$

where the subscript i represents stars or gas, respectively. So for similar fractional wave amplitudes between stars and gas, the mass component that has higher phase shift will have higher mass flow rate per unit surface density.

For galaxy NGC 0628, the stellar and gaseous accretion efficiencies are similar (Figure 4.16) as are the respective total mass accretion rates (Figure 4.15), since this is a late-type spiral galaxy and is gas rich. The alignment of the stellar and gaseous phase shifts is not consistent, especially for the outer disk, indicating that the galaxy is yet to evolve into a state of dynamical equilibrium.

NGC 4321 is relatively quiescent and of intermediate Hubble type. From Figure 4.16, we see that for much of the central region (except the very center) the gaseous phase shift with respect to their common potential is much larger than the stellar phase shift, indicating that gas leads in phase in this region compared to stars, revealing the higher dissipation rate and thus mass redistribution efficiency of the gas compared to stars for the central region of this galaxy. The values of phase shifts of stars and gas are comparable for the outer region indicating that the two mass components have similar mass-redistribution efficiency there. The shapes of the positive and negative humps of phase shifts for stars and gas have more consistent radial

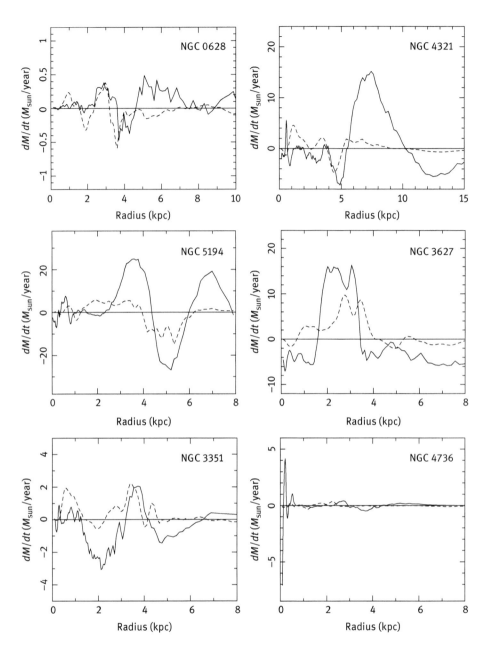

Figure 4.15: Stellar and gaseous mass flow rates for the six sample galaxies, calculated from the IRAC 3.6 μm for NGC 0628, an average of IRAC 3.6 μm and SDSS i-band for NGC 4321, 5194, 3627, 3351, and SDSS i-band for NGC 4736, plus VIVA, THINGS, and BIMA SONG data. The total potentials used for these calculations were the same as previously derived using IRAC and/or SDSS for the stellar contributions, with appropriate averaging, plus the gas contributions (adapted from Zhang & Buta 2015).

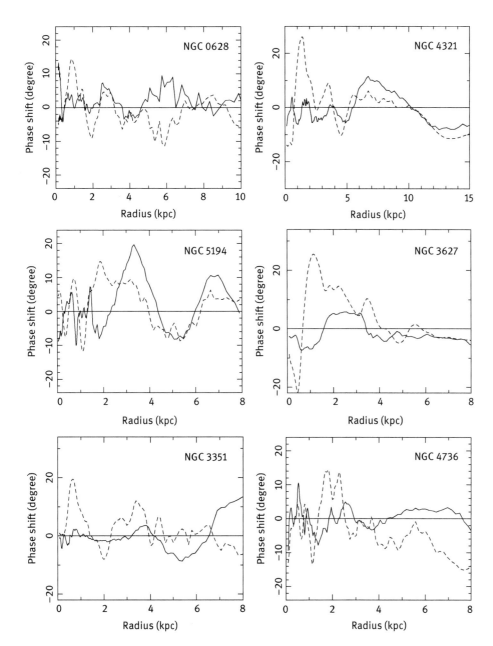

Figure 4.16: Stellar and gaseous phase shift with respect to total potential for the six galaxies. *Solid lines:* Stellar phase shifts. *Dashed lines:* Gaseous phase shifts. The stellar maps were derived using the IRAC 3.6 μm data for NGC 0628, using an average of IRAC and SDSS data for NGC 4321, 3351, 3627, 5194, and using SDSS i-band data for NGC 4637. The maps were from VIVA, THINGS, and BIMA SONG observations. The total potentials used for these calculations were the same as previously derived using IRAC and/or SDSS for the stellar contributions, with appropriate averaging, plus the gas contributions (adapted from Zhang & Buta 2015).

distributions for this galaxy compared to NGC 0628, especially for the outer region, indicating a higher degree of dynamical equilibrium. The overall contribution of the stars to mass redistribution, however, is much higher for stars than for gas in the outer region (Figure 4.15) because of the higher overall stellar surface density there.

Next the stellar and gaseous mass flow rates and phase shifts for NGC 5194 (M51) are presented. For this galaxy, unlike for NGC 4321, the stellar and gaseous phase shifts have significantly different radial distribution, even though both galaxies were of intermediate Hubble type. This indicates the non-dynamical-equilibrium state of M51 due to the tidal pull of the companion, and the inevitable evolution toward forming a new set of nested resonances, with the gas playing a leading role in seeking the new dynamical equilibrium because of its more dissipative nature, and the stellar component lagging somewhat behind in this action. But at every moment of this re-establishment of the dynamical equilibrium the overall density (i. e., the sum total of stellar and gaseous) still has a much more coherent phase shift distribution with respect to the total potential than each component considered separately. The overall mass flow rate of stars is much higher than gas for this galaxy.

Next the stellar and gaseous mass flow rates and phase shifts for NGC 3627 are presented. For this galaxy, even though the phase shift in the central region shows that gas has a higher accretion efficiency than stars, the overall accretion rate of stars still exceeds those of gas. The second CR location is not shown here because of more limited radial range plotted, but is present in the total gas phase shift curve when examined further outward.

For the earlier-type galaxy NGC 3351, we see that the central region gas-star relative phase shift for this galaxy is even more severe than for NGC 4321. This is likely due to the fact that the straight bar potential in the central region of this galaxy is mostly contributed by stars, which has small phase shift with respect to the total potential, and gas thus has a larger contribution to the total phase shift (through its dissipation in the bar potential and the phase offset of its density peak from the stellar density peak). The overall mass flow rates are contributed similarly by stars and gas for this galaxy. Note that the somewhat chaotic appearance of phase shifts in the outer region of this galaxy is due to the low surface density there, and thus noise begins to dominate.

For NGC 4736, gas leads stars in part of the intermediate radial range, but the overall contribution to mass flow is mostly due to stars, especially for the central region, due to the fact that in this early-type galaxy the stellar surface density much exceeds that of gas. Also once again the low surface density in the outer disk of this early-type galaxy leads to the more chaotic phase shift distribution there.

We see from this set of plots that one of the old myths, such as that the gas always torques stars inside/outside CR in the leading/trailing sense, and from the change of sign of the relative phase between stellar and gaseous density distributions one can tell the CR location, is only true in small number of instances. To obtain a reliable CR estimate, one really needs to use the total density (star plus gas) and total potential, and the calculation of the phase shift positive-to-negative zero crossings between these two components to determine the CR locations. In the absence of the gas surface densities, stellar surface density alone and the potential calculated from it (as is done in Zhang & Buta 2007) is the next best compromise. Both of these approaches are much more reliable than if using the phase shift between the stellar density and gas density distributions. We also see that in the majority of the galaxies in the local universe secular mass flow is dominated by the stellar mass redistribution rather than by gas redistribution, unlike what has been emphasized by many of the earlier works on secular evolution.

We comment once again that even for the gas accretion in disk galaxies the mechanism responsible for its viscosity is still the collective gravitational instabilities (which manifests as the phase shift between the gas density and total potential), rather than the microscopic gaseous viscosity, which was long since known to be inadequate both for the stellar accretion disks and for the galactic gaseous accretion disks – thus the well-known need for anomalous viscosity in generalized accretion disks (Lin & Pringle 1987). The current work shows that the large-scale density-wave-induced gravitational viscosity is likely to be the source of anomalous viscosity in both the stellar and gaseous viscous accretion disks of galactic and stellar types.

4.4 Secular Heating and the Age–Velocity–Dispersion Relation[†]

The secular radial migration of mean stellar orbit is not the only effect produced by the wave/basic-state energy and angular momentum exchange. There is also an associated heating effect due to the fact that the circular speed Ω of stars at a given galactic radius is in general different from the pattern speed Ω_p. Since the energy and angular momentum of the basic state stars are related through a factor Ω, the angular speed of the orbiting stars, and those for the wave are related through a factor Ω_p, the pattern speed of the wave, the complete angular momentum exchange between the wave and the basic state implies that there will be certain amount of the energy left, which is used to heat the disk stars. We have derived in Chapter 2, as well as quantitatively verified through N-body simulations in Chapter 3, that the rate of increase of the mean square random velocity of the basic state stars due to their interaction with a quasi-stationary spiral structure is given by

$$\frac{d\sigma^2}{dt}(r,t) \equiv D^{(3d)}(r,t) = (\Omega - \Omega_p)F^2 v_c^2 \tan i \sin(m\phi_0), \qquad (4.14)$$

where σ and D^{3d} are the dispersion and the effective diffusion constant of the random velocity of basic state stars; and the rest of the symbols have their meanings as given in

---

[†] Portions of this section used material previously published in Zhang (1998, 1999), reproduced with modifications from The Astrophysical Journal @ AAS. Reproduced with permission.

Chapter 2. The above results could explain a number of observational facts pertaining to the magnitude and distribution of the random velocity of stars in the galactic disks.

4.4.1 The Age–Velocity Dispersion Relation of the Solar Neighborhood Stars

The number density f_0 of stars at the solar neighborhood was found to follow the Schwarzschild distribution in peculiar velocity $\vec{v}$

$$f_0(\vec{v})d^3\vec{v} = \frac{n_0 d^3\vec{v}}{(2\pi)^{3/2}\sigma_r\sigma_\phi\sigma_z} \exp\left[-\left(\frac{v_R^2}{2\sigma_r^2} + \frac{v_\phi^2}{2\sigma_\phi^2} + \frac{v_z^2}{2\sigma_z^2}\right)\right], \qquad (4.15)$$

(Schwarzschild 1907), where n_0, σ_r, σ_ϕ, and σ_z are constants. Furthermore, it has been found that the velocity dispersion components σ_r, σ_ϕ, and σ_z of the solar neighborhood stars increase smoothly with the mean age of a stellar group (Gliese 1969), following roughly a $t^{1/2}$ law (Wielen 1977). Starting with the early attempt of Spitzer and Schwarzschild (1951, 1953), a spectrum of mechanisms has been proposed to account for the secular heating of disk stars (Barbanis & Woltjer 1967; Carlberg & Sellwood 1985; Wielen & Fuchs 1990). However, none of the previously proposed mechanisms was able to satisfy the requirement of being physically realistic on the one hand, and being able to account for both the form of the age–velocity dispersion law and the shape of the velocity ellipsoid on the other (Binney & Lacey 1988; Gilmore et al. 1990). As Ivan King pointed out in his Saas Fee lectures, "It is clear that we do not yet fully understand this intriguing problem" (Gilmore et al. 1990, p. 175).

In the past few decades, it was widely believed that a quasi-stationary spiral structure could not serve to heat the disk stars. However, we have presented in this work the dynamical mechanism by which a quasi-stationary spiral density wave mode could interact and exchange energy and angular momentum with the basic state of the galactic disk. We now calculate the form of age–velocity dispersion relation due to the spiral-heating mechanism, as well as the magnitude of this heating effect for the solar neighborhood stars.

Since heating by the spiral structure is due to the action of a local gravitational instability, or a collisionless shock, the energy injection into the stars is sudden and is local. We can thus employ the impulse approximation. This implies that the successive increases of stellar random velocity at each crossing of the spiral arms are independent events and are also independent of the magnitude of the peculiar velocity a given star already has. Therefore, after n crossings of the spiral arms, the total increase of stellar random velocity for an average star is

$$\Delta\sigma \sim \sqrt{n(\delta v)^2}, \qquad (4.16)$$

4.4 Secular Heating and the Age–Velocity–Dispersion Relation

where δv^2 is the increase in mean square stellar random velocity per crossing of the arm. Since over a long time span, $n \propto t$, the age of the star, we see that heating by a quasi-stationary spiral structure can produce a $t^{1/2}$ law for the age–velocity dispersion relation. In general, the stellar velocity dispersion at the current epoch depends also on the average random velocity of stars acquired at birth, so we can write the age–velocity dispersion law in general as

$$\sigma(t) = \sqrt{\sigma_0^2 + D^{(3d)} t}, \qquad (4.17)$$

where σ_0 is the initial velocity dispersion of stars at zero age, and with $D^{(3d)}$ given by eq. (4.14).

A couple of points need to be noted in calculating the numerical values in the coefficients in the age–velocity dispersion relation due to the spiral structure. First of all, the correct values to use in (4.14) for deriving the age–velocity dispersion relation over the life span of the Milky Way are not those for the *current epoch* since the secular evolution effect induced by the Galactic spiral structure changes both the properties of the basic state as well as the properties of the wave mode. A further complication is introduced by the secular radial migration of stars during the same evolution process. Therefore, the overall effect is that the exact form of the age–velocity dispersion relation for any given spatial location is expected to be modified from that of (4.17).

However, the result of N-body simulations shows that the actual age–velocity dispersion curve is still quite close to (4.17) despite the above-mentioned complications. In Figure 4.17, we plot the evolution of the mean-square peculiar velocity of one of the N-body spiral mode obtained in Zhang (1998, 1999), which is a run with a large number of particles ($N = 300{,}000$), so the contribution of Poisson noise to heating is small, and the acceleration of particles witnessed there can be entirely attributed to spiral heating. On the same figure, we also plot a line which corresponds to $t^{1/2}$ evolution

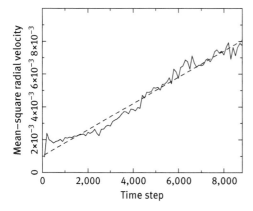

Figure 4.17: Comparison of the N-body evolution of the mean-square velocity of a group of stars inside corotation with the expected linear evolution function (adapted from Zhang 1999).

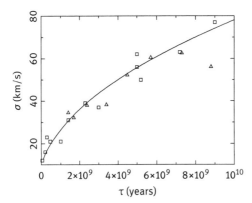

Figure 4.18: The dependence of the stellar velocity dispersion on the mean stellar age. In the figure, the squares are from Wielen (1977), the triangles are from Carlberg et al.'s (1985) tangential velocity measurement and scaled to the space velocities. The solid line is a fit of the form (9) with $\sigma_0 = 10$ km s^{-1}, $D^{(3d)} = 6 \cdot 10^{-7}$(km s^{-1})^2 year^{-1} (adapted from Zhang 1999).

with an effective diffusion coefficient $D^{1d} = 0.8 \cdot 10^{-6}$, calculated by eq. (4.14) using parameters appropriated for the spiral mode in that simulation. A simple explanation for the observed behavior of the N-body heating is that first of all, the factor $F^2 v_c^2 \tan i \sin(m\phi_0)$ in (4.14) can be shown to be approximately independent of galactic radius for the inner disk region. Furthermore, the factor $\Omega - \Omega_p$ is also approximately constant for the short radial range that this particular group of particles traverses between the time step of 0 and 8,000 (or about 25 rotation periods at radius 20), since this mode is slow evolving. However, over the time span of a Hubble time, a more fast-evolving mode in a fast-changing basic state would result in some departure from the $t^{1/2}$ law.

Having established the approximate validity of the age–velocity dispersion law (4.17), we now estimate an average value of the *apparent* diffusion coefficient D^{3d} for the solar neighborhood stars, appropriate as a mean value for the past 10^{10} years. For a 20° pitch-angled spiral with $1/r$ density modulation, which results in an averaged phase shift of 3° in the inner disk; and for an average $\Omega - \Omega_p = 8$ km/s/kpc, which is two-thirds of its current value at the solar neighborhood; and using for the fractional amplitude of the wave $F = 0.15$ as an average over the past 10^{10} years, we obtain that the mean velocity diffusion coefficient for the current solar neighborhood stars over the past 10^{10} years is $\bar{D}^{(3d)} = 5.7 \cdot 10^{-7}$ (km/s)^2/year, which is quite close to the expected value of $\bar{D}^{(3d)} = 6 \cdot 10^{-7}$ (km/s)^2/year (Wielen 1977). In Figure 4.18, we plot the age–velocity dispersion curve with this mean diffusion coefficient, together with some of the published data for the actually stellar velocity dispersion and age group correlation. The agreement between the $t^{1/2}$ curve and the observed data, as was already found by Wielen (1977), was excellent. Note that we have omitted plotting the data beyond the age of the Galaxy, which if done would show the well known Parenago's discontinuity, as also confirmed in the analysis of Hipparcos data by Dehnen & Binney (1998). This discontinuity could be associated with the emergence of density wave modes in the primordial Galactic disk.

4.4.2 The Cause of Isotropic Velocity Diffusion in Three Dimensions and the Preservation of the Gaussian Velocity Distribution through Time

Two further aspects of the distribution of stellar random velocities in the solar neighborhood which require explanation are (1) the observed axis ratios of the velocity ellipsoid and (2) the preservation of the gaussian (Schwarzschild) velocity distribution in time.

From observations of the representative K+M dwarfs in Gliese's catalog, the velocity dispersions in the r, ϕ, and z directions are found to be $\sigma_r = 48$ km s^{-1}, $\sigma_\phi = 29$ km s^{-1}, and $\sigma_z = 25$ km s^{-1}, respectively. Wielen (1977) demonstrated that the apparently different rates of the velocity diffusion in the three spatial dimensions is in fact an illusion. The ratios of the apparent diffusion constants C_r, C_ϕ, and C_z can be obtained from the epicycle motion of the stars in the galactic plane, and from the stellar vertical oscillations, using an *isotropic* one-dimensional diffusion coefficient $D^{(1d)}$.

Both of these aspects can be naturally explained by the spiral heating mechanism. Since the operation of the spiral-arm gravitational instability is sudden and is *local*, it naturally leads to isotropic velocity diffusion in the three spatial dimensions, which explains the observed correlation between the z-component and the r- and ϕ-components of the stellar random velocities (Wielen 1977). This correlation on the other hand is difficult to explain in the transient spiral wave, or else in the GMC complex scattering scenarios.

Furthermore, in a general analysis of the stellar diffusion through phase space, one of the most important conclusions drawn by Binney & Lacey (1988) was that the *only* stellar acceleration mechanism which could preserve the gaussian distribution of random velocity and also produce a $t^{1/2}$ evolution is one in which the potential perturbation lasts significantly less than an epicycle period, so that the impulse approximation is valid. The spiral heating through a local gravitational instability certainly satisfies this criterion. And it differs from the stochastic spiral heating mechanism not only in terms of the effective 3D isotropic diffusion of random velocity, but also in that the effect of heating by a quasi-stationary spiral structure can last persistently and smoothly on the order of a Hubble time, which is what is needed to explain the smoothness of the age-velocity dispersion curve (Figure 4.18) over the past 10^{10} years.

4.4.3 The Origin of Radial Variation of the Stellar Velocity Dispersion with Galactocentric Distance

So far there is no complete measurement of the radial variation of the stellar velocity dispersion with the Galactocentric distance for our own Galaxy. However, indirect evidence indicates that stellar velocity dispersion decreases with the Galactic radius (Gilmore et al. 1990, p. 194). The observation of external galaxies has further confirmed the radial decrease of the stellar velocity dispersion with galactocentric distance (van der Kruit & Freeman 1986). This radial decrease in stellar velocity dispersion can also be inferred from the observed constant scale height of the old-disk population versus galactic radius (van der Kruit & Searle 1981, 1982; Gilmore et al. 1990, p. 193, p. 373).

The variation of stellar velocity dispersion with galactic radius has a simple explanation from the current theory. We have shown previously that if the spiral heating is the main contributor to the secular increase in stellar velocity dispersion, then the radial dependence of stellar velocity dispersion will likely to have a form of

$$\sigma \propto \sqrt{(\Omega - \Omega_p) F^2 v_c^2 \tan i \sin(2\phi_0)}. \tag{4.18}$$

This can be further simplified to $\sigma \propto \sqrt{\Omega(r) - \Omega_p}$ for the inner disk stars since the product of the other factors is approximately constant for the inner disk region. Therefore for a nearly constant circular velocity v_c, we can express the radial dependence of the velocity dispersion for the *inner disk region* as

$$\sigma(r) \approx \sigma_\odot \sqrt{\frac{\frac{v_c}{\Omega_p r} - 1}{\frac{v_c}{\Omega_p r_\odot} - 1}}, \tag{4.19}$$

where the subscript ⊙ denotes the solar neighborhood values.

Similar to the case inside corotation, there is heating outside corotation radius as well, and the effective diffusion constant is again described by (4.14). Furthermore, the heating at the corotation radius is not expected to be zero, but rather smoothly joins the values at the inner and outer disk through the process of conduction (i. e, regular diffusion). Therefore, the overall $\sigma(r)$ profile is expected to resemble a single $r^{-1/2}$ function, that is,

$$\sigma(r) \approx \sigma_\odot \sqrt{\frac{r_\odot}{r}}. \tag{4.20}$$

In Figure 4.19, we plot the result of the radial velocity dispersion for the N-body simulation run of Zhang (1998) with 300,000 particles, at the beginning ($n = 0$) and the end ($n = 8,000$) of the run. Also plotted was a $r^{-1/2}$ fit to the velocity dispersion curve at the end of the run. We see that the $r^{-1/2}$ curve fits very well the $n = 8,000$ curve, and the tendency of the variation of the velocity dispersion distribution is such that as the simulation progresses the curve looks more and more like a $r^{-1/2}$ curve throughout the galactic radii (i. e., the tail end at larger radii gradually lifts up). Of course, the initial condition for the stellar velocity dispersion already has a nonuniform radial dependence, so the final radial dependence in the N-body simulated stellar velocity dispersion is not entirely contributed by the spiral heating, and there is indeed also the impact of the Poisson noise which produces excess heating. Still, the least we can say is that the spiral activity appears to be *compatible* with the maintenance of the $r^{-1/2}$ radial dependence of the velocity dispersion.

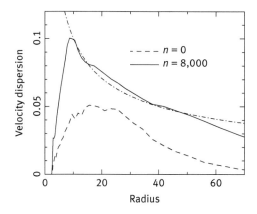

Figure 4.19: Radial velocity dispersion of the disk stars at the beginning and the end of an N-body simulation with $N = 300{,}00$ particles. The dot-dashed line is an $r^{-1/2}$ function fit to the $n = 8{,}000$ curve (adapted from Zhang 1999).

The $r^{-1/2}$ function fits very well the observed radial variation of stellar velocity dispersion for certain external galaxies (van der Kruit & Freeman 1986). Furthermore, if we fit an exponential to an $r^{-1/2}$ function, we obtain a fitted exponential length scale which is twice that for fitting a r^{-1} function (the approximate functional form for the surface density of many disks which can also be approximated by exponential disks). This explains the well-known observation that the exponential length scale of radial velocity dispersion for many disk galaxies is twice that of the exponential length scale for the surface density (Gilmore et al. 1990).

4.5 Secular Heating and the Size–Line-Width Relation[†]

The same energy injection process mediated by the spiral density wave, apart from being responsible for producing the age–velocity dispersion relation of the solar neighborhood stars, is also responsible for producing the size–line-width relation of the ISM (or the so-called Larson Law) when the injected energy is channeled into the random motion of the Galactic cloud complexes (Zhang et al. 2001; Zhang 2002).

4.5.1 History and Motivation

More than half a century ago, von Hoerner (1951) and von Weizsacker (1951a, 1951b, 1951c) were among the first to point out that the Galactic molecular clouds and complexes appear to be in a well-organized hierarchy. It was later demonstrated that this hierarchy can be quantified as a power-law correlation between the cloud sizes and their velocity dispersions, spanning the size range from that of the giant molecular cloud (GMC) complexes, or ∼ 1 kpc, to that of the cores of low-mass star-forming

---

[†] Portions of this section used material previously published in Zhang et al. (2001), reproduced with modifications from The Astrophysical Journal @ AAS. Reproduced with permission.

regions, or ~ 0.1 pc (Larson 1979, 1981; Solomon, Sanders, & Scoville 1979; Fleck 1980; Leung, Kutner, & Mead 1982; Myers 1983; Dame et al. 1986; Elmegreen 1989; Elmegreen & Scalo 2004; Falceta-Gonçalves et al. 2015 and the references therein).

This coherent organization of cloud sizes and velocity dispersions has been speculated to be produced by a hierarchy of turbulent energy cascade (Larson 1981), with the smaller-scale motions produced by the turbulent delay of larger-scale ones. The topmost level of this energy injection appears to be on the order of several hundred parsec to a kiloparsec (Larson 1979), and stars and many forms of ISM appears to be on the *same* hierarchy, as reflected in the observed correlation lengths of a variety of processes (Fleck 1981), including the brightness fluctuations of starlight (Ambartzumian & Gordeladse 1938; Chandrasekhar & Munch 1952), cosmic ray transport (Jokipii & Lerche 1969a, 1969b), and fluctuations in Faraday rotation and pulsar dispersion measures (Jokipii & Lerche 1969). A similar hierarchy in the initial mass function of stars, also over four orders of magnitude of mass range, is thought to be related to the turbulent hierarchy of the ISM out of which the stars form (e. g., see Elmegreen 1998).

It is well known that the physical state of the galactic ISM satisfies conditions for both compressibility and turbulence (Elmegreen & Scalo 2004 and the references therein). Due to the natural tendency of turbulence to cascade energy from large to small scales, and to dissipate this cascaded energy at the smallest scales, the maintenance of the turbulent motion requires the continuous injection of kinetic energy at larger scales. The rate of energy injection required for maintaining a quasi-steady state of the turbulent motion depends of course on the rate of energy cascade, which is usually on the order of the eddy turnover time at the largest scale (i. e., L_1/v_1), which for the interstellar cloud complex of a half-kpc size is on the order of several times 10^5 years, whereas the lifetime of the molecular clouds themselves is on the order of 10^7 years.

It had been hoped that magnetohydrodynamic (MHD) turbulence would prolong the decay time compared to that of the pure hydrodynamical turbulence. However, 3D simulations of the compressible MHD turbulence show that the decay timescale for the MHD turbulence is comparable to that for pure hydrodynamical turbulence, that is, on the order of the eddy turnover time at the largest eddy scale (Stone, Ostriker, & Gammie 1998). Therefore external driving mechanism still needs to be identified for producing the observed long lifetime turbulent motion of the Galactic molecular clouds.

Many candidate energy-injection mechanisms for feeding the interstellar turbulence have been envisioned over the past few decades, covering both the large-scale and small-scale mechanisms (e. g., see von Weizsacker 1951a, 1951b, 1951c; Larson 1979, 1981; Fleck 1981; Myers 1983; Zhang 2002; Falceta-Gonçalves et al. 2015). It is generally agreed that the relevant mechanism has to involve the injection from the largest scales of the hierarchy, even though it does not have to exclude the injection at smaller scales. This is both because any small-scale energy injection mechanism alone (such as outflow, angular momentum cascade, magnetic field, etc.) would have difficulty producing a tight correlation upward to the kiloparsec scale, especially since the resulting velocities would be largest on the smallest scales, contrary to what is observed (Larson 1981); and also because the failure of these mechanisms to provide sufficient amount of total energy input to power the level of turbulence observed in the Galactic ISM. The small-scale mechanisms also lack the uniformity to explain the universality of the size-line-width correlation across the Galactic disk in regions of varying physical conditions (Falceta-Gonçalves et al. 2015 and the references therein).

The most obvious reservoir of turbulent energy on the largest scale is of course the galactic rotation. The kinetic energy contained in this motion (about 500 eV/nucleon) is much larger than that of the other energy sources, such as cosmic rays, magnetic field, or gas pressure (each about 1 eV/nucleon) (Pfenniger 1998). If the entire energy reserve of the orbital motion of the ISM can be used to feed the interstellar turbulence, Fleck (1981) estimated that this would supply enough energy to sustain the turbulence at the observed level for about 5×10^{10} years. One problem with the previously proposed mechanism for tapping into this energy reserve, that is, through the shear in the differential rotation (von Weizsacker 1951a, 1951b, 1951c; Fleck 1981), is that galactic shear when coupled to the cloud-complex length scale will cause these complexes and clouds to rotate rapidly. On the other hand, most of the clouds which possess turbulent line widths are not found to be fast rotators. One further problem with this method of tapping into the rotational kinetic energy is that galactic shear alone does not allow stars to convert orbital energy into turbulent energy, whereas stars appear to be in the same hierarchy of turbulent motion as the GMCs and the diffused HI clouds (Larson 1979). Furthermore, detailed numerical simulations show that it is in fact rather difficult to couple the galactic rotation energy into internal motion energy of the cloud (Das & Jog 1995).

The collective interaction mechanism between the density wave modes and the basic state, in this case the ISM which circles the galaxy disk just like stars, can effectively tap into the energy reserve of galactic rotation without causing significant vertical motion of the galactic clouds or causing the clouds to rotate rapidly (Zhang et al. 2001; Zhang 2002). Like the mechanism that produced the age–velocity dispersion relation for the solar neighborhood stars, the cloud-heating mechanism once again is a byproduct of the energy and angular momentum exchange process between the density wave and the disk matter. In addition, the spiral-induced energy injection mechanism, being global and top-down, possesses the university characteristic needed to account for the uniformity of the size-line-width relation across different regions and size ranges observed for the Galactic interstellar medium.

4.5.2 Energy Injection into the Star-Gas Two-Fluid through the Spiral Collisionless Shock

We have previously derived that the rate of random energy gain per unit area of the disk matter, due to its interaction with the density wave mode at the quasi-steady state, is related to its angular momentum exchange rate per unit area through

$$\frac{d\Delta E}{dt} \equiv \frac{d(E_{\text{basic state}} - E_{\text{wave}})}{dt} = (\Omega - \Omega_p)\frac{dL_{\text{wave}}}{dt}, \quad (4.21)$$

where L_{wave} is the angular momentum density of the wave. Note that this expression is true (and has a positive sign) both inside and outside corotation, since both $\Omega - \Omega_p$ and dL_{wave}/dt change sign across corotation.

The above expression can be further written in terms of the spiral parameters by

$$\frac{d\Delta E}{dt} = \frac{1}{2}(\Omega - \Omega_p)F^2 v_c^2 \tan i \sin(m\phi_0)\Sigma_0, \quad (4.22)$$

where i is the pitch angle of the spiral, m is the number of spiral arms, ϕ_0 the potential and density phase shift, F the fractional amplitude of the spiral, v_c the circular speed of the galaxy, and Σ_0 the surface density of the disk. Equation (4.22) gives the rate of random energy increase of matter per unit area, valid for *both* the stellar *and* the gaseous components.

4.5.3 The Rate of Energy Injection and Rate of Energy Cascade

In the case of the ISM of our own Galaxy, we verify below that the above rate of energy injection is of the same order of magnitude as the rate of downward turbulent energy cascade. The source of the top-level injected energy is the orbital energy converted into random energy by the spiral density wave collisionless shock.

The average rate of energy injection per unit mass into the orbiting disk matter, using the fitted value of the stellar velocity diffusion coefficient D^{3d} of $D^{3d} = 6 \times 10^{-7}$ (km/s)^2 yr^{-1} (Wielen 1977; Zhang 1999), is

$$\left(\frac{d\Delta E}{dt}\right)_{\text{injection}} = \frac{1}{2}\frac{d\Delta v^2}{dt} = \frac{1}{2}D^{3d} = 3 \times 10^{-7} (\text{km/s})^2 \text{yr}^{-1}. \quad (4.23)$$

On the other hand, the rate of energy cascade can be calculated using the theory of fully developed turbulence of Kolmogorov (Shu 1992; Frisch 1995; Elmegreen & Scalo 2004). Using the standard values of the Galactic molecular cloud velocity dispersion of 10 km/s, and the cloud complex size scale of 1 kpc, we have that for the rate of energy cascade

$$\left(\frac{d\Delta E}{dt}\right)_{\text{cascade}} = \frac{\Delta v^3}{L} = \frac{(10 \text{ km/s})^3}{1{,}000 \text{ pc}} = 8.7 \times 10^{-7} (\text{km/s})^2 \text{yr}^{-1}. \quad (4.24)$$

The two rates are quite comparable, taking into account of the fact that the average energy injection rate calculated above is an underestimate of the *instantaneous* energy injection rate during the crossing of the ~1 kpc width spiral arm shock since this average is taken over the time period of the entire orbital cycle, which includes the long inter-arm migration period during which there is no energy injection from the spiral shock.

The spiral density wave mechanism is capable of injecting energy to the disk matter on all scales since it is operated through the gravitational potential on the disk matter, which eliminated the need of a direct coupling of energy from the size scale of 1 kpc to a few scales downward. This direct coupling aspect of the gravitational energy injection could change the slope of the scaling law from that of a pure cascade process.

4.5.4 Application: The Carina Molecular Cloud Complex

Zhang et al. (2001) presented large area, fully sampled maps of the Carina molecular cloud complex in the carbon monoxide CO ($J = 4 \to 3$) and neutral carbon CI ($^3P_1 \to {}^3P_0$) transitions. These data were collected in the austral winter of 1998, when the author served as a winter-over scientist at the South Pole, using the 1.7-meter diameter radio telescope of the Antarctic Submillimeter Telescope and Remote Observatory (AST/RO), with a northern base at the Harvard-Smithsonian Center for Astrophysics (Stark et al. 1997).

Analysis of the AST/RO data, in conjunction with CO ($J = 1 \to 0$) data from the Columbia CO survey and the IRAS HIRES continuum maps for the same region, suggests that the spiral density wave shock associated with the Carina spiral arm played an important role in the formation and dissociation of the cloud complex, as well as in maintaining the internal energy balance of the clouds in this region. Massive stars form at the densest regions of the molecular cloud complex. The winds and outflows associated with these stars have a disrupting effect on the complex and inject mechanical energy into the parent clouds, while the UV radiation from the young stars also heats the parent clouds. However, the new data suggested, contrary to previous beliefs, that massive stars alone may not account for the energetics of the clouds in the Carina region. The details of the data and the correlation among the various data sets hint at the important role that the spiral density wave shock plays in feeding interstellar turbulence and in heating molecular clouds.

4.5.4.1 Background

Studies of the ISM on large scales are often pursued through radio surveys because the Galaxy is transparent at radio wavelengths and because interstellar gas in its various phases emits radio lines which can be observed in emission over large areas of sky. The line shapes are easily resolved by radio spectroscopic techniques and reflect the motion of the ISM on a galactic scale as well as the internal dynamics of clouds.

Zhang et al. (2001) observed a region of roughly 150 pc × 100 pc in linear extent, surrounding the bright southern peculiar star η Carinae in the mid-submillimeter lines of CO ($J = 4 \to 3$ at 460 GHz) and neutral carbon CI ($^3P_1 \to {}^3P_0$ at 492 GHz). The CO ($J = 4 \to 3$) line is a tracer of warm (~55 K), dense ($n \sim 10^5$ cm^{-3}) cores in molecular clouds (Viscuso & Chernoff 1988), whose average properties as seen in CO $J = 1 \to 0$

line studies is usually rather colder (~10 K) and more diffuse ($n \sim 10^3 \, \text{cm}^{-3}$). Like density-sensitive species such as H_2CO (e. g., Magnani, LaRosa, & Shore 1993) and CS $J = 2 \rightarrow 1$ (e. g., Lada, Bally, & Stark 1991), it is a tracer of molecular cloud cores, but unlike those species it is not excited into emission unless the gas is warm. The CI line, on the other hand, is expected to trace the photon-dominated regions (PDR) in the outer layers of molecular clouds (Tielens & Hollenbach 1985), the boundary layer between low-density regions where carbon is mostly ionized by the interstellar ultraviolet radiation field and the interior of molecular clouds where chemical processes drive most of the carbon into CO. The CI $^3P_1 \rightarrow {}^3P_0$ transition has a minimum excitation temperature of 24 K and critical density $n \sim 10^3 \, \text{cm}^{-3}$ for collisions with H_2 (e. g., Schroder et al. 1991). It is, therefore, easily excited in dense interstellar gas. Actual observations show, however, that the CI emission is surprisingly well-mixed and well correlated with ^{12}CO and ^{13}CO emissions (Plume 1995; Keene et al. 1997). Stutzki et al. (1988) suggest that this effect results from the clumpiness of molecular material, so that the "surface" layers are distributed throughout the volume of the cloud.

It is a well-known fact that the most massive molecular cloud complexes are concentrated in spiral arms (Stark 1979; Elmegreen 1979; Dame et al. 1986). Indeed, massive stars, HII regions, and dust lanes, which are the visual tracers of spiral arms, are all manifestations of concentrations of GMCs. The processes leading to the formation of these complexes are not completely understood, but the observational evidence suggests they are strongly linked to the passage of the gas through the spiral density wave shock. The role of spiral density wave shock in the formation and evolution of the galactic GMC complexes has been speculated upon and investigated since the advent of the density wave theory, and this effort has continued for many decades (Roberts 1969; Elmegreen 1979; Balbus & Cowie 1985; Dame et al. 1986; Heyer 1998; Zhang 1998, 1999).

The problem of cloud complex formation and dissociation is closely related to two other issues:

(i) The source of the turbulence energy injection into the clouds (e. g., see Larson 1981; Myers 1983 and the references therein). Turbulence has a natural tendency to cascade downward and dissipate into heat and line radiation at the smallest scales. Therefore, turbulent energy must be constantly injected into the ISM to sustain the line-widths observed in galactic molecular clouds. The small-scale energy injection mechanisms considered (e. g., stellar winds and outflows) usually fail to either generate sufficient energy injection or reproduce the correct form of the size–line-width relation. Associated with the issue of the source of turbulence energy input is the issue of how and where the turbulence energy is dissipated.

(ii) The processes dominating the overall energy balance in the ISM. The theory of PDRs (believed to constitute more than 90% of the galactic ISM) has gradually confronted serious challenges as observational data accumulate. In their review article, Hollenbach (1999) cite several instances of observations of Galactic and extragalactic star-forming regions where the current theory of the PDR often produces a much a lower temperature than that measured in the rotational quadrupole transitions of H$_2$, and a much higher ratio of [CII]/L_{FIR} than is seen in infrared luminous regions. These authors conclude that these regions must have additional sources of energy input in order to account for the energy balance of the PDR. These sources may include the dissipation of magnetohydrodynamic turbulence.

The Carina molecular complex is a segment of the Carina spiral arm surrounding the extraordinary Luminous Blue Variable (LBV) star η Carinae. It is located between Galactic longitudes 284 and 289° and latitudes −2 and 1°. Situated near the center of the Carina molecular complex is the Carina nebula, which contains an extremely bright and extended OB association (Car OB1) and a bright HII region, NGC 3372. Many spectroscopic and morphological studies of the Carina region have been made, covering the entire spectral range from centimeter to X-rays.

Past study of the Carina region has focused mainly on the peculiar star η Carinae and on the HII region nebula surrounding it. The larger molecular complex is mapped through the Columbia CO survey with a 8.8' beam, and by the IRAS satellite in its four spectral bands. All of the above observations are large surveys which are not particularly focused on the Carina region. We have chosen to map the entire complex with the AST/RO telescope, since its primary beam size of ~3' as well as its main function as a dedicated survey telescope make it an ideal instrument for the purpose of mapping out the large-scale physical conditions in the Carina region, in order to study of the role of spiral density wave to the molecular cloud complex formation and dissociation.

The Carina molecular complex is well-suited for the study of the interaction of spiral density wave and the ISM. This complex is located at the sub-tangent region of the sun and has a clear line of sight, with a mean color excess E_{B-V} = 0.5 at a distance of 2.5 kpc (Feinstein 1995). The various clouds and sub complexes have a sequential distribution along the galactic plane, and their kinematics indicate that locations of decreasing longitude correspond to advancement in the spiral-arm-crossing phase. This correspondence of the galactic longitude with the spiral-crossing phase is further confirmed by the observed age gradient for the various star clusters across the complex, with the Tr14 and Tr16 clusters within the Carina nebula NGC 3372 at (287.6°, −0.65°) being the youngest (age ~10^6 year) and IC 2581/NGC 3293 at (284.7°, 0.1°) being the oldest (age ~5×10^6 yr). They are separated by a projected distance of ~130 pc. This sequential arrangement is advantageous to the study of the evolution of physical conditions in the clouds as they stream across the spiral arm.

The Carina nebula itself contains an archetypal outflow (Duschl et al. 1995) centered on the LBV star η Carinae at (287.6°, −0.64°), and the highest concentration of early type stars known in the Galaxy in the two ionizing clusters Tr 16 (which is centered on η Carinae, and which also includes a smaller cluster Cr 228 to the south) and Tr 14 (about 10' to the north of η Carinae). The region offers the opportunity to study and possibly disentangle the effects of energy input to the molecular clouds by massive stars and by spiral density wave shocks.

4.5.4.2 Observational Results and Analyses

Figure 4.20 shows an overlay map of the AST/RO CI and CO (4 → 3) observations. The maps were obtained by integrating the calibrated data cube over the entire velocity range of the Carina complex, that is, between −50 and −9km s^{-1}. This velocity range is chosen to coincide with that used by Grabelsky et al. (1988) for the CO (1 → 0) data. Each map is a composite of three individual, partially overlapping submaps for each transition. While the southernmost and northernmost submaps have identical size in CI and CO (4 → 3), the central submap in CI is significantly smaller than its CO (4 → 3) counterpart (as it is evidenced by the color and contour boundaries in Figure 4.20).

From this figure, it is immediately evident that the CO (4 → 3) and CI transitions are approximately coextensive throughout the whole Carina region. This is the first instance in which a higher transition of CO and the CI emission are found to be coextensive over such a large area, spanning approximately 150 × 100 parsecs. Previous studies have found a similar result for the lower transitions of CO and ^{13}CO with CI (e. g., Phillips & Huggins 1981; Keene et al. 1985, 1997; Plume 1995, and the references therein). This result is noteworthy, as CO (4 → 3) requires warm, high-density conditions to be excited. Cuts in Galactic longitude observed by AST/RO's southern survey project find CO (4 → 3) emission in most cases only in the high-density and high-temperature cores (presumably actively forming stars) embedded in the more extended CI emission. The η Carina region is special in this regard probably because it is situated right at a spiral density wave shock crest.

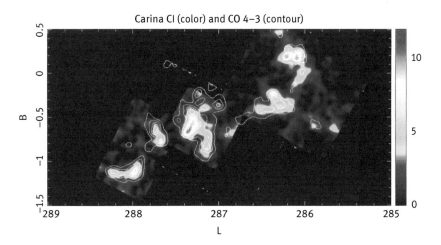

Figure 4.20: AST/RO CO (4 → 3) and CI maps. The contour levels for CO (4 → 3) are 10–90% of 80 K kms^{-1}. The half tone for CI is from 0 to 12 K kms^{-1}. The extent of the mapping area for each transition is indicated by the boundaries of the contour or halftone images (adapted from Zhang et al. 2001).

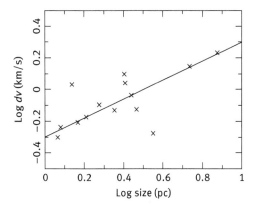

Figure 4.21: Size–line-width relation for the clumps in the Carina molecular cloud complex, derived from the CO (4 → 3) data cube (adapted from Zhang et al. 2001).

In Figure 4.21 we show the cloud-size and cloud–velocity–line–width correlation plot for the molecular clumps derived from the CO (4 → 3) data cube, using the clump-finding algorithm developed by Y. Lee (one of the coauthors on the Zhang et al. 2001 paper). The boundary of the clumps is defined to be three times the rms noise level of the data cube pixels. The fitted size–line-width relation has a slope of 0.6, similar to that found in other studies of the galactic molecular clouds (cf. Myers 1983 and the references therein). This slope (0.6) is steeper than that (0.33) expected for homogeneous isotropic turbulence in an incompressible fluid (Kolmogorov 1941a, 1941b, 1941c; Frisch 1995), likely due to the compressibility of the interstellar clouds which would raise the slope of the turbulence scaling relations (Fleck 1996; Kowal & Lazarian 2007 and the references therein).

The major trend of the correlation in Figure 4.21 is a single linear relation across the entire complex, regardless of whether a particular clump lies near or farther away from the η Carinae outflow (near the most prominent peak in Figure 4.20) – showing that the fundamental cause of the size–line-width relation is not the disruptive influence of the outflow. The role of this outflow appears to be mainly in perturbing the velocities of several clumps. The outflow also perturbs the size–line-width relation from a perfect linear correlation – that is, it adds noise into the relation. In fact, the two extreme outliers on the correlation plot are clumps which are most affected by the outflow, as the more detailed analysis in Zhang et al. (2001) has shown. In view of the basic uniformity of the correlation law across the whole complex, it was concluded that the outflow is not the cause of the size–line-width correlation, but is rather a cause for departure from a perfectly linear relation.

We are thus still in need of a mechanism which is capable of injecting energy into the interstellar clouds on spatial scales of hundreds of parsecs. For the particular region of the Carina molecular complex at least, many of the proposed large-scale processes, such as supernovae and superbubbles (Kornreich & Scalo 2000), do not appear to be applicable. Another proposed mechanism operating on the galactic level is the coupling of galactic rotational energy (von Weizsacker 1951a; Fleck 1982). However, detailed numerical simulations have already shown that it is in fact rather difficult to couple this energy into the internal motion energy of the cloud (Das & Jog 1995).

An alternative candidate mechanism is the spiral density wave. As we have shown earlier, spiral density waves constantly inject energy into the ISM during spiral arm crossings, at size scales from 1 kpc down to a few parsecs. Since this energy is injected through the mediation of the gravitational potential, it happens simultaneously on large and small spatial scales. Using average Galactic spiral parameters, the orbit-averaged rate of energy injection per unit mass due to the interaction of a spiral density wave with the disk matter is calculated to be (see the earlier discussion in Section 4.5.3)

$$\left(\frac{d\Delta E}{dt}\right)_{injection} = 3 \times 10^{-7} (\text{km s}^{-1})^2 \text{ yr}^{-1}. \tag{4.25}$$

Using an average line-width of $\Delta v = 2 \text{ km s}^{-1}$ at size scale of 10 pc from Figure 4.21, the rate of the energy cascade can be found from the following equation

$$\left(\frac{d\Delta E}{dt}\right)_{cascade} = \frac{\Delta v^3}{L} = \frac{(2 \text{ km s}^{-1})^3}{10 \text{ pc}} = 8.1 \times 10^{-7} (\text{km s}^{-1})^2 \text{ yr}^{-1}. \tag{4.26}$$

These two numbers are quite comparable, especially considering, as we have done previously, that the energy injection rate during the period of spiral arm crossing is expected to be several times larger than its value averaged over the entire orbital period. Energy injection due to the spiral density wave is therefore a plausible source for maintaining the degree of turbulent motion and producing the basic trend of size–line-width correlation observed in the Carina arm region.

In Chapter 5, we will present further evidence that both the stellar and the gaseous motions are on the same cascade hierarchy due to the collisionless-shock energy injection. How successfully this hierarchy is maintained in fact can serve as a practical criterion for judging the modal quasi-steady-state itself.

4.6 Other Characteristics of the Milky Way Galaxy and External Galaxies

Besides the ability to quantitatively account for the age–velocity dispersion relation of the solar neighborhood stars, the size–line-width relation of the Galactic molecular cloud complexes, as well as the growth of the Galactic Bulge within a Hubble time, all using the same set of Galactic disk and Galactic spiral parameters that are well supported by the most recent observations[†], the secular evolution process also provides a natural explanation for a host of other known characteristics pertaining to

---

[†] Note that the uncertainty on the number of spiral arms of the Galactic spiral structure is matched by the uncertainty on their pitch angles and the resulting PDPS across the entire Galactic disk and

the Milky Way Galaxy, as well as other external galaxies where such characteristics can be measured.

4.6.1 Mass Distribution of the Different Galactic Components

Like many external galaxies, the Milky Way is known to have a nearly flat rotation curve. From the analysis of the dynamical balance between mass distribution and rotation characteristics (see the discussions in Section 4.7), we know that this implies a distribution for the total enclosed mass within galactocentric distance R, of $M(R)$ proportional R, or a volume density proportional to $1/R^2$, for a total mass distribution of nearly spherical shape. This distribution is the same as the outer envelope of an isothermal sphere.

For the outer part of the Galaxy where the dark halo mass dominates, the above mass distribution was considered reasonable based on galaxy formation theories. However, for decreasing galactocentric distance, especially for $R < 10$ kpc where the disk mass begin to dominate that of the halo mass, the flat rotation curve remains. This is the well-known phenomenon of "disk-halo conspiracy" where the different mass components (stellar halo, dark halo, and stellar disk) conspire to maintain the flat rotation curve, and its origin was not apparent from the current theories of galaxy formation (Gilmore et al. 1990). Due to the prevalence of the flat or nearly flat rotation curves among observed external galaxies, such disk-halo conspiracy is a universal feature in need of explanation.

In the secular evolution scenario, especially after taking into account the possibility that the composition of at least the galactic part of the dark matter could be primarily baryonic (see Section 5.2.2.2), this phenomenon of the conspiracy of the different mass components to form flat rotation curves in not only our Galaxy, but in a great many external galaxies as well, becomes easy to understand. The disk forms through the dissipative vertical infall of material from the halo region, starting from the original isothermal primordial clump, and the nearly flat rotation curve can be maintained throughout the disk formation and secular evolution process because of the detailed manner this evolution process proceeds (see Section 4.7).

The so-called bulge-disk connection in disk galaxies (Courteau et al. 1996), wherein many of the characteristics of the disk and bulge in galaxies are found to correlate, is also most naturally understood under the secular evolution scenario, in which the bulge grows gradually from the radially accreted disk mass, including both stars and gas. Shen et al. (2010) discussed evidence for the pseudo-bulge nature of the Galactic Bulge, making our Galaxy another likely candidate galaxy built primarily through the secular evolution process, rather than through mergers.

---

throughout the age of the Milky Way Galaxy. So we emphasize here that it is the same *set* of parameter choices, in the same combination, that enters into most of the calculations of wave/basic state interaction that produced the aforementioned excellent agreement.

4.6.2 Stellar Population and Kinematics in the Thin and Thick Disks

The above scenario about the disk formation and evolution also naturally accounts for the close relation between the thin and thick disks (Gilmore et al. 1990 and the references therein), including the similarity in scale lengths between these two components, and the close relation between the stellar populations in these two components. It also provides a natural mechanism for effectively heating the disk stars to populate the thin and thick disks in a continuous manner.

Gilmore et al. (1990), especially their Chapter 11, showed evidence that in the Milky Way the stellar population is consistent with the thick disk's population being a continuation of the thin disk's older stellar population, rather than being a continuation of the halo field population. This shows that the formation and evolution of the thick disk is likely to have been affected by the spiral density waves in the same manner that the thin disk had been affected.

As we have shown previously, the past proposed mechanisms for stellar velocity diffusion in phase space include the perturbation by transient spiral waves in the disk plane, as well as the redirection of some of the transverse random velocity into the vertical direction through the scattering with GMC complexes (see Sellwood 2013 and the references therein). A known problem with this heterogeneous stellar acceleration mechanism is that it cannot account for the smooth trend of the random velocity increase (while keeping the velocity ellipsoid of approximately the same shape) across the galactic radii, as well as across the different Hubble types of galaxies (Gilmore et al. 1990 and the references therein) since the inner galactic regions and the early type galaxies have much fewer GMC complexes, yet the smooth trend of acceleration of stellar random velocity continues from the outer region into the central region of a galaxy, and from the late-type into early-type galaxies.

Given that the stellar population and the kinematics of the thick disk stars appear to connect smoothly with the thin ones, the acceleration mechanism has to involve a kind of local instability that is inherently three-dimensional, rather than a 2D mechanism such as the transient waves aided by GMC complex scattering – the scattering mechanism was already stretched to explain the most extreme of the random velocities for thin disk old star population of 20 km/s, and it is simply not possible to stretch it even further to explain the acceleration of the thick disk stars as well, which has vertical velocity dispersions on the order of 40 – 50 km/s (Gilmore et al. 1990, p. 369). This difference between the 2D and the 3D heating is one of the central differences in the mechanisms of heating either by transient waves on overstable disks, or by collective instabilities of self-organized modes on unstable disks: The transient waves sweep across the originally stable disk, they affect the velocity dispersions of the disk stars in a superficial manner, whereas a true self-organized global instability exercises the inherent randomizing capability of the participating particles in a thorough, inside-out manner. The latter is thus able to cause the coherent interaction of particles that participate in such self-organized instability in regions well above the thin disk (i. e., the skewed density wave perturbation can include the thick disk stars, and possibly even the skewed halo components).

De Simone et al. (2004) used numerical simulations to study the heating of solar neighborhood stars using recurrent transient waves. They conducted a local analysis and used test particles initial on circular orbits in a sheared sheet, and subject them to the enforced transient spiral perturbations. They attempted to use this mechanism to explain the presence of small-scale structure in the velocity distribution, or the so-called "moving groups" found by the Hipparcos satellite, as well as the fact that stars in a single velocity–space moving group can have wide range of ages.

De Simone et al. (2004) found that the duration of the transient waves that were effective in heating the stars was very short and with regular repetition, a conclusion corroborating that found also in Carlberg & Sellwood (1985). Such short and repeated transients were not observed in physical galaxies. As a matter of fact, if we consider the shock effect produced by the dynamical mechanism we presented in this work, each time as a spiral-arm collisionless shock sweeps across a disk area, the effect this has on stars is actually very similar to what de Simone et al. investigated using short duration, repeated transients (i. e., the spiral collisionless shock repeats on a dynamical time scale, and the duration is a fraction of the orbital period). What is more, the random transients heat disk stars but also disperse them, whereas spiral collisionless shocks not only heat the stars but also organize them into systematic patterns of mass inflow inside corotation and mass outflow outside corotation, as we have shown before.

The smoothness of the age–velocity dispersion curve of the solar neighborhood stars is easily reproduced through spiral collisionless shocks, but very difficult for the short and repeated transient scenario because this requires a very regular transient generation mechanism, in contradiction to the random origins of these transient waves. The spiral collisionless shock, since it is a local gravitational instability, also naturally produces the small scale velocity structure and mixes stars of different ages.

4.7 Universal Rotation Curve

An important descriptor of the structural characteristics of disk galaxies is the rotation curve, which reflects the galaxy mass distribution on the large scale. After assembling hundreds of observed rotation curves for nearby disk galaxies, Persic, Salucci, and Stel (1996) found that these rotation curves fall onto a 2D surface in the three-parameter space of normalized galactic radius, velocity, and the absolute magnitude. Since for nearby galaxies the variation of absolute magnitude corresponds to the variation of Hubble type (i. e., bigger and brighter galaxies generally have earlier Hubble types), this means that the rotation curve shapes also vary systematically for galaxies of varying Hubble types. In Figure 4.22, we plot three typical rotation curves generated using the data from Persic et al. (1996).

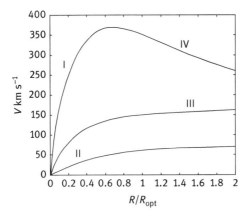

Figure 4.22: Typical rotation curves for disk galaxies of varying absolute magnitudes, as well as four characteristic regimes which are analyzed in the main text. Top: $M = 18$; middle: $M = 21$; bottom: $M = 24$, where M is the absolute magnitude of the representative galaxy, and R and R_{opt} are the galactic radius and optical radius, respectively (data from Persic et al. 1996) (adapted from Zhang 2004).

Following Zhang (2004), we employ dynamical relations for equilibrium axisymmetric/spherically symmetric mass distributions, that is,[†]:

$$V(R)^2 = \frac{GM_{\mathrm{dyn}}(R)}{R}, \qquad (4.27)$$

as well as

$$M_{\mathrm{dyn}}(R) = \int_0^R \Sigma_{\mathrm{proj}}(r) 2\pi r \, dr = \int_0^R \rho(r) 4\pi r^2 \, dr, \qquad (4.28)$$

where $M_{\mathrm{dyn}}(R)$ is the dynamical or total mass within a galactic radius R (including the contributions from both the luminous and the dark components), and $\Sigma_{\mathrm{proj}}(R)$ is the total projected surface density (disk plus spheroidal) at radius R, we can infer the underlying mass and surface density distributions corresponding to the different regimes for each type of rotation curves (c.f. Figure 4.22).

(I) Steeply rising solid body rotation curve, as observed for the inner regions of early-type galaxies.

This regime is characterized by solid-body rotation ($\Omega = \Omega_0 = $ constant, or a nearly linear rise of rotation velocity $V(R) = \Omega_0 R$). Using

$$V(R)^2 = \Omega_0^2 R^2 = GM_{\mathrm{dyn}}/R, \qquad (4.29)$$

we obtain

$$M_{\mathrm{dyn}} = \frac{\Omega_0 R^3}{G}, \qquad (4.30)$$

[†] The following results will be exact for a perfectly spherical mass distribution and will be off by a factor of 2 for axisymmetric razor-thin disky mass distribution. If the mass distribution is in between of these two extremes, then the scaling factor is between 1 and 2. So the "=" sign in the following derivations are more appropriately interpreted as "~".

which implies a constant volume density

$$\rho = \rho_0 = \frac{\Omega_0}{G}\frac{3}{4\pi}. \qquad (4.31)$$

Therefore the projected surface density can be obtained by equating

$$M_{\mathrm{dyn}}(R) = \int_0^R \Sigma_{\mathrm{proj}}(r) 2\pi r\, dr \qquad (4.32)$$

and

$$M_{\mathrm{dyn}}(R) = \frac{3}{4}\pi R^3 \rho_0, \qquad (4.33)$$

which gives

$$\Sigma_{\mathrm{proj}}(R) = 2\rho_0 R, \qquad (4.34)$$

that is, the solid-body regime corresponds to a constant column density ρ_0 and linearly rising projected surface density $\Sigma_{\mathrm{proj}}(R) \propto R$.

(II) Gently rising rotation curve, as representative of the very late-type disks, including most of the dwarf galaxies and low-surface-brightness (LSB) disk galaxies.

This regime can be shown to correspond to approximately constant projected surface density, as found to be the case for many LSB disks. Assume $\Sigma_{\mathrm{proj}}(R) = \Sigma_0 = $ constant, we obtain

$$M_{\mathrm{dyn}}(R) = \int_0^R \Sigma_0 2\pi r\, dr = \Sigma_0 \pi R^2, \qquad (4.35)$$

which we equate to $V(R)^2 R/G$ to obtain

$$V(R)^2 = \sqrt{\Sigma_0 \pi R G}, \qquad (4.36)$$

which indeed corresponds to a gently rising rotation curve.

Instead of constant projected surface density, many LSBs are found to have constant volume density (de Blok 2003). These two views (constant projected surface density or constant mass volume density) can be consistent if the scale height in the central regions of these LSBs are nearly constant, as for a pancake-type proto galaxy cloud collapse.

(III) Flat rotation curve, as is representative of the outer regions of most Sb/Sc galaxies.

In this regime, the projected surface density profile should be close to $1/R$, as shown below. Assume

$$\Sigma_{\mathrm{proj}}(R) = \frac{\Sigma_0 R_0}{R}, \qquad (4.37)$$

where $\Sigma_0 = \Sigma_{\text{proj}}(R = R_0)$, we have

$$M_{\text{dyn}}(R) = \int_0^R \Sigma_{\text{proj}}(r) 2\pi r \, dr \qquad (4.38)$$

$$= \Sigma_0 R_0 2\pi R.$$

Using again $V(R)^2 = GM_{\text{dyn}}/R$, we obtain

$$V(R)^2 = G\Sigma_0 R_0 2\pi R/R, \qquad (4.39)$$

or

$$V(R) = \sqrt{G\Sigma_0 R_0 2\pi} = \text{constant}. \qquad (4.40)$$

Therefore the flat rotation curve regime corresponds to $1/R$ projected surface density distribution.

(IV) The falling rotation curve regime, as representative of the outer part of the early-type galaxies.

In this regime the projected surface density $\Sigma_{\text{proj}}(R)$ falls off faster than $1/R$, and can be up to $\Sigma_{\text{proj}} = 0$ (Keplerian). For the extreme case of Keplerian rotation curve due to a concentrated central mass distribution, $\Sigma_{\text{proj}}(R_{\text{outer}}) \approx 0$, and

$$V(R)^2 = \frac{GM_{\text{dyn}}}{R}, \qquad (4.41)$$

or

$$V(R) = \sqrt{\frac{GM_{\text{dyn}}}{R}}, \qquad (4.42)$$

which is indeed falling.

In the secular evolution scenario a typical galaxy starts its life with properties similar to that of an LSB galaxy. The high-z correspondence of the local LSBs appears to be closely approximated by the damped L_α system (Wolfe 2001), which also possesses disk-like morphology with large-scale length, and a constituting mass component which is gas rich and metal poor.

As the proto galaxy disk contracts and condenses, the global spiral instability develops, which begins the mass-redistribution process that transforms the flat surface density distribution to a more and more centrally concentrated mass distribution. The intermediate stage of this evolution resembles Freeman's type II disks (Freeman 1970), and the later stage with a $1/R$ projected surface density resembles a Freeman's type I disk.

As the evolution progresses, the inward-accreted disk stars rise out of the disk plane to become bulge stars. The thermalized 3D bulge mass will have nearly constant volume density and undergoes solid-body rotation. The outer part of the visible disk starts to be drained of matter, producing the falling rotation curve.

This entire evolution sequence therefore corresponds to a galaxy evolving from the lower rotation curve to the upper rotation curve in Figure 4.22. Therefore secular evolution provides the underlying dynamical process responsible for the formation of the universal rotation curve (URC). The fact that the URC is a thin plane in the available 3D parameter space shows that the density-wave-assisted galaxy morphological transformation can be regarded on a global sense as an extended phase transition process (i. e., not just the emergence of the wave mode, but also the entire Hubble sequence formation and evolution process, can be regarded as a slow phase transition process), with its trajectory determined by the dynamics of this process.

The twisted isophotes often found for the Hubble Deep Fields galaxies are expected to evolve into the observed variation of major axis with radius for elliptical galaxies (Faber & Jackson 1976), possibly bypassing the disk formation phase, at least for some galaxies. Like the spiral structure, the twisted isophotes are results of dissipation and differential rotation, and just as in the case of spiral structure, these 3D skewed structures will also lead to a phase shift between the potential and density distributions, resulting in accelerated secular morphological changes.

4.8 Formation and Maintenance of Galaxy Scaling Relations

In this section we address possible connections between the secular evolution of galaxies and the formation and maintenance of several well-known galaxy scaling relations (Zhang 2004).

4.8.1 General Considerations

The earliest known versions of the scaling relations link the radii and internal velocities of galaxies to their luminosity: these are the Tully–Fisher relation for spiral galaxies (Tully & Fisher 1977) and the Faber–Jackson relation for elliptical galaxies (Faber & Jackson 1976). These two relations are very similar in form, which itself is a remarkable fact in need of explanation.

In the secular evolution picture, we know that during the radial mass accretion process approximately one-half of the orbital energy of the accreting stars becomes their random motion energy in the final configuration (the other half being transported outward by the spiral density wave), and the final radius of the evolved galaxy is also simply a fraction of its initial radius. Therefore the scaling law for a galaxy in the spiral phase should be mostly preserved as it evolves from late to early, besides a change of the constant multiplication factor. Observationally, it was indeed found that the combined scaling law for elliptical and spiral galaxies can be written in the following form of a generalized Tully–Fisher/Faber–Jackson relation (Williams 1997),

$$L_{\text{galaxy}} = K \cdot V_{\text{int}}^{x} \qquad (4.43)$$

where L_{galaxy} is the total intrinsic luminosity of a galaxy, V_{int} is the internal velocity dispersion of the galaxy, the exponent x depends on the band of observation, and the proportionality constant K depends on the Hubble types of galaxies.

The secular evolution scenario provides a natural explanation also for the existence of the fundamental-plane relation of elliptical galaxies (Dressler et al. 1987; Djorgovski & Davis 1987), at least for the disky ones. The fundamental plane of elliptical galaxies describes a set of bivariate correlations connecting a set of global observables, such as radii, luminosities, surface brightness, velocity dispersions, masses, densities, etc. In a parameter space defined by at least three independently measured variables from this set, ellipticals occupy a 2D surface, which is close to a plane if logarithmic variables are used.

Although physical relation such as the virial theorem appears to underlie the fact that elliptical galaxies of various characteristics can find their quasi-equilibrium configurations along the fundamental plane (see following derivations in Section 4.8.2), a question remains as to why, for a proto galactic clump of a *given* mass, considerably different equilibrium configurations are chosen (which the virial theorem is not able to explain). From the standpoint of the secular evolution scenario, the reason for the existence of varying configurations for a given mass is that this indicates *the varying degree of advancement of secular angular momentum evolution*. Furthermore, the fact that for a given mass range, there could be a whole spectrum of the degrees of advancement of secular evolution is likely to be due to the influence of *environment* on the speed of secular evolution. In other words, in the single independent parameter case, such as the Tully–Fisher or Faber–Jackson relation with a constant proportionality K, the initial condition alone would determine how fast a galaxy evolves. Whereas in reality, the factor of the environment also comes into play in determining the rate (and the current degree) of secular evolution.

Viewed in this light, we see that the variation of the "constant" K factor with Hubble types in the generalized Tully–Fisher/Faber–Jackson relation (4.43) is consistent with the need for a second independent parameter such as surface brightness in the fundamental plane relation; and both the fundamental plane relation and the generalized Tully–Fisher/Faber–Jackson relation are a reflection of galaxy evolution under the influence of both nature and nurture.

4.8.2 Origin and Evolution of the Scaling Relations

Galaxy scaling relations own their existence first and foremost to the fact that galaxies are equilibrium configurations and satisfy the virial theorem relation (e. g., see Binney & Tremaine 2008). This relation dictates that during any quasi-equilibrium evolution process of a gravitational system, $-E \sim T \sim -V/2$ where T is the kinetic energy, V is the potential energy, and $E = T + V$ is the total energy of the system. The fact that spiral galaxies can remain on the Tully–Fisher relation during secular morphological evolution despite the fact that stars cannot radiate about 1/2 of the converted potential energy away is due to the fact that spiral density wave actually carries about 1/2 of the dissipated orbital energy (equal to potential energy) away to the outer disk to be deposited there. The existence of a collective dissipation process at the spiral arms also makes possible the regulated conversion of orbital velocity to random velocity, so a single kinetic energy (the half that is not being carried away) can enter the virial relation no matter how it is apportioned between the circular and the random motions.

According to the virial theorem (we set $G = 1$ in the following derivations of scaling relations)

$$M_{\text{dyn}} = V_e^2 R_e, \tag{4.44}$$

where M_{dyn} is the dynamical mass of a galaxy, V_e is the effective velocity spread which corresponds to the total velocity dispersion for an elliptical galaxy or the maximum rotation velocity for a spiral galaxy, and R_e is an effective radius which makes the virial relation hold.

Define the mean surface brightness of a galaxy SB as

$$SB \equiv \frac{L}{R_e^2}, \tag{4.45}$$

where L is the luminosity of the galaxy, and the dynamical mass-to-light ratio M_{dyn}/L can be written as

$$M_{\text{dyn}} \equiv (M_{\text{dyn}}/L) * L. \tag{4.46}$$

Therefore it follows that

$$(M_{\text{dyn}}/L)^2 * L^2 = M_{\text{dyn}}^2 = V_e^4 R^2 = V_e^4 \frac{L}{SB}, \tag{4.47}$$

or

$$(M_{\text{dyn}}/L)^2 L = \frac{V_e^4}{SB}. \tag{4.48}$$

Therefore we finally have

$$L = V_e^4 \frac{1}{SB} \frac{1}{(M_{\text{dyn}}/L)^2}. \tag{4.49}$$

We thus see that in order to obtain the classical Tully–Fisher relation, which has $L \propto V_e^4$, we must have $SB \cdot (M_{\text{dyn}}/L) \sim$ constant. This can be accomplished in two ways: to have SB and M_{dyn}/L each being constant; or, to have the variations of the two factors offset each other during an evolution process.

Traditionally, the former was assumed to be the case (e. g., see the discussions and references in Shu 1982). It is now known that both surface brightness and dynamical mass-to-light ratio vary considerably for galaxies along the Hubble sequence. The surface brightness is found to be higher for early-type galaxies (McGaugh & de Blok 1998 and the references therein), whereas the dynamical mass-to-light is found to be lower for the earlier Hubble types (Zwaan et al. 1995; Bell & de Jong 2001). The observed opposing trends of variation of surface brightness and dynamical mass-to-light ratio is naturally explained as the outcome of the secular baryonic mass accretion and the increased fraction of luminous baryon mass (see Section 5.2.2.2), especially toward the central region of a galaxy as its Hubble type evolves from late to early. The Tully–Fisher relation can be maintained as long as the increase in surface brightness is accompanied by a corresponding decrease in the dynamical mass-to-light ratio during the secular evolution process.

In the secular evolution scenario, since most of the elliptical galaxies (i. e., those with lower luminosity, or the so-called disky ellipticals) are the end results of evolution from the initial condition of disks, we speculate that spiral galaxies may satisfy similar kind of fundamental plane relation just as ellipticals. This has been found to be so (e. g., see Pharasyn, Simien, & Heraudeau 1997).

The fundamental plane relation for spirals can likewise be derived from the virial theorem. Starting from

$$V_e^2 = G \frac{M_{\mathrm{dyn}}}{R_e}, \tag{4.50}$$

we have

$$V_e^2 = G \frac{M_{\mathrm{dyn}}}{L} \frac{L}{R_e} = L^{1/2} \left(\frac{L}{R_e^2}\right)^{1/2} \frac{M_{\mathrm{dyn}}}{L}, \tag{4.51}$$

which leads to

$$10 \log V_e = -(1 + 2\beta) M_t - \mu_e + \text{constant}, \tag{4.52}$$

where M_t is the absolute magnitude and μ_e is the face-on average surface brightness in mag/arcsec^2, and where we have assumed $M_{\mathrm{dyn}}/L \propto L^\beta$. Pharasyn et al. (1997) found that fitting I band and K band data of a sample of spiral galaxies to this relation resulted in $\beta \approx -0.15$, that is, the mass-to-light ratio slightly decreases with increasing luminosity, which is consistent with what we would expect from a secular evolution picture, that is, as the secular evolution advances, L increases and M_{dyn}/L decreases.

The fundamental plane relation for elliptical galaxies and bulges were first obtained by Faber et al. (1987), Djorgovski and Davis (1987), and Dressler et al. (1987). One type of the fundamental plane relation for ellipticals can be written as (c.f. eq. 2a and 2b of Djorgovski & Davis 1987)

$$L \sim V_e^{3.45} SB^{-0.86}. \tag{4.53}$$

This relation can be rewritten into the form

$$M_t(R_e) = -8.62(\log V_e + 0.1\mu_e) + \text{constant}, \tag{4.54}$$

which can be further written as

$$10 \log V_e = -1.25 M(R_e) - \mu_e + \text{constant}. \tag{4.55}$$

Comparing eqs (4.52) and (4.55), we see that apart from a possible difference in the "constant" term (accounting for the change in total luminosity), the only difference between the spiral and elliptical fundamental plane relations is in the different sign of the mass-to-light ratio exponent β: $\beta = -0.15$ for spirals and $\beta = 0.13$ for ellipticals and bulges, where β is defined through $M_{\text{dyn}}/L = L^\beta$. This difference is apparently brought about by the fact that spiral galaxies still have varying reserves of baryonic dark matter to form stars (see Section 5.2.2.2), therefore as the secular evolution proceeds (and therefore L increases) the mass-to-light ratio decreases; whereas elliptical galaxies have essentially exhausted their central baryonic dark matter supply, thus the ellipticals in more advanced stage of evolution (which generally have larger L) will experience greater degree of dimming, which is reflected in the increase of mass-to-light ratio with L.

The observed elliptical galaxies come roughly in two types. While the boxy giant ellipticals have characteristics which indicate that they are likely to be the product of mergers, that is, being pressure supported, having multiple nuclei, having cuspy cores in their central density profile, having two populations of globular clusters of different colors, and are radio and X-ray loud, etc., the more numerous disky ellipticals, on the other hand, are mostly rotationally supported, having power-law central density profiles, show little evidence of merger, having only one globular cluster population, and are radio and X-day quiet (Bender et al. 1989; Faber et al. 1997). Furthermore, the stellar populations in the lower-mass disky ellipticals are found to be systematically younger, and have more extended history of star formation, than that in the high-luminosity boxy ellipticals (Bender & Saglia 1999). These results suggest that boxy ellipticals are the true merger products, whereas disky ellipticals are produced mostly by secular evolution. Bender et al. (1989) also noted that the average mass-to-light ratios are higher in boxy ellipticals than in disky ellipticals. The secular evolution scenario naturally accounts for the smaller average values of M/L in disky ellipticals as compared to boxy ellipticals: secular mass inflow would lead to an increase in the central phase-space density of galaxies, as well as an increase in both the ignition of baryonic mass and the displacement of smaller-mass dark baryons (see Section 5.2.2.2); whereas the merger process at most keeps the phase space density the same.

4.9 Butcher–Oemler Effect and Evolution of Cluster Galaxies

The previous sections of this chapter focused on the general physical processes pertaining to galaxies in *any* environment. In the next three sections, we will turn our attention to the environmental influence on the speed of secular evolution of galaxies. Most of the content in Section 4.9 and 4.10 was first published in Zhang (2008).

The morphological Butcher–Oemler (BO) effect, or the rapid transformation of a large population of late-type galaxies to earlier Hubble types in rich cluster environment between intermediate redshifts and the local universe (Butcher & Oemler 1978a, 1978b; Dressler et al. 1997), was among the earlier evidence that galaxy environment plays an important role in the secular morphological transformation of galaxies. The determination of the dynamical causes of the morphological BO effect was an important unsolved problem central to our understanding of the general problems of galaxy formation and evolution. In this section, we first survey the existing proposed mechanisms for cluster galaxy transformation, and then discuss their relevance and limitations to the explanation of the morphological BO effect. A new infrared diagnostic approach is devised to disentangle the relative importance of several major physical mechanisms to account for the BO effect, and an example of the first application of this procedure to a single rich, intermediate-redshift galaxy cluster is given to demonstrate the viability of this approach. The preliminary result of this analysis favors the interaction-enhanced secular evolution process as the major cause of the cluster-galaxy morphological transformation. This conclusion is also supported by a wide range of other published results, which are assembled here to highlight their implications on a coherent physical origin for the morphological BO effect.

4.9.1 Observations and Candidate Mechanisms of the Morphological BO Effect

At the heart of the BO effect is the spiral to early-type transformation (Dressler et al. 1997). This transformation not only involves the extinguishing of the star-formation activity, but also the gradual increase of bulge-to-disk ratio (Franx 2004; Treu 2004 and the references therein). Furthermore, whichever physical mechanism(s) produced the BO effect, in addition to producing the observed number density evolution of the different types of cluster galaxies (i. e., the transformation of spirals into S0s and ellipticals), must simultaneously be able to preserve the relatively tight color–magnitude relation observed throughout the redshift range $z = 0 - 1$ (van Dokkum & Franx 2001; see also the discussion in Section 4.10). The relevant process for galaxy transformation must also preserve the outer disks and spiral structures observed for many of these galaxies during the quenching of star formation and the morphological transformation (Couch et al. 1998; Poggianti et al. 1999; Goto et al. 2003b).

Several candidate mechanisms have been proposed to account for the morphological BO effect. These include major and minor mergers (Icke 1985; Kauffmann 1995), galaxy harassment through tidal shocks (Moore et al. 1996, 1998, 1999), galaxy infall onto the forming cluster and the resulting stripping of either the disk or halo gas (with the latter having the name of "starvation") due to the ram pressure of the intracluster medium (Gunn & Gott 1972; Larson, Tinsley, & Caldwell 1980), as well as the secular dynamical process discussed in the current work which is mediated by the large-amplitude density wave patterns excited during the tidal interaction of galaxies in clusters.

Most of the previously proposed mechanisms run into contradictions with one or another of the observed characteristics of cluster galaxies upon careful analyses. Due to the high-speed nature of galaxy–galaxy encounters within clusters which prevents the encountering galaxies from "sticking" to one another, major mergers are not believed to have a significant probability in clusters (Dressler et al. 1997), though an exception to this conclusion has been observed for one high-z cluster (van Dokkum et al. 1998). Moreover, there does not seem to be a large reservoir of dwarf spheroidal (dE) satellites in these clusters (Trentham 1997) to cause the simultaneous morphological transformation of the large number of BO galaxies by minor mergers. The tightness of the color–magnitude relation of cluster early-type galaxies (Franx 2004 and the references therein) and the well-preserved disks throughout the BO evolution process (Moran et al. 2007b and the references therein) also argue against a significant role played by mergers.

Ram pressure stripping of the ISM for infalling cluster galaxies (Gunn & Gott 1972) by itself appears not to be sufficient to account for the observed level of cluster galaxy transformation. Stripping of gas generally does not lead to a significant change in the bulge-to-disk ratio, though Larson et al. (1980) had envisioned the reduction of disk size due to gas stripping as one way of increasing the B/D. However, observations and analysis by Moran et al. (2007a) showed that cluster spiral galaxies have systematically higher central density than their field counterparts, an effect which is difficult to produce if the outer-disk stripping is the major cause of the cluster BO effect.

There is also the issue of time scale with any stripping/starvation mechanism which relies on the infall of galaxies from the field region into the centers of clusters: If an infalling galaxy starts its journey from the field at 5 Mpc distance to the cluster center and has an average velocity of 500 km/s[†], it takes the galaxy 4×10^9 years to reach the central region. This is a significant fraction of the Hubble time and

---

[†] Even though the cluster velocity dispersion is generally on the order of 1000 km/s, the velocities of galaxies at the outskirts of the clusters, especially their radial components, are generally quite low since these outer galaxies have not been virialized and thus have not gained the kind of high-velocity dispersion that the cluster-center galaxies have.

comparable to the cluster formation/evolution time scale. So the late-type galaxies observed in the central regions of clusters as found by Dressler et al. (1997) are likely to be primordial instead of new in-fallers from the field region. The existence of the morphology–density relation over several orders-of-magnitude variation of local surface density (Dressler 1980; Dressler et al. 1997; Goto et al. 2003c), from the densest region of cluster center all the way to the group environment in the field, also supports the segregation of the varying-density regions of a cluster during cluster-galaxy morphological transformation, and argues against significant mixing of field populations into the central regions of clusters.

The galaxy harassment mechanism (Moore et al. 1996, 1998, 1999) has gained some popularity in recent years. When first proposed, harassment was meant as a process whereby the gravitational tidal shock from a large cluster galaxy tears away the outer disk of a small late-type neighboring galaxy, thus transforming the small galaxy into an early-type dE galaxy. Numerical simulations by Moore et al. (1998) showed that the harassment process generates remnants which are predominantly prolate in shape, as opposed to the oblate shape that an S0 galaxy (which is the typical end product of BO transformation) usually has. The follow-on exploration of the harassment mechanism by Gnedin (1999, 2003a, 2003b) showed that this mechanism is rather inadequate in transforming the morphology of large disk galaxies: even though a moderate puffing-up of the entire disk was observed, there is not a significant increase in the bulge-to-disk ratio which is a required element for the spiral-to-S0 transformation. The simulations conducted by Moore et al. (1999) showed that the low-surface-brightness galaxies in their tests were transformed to dE galaxies due to tidal shocks, whereas for high-surface-brightness (HSB) galaxies "the scale height of the discs increases substantially and no spiral features remain." This is in marked contrast to the post-interacting galaxies observed by, for example, Poggianti et al. (1999) and Moran et al. (2007b), which showed prominent spiral structures and well-preserved outer disks after whatever interaction processes which had resulted in the star-burst episodes in these galaxies.

Another proposed mechanism for producing the BO effect is the secular dynamical evolution process in galaxies discussed in this work, which is further enhanced by the tidal interactions in the cluster environment. In this scenario, tidal forces between neighboring galaxies and between a galaxy and the cluster potential serve to excite large-amplitude, open spiral or bar patterns. Most of the morphological transformation is realized through the post-interaction mass redistribution process due to the presence of tidally enhanced density wave patterns. In this process the radial mass accretion and the vertical heating of stars occur simultaneously over the time span of a few Gyrs due to the irreversible energy and angular momentum exchange between the density wave and the disk matter and the outward transport of these exchanged energy and angular momentum by the density wave to the outer disk, which lead to the gradual build-up of the bulge, in addition to the consumption of gas through density-wave-induced star formation. This mechanism is able to preserve not only the tight scaling relation of early-type galaxies during the secular mass accretion process (Section 4.8), but also the color–magnitude relation of early-type galaxies (Section 4.10), including those galaxies in clusters.

4.9.2 An Infrared Diagnostic Approach for the Star-Formation States of Cluster Galaxies

In order to assess the relevance of the different proposed mechanisms to the explanation of the BO effect, an infrared (IR) diagnostic approach was developed (Zhang 2008) which has the potential to shed new light on the BO problem using the data offered by the new generation of the space IR telescopes such as Spitzer and Herschel.

The method makes use of the fact that most of the proposed mechanisms for BO transformation will involve varying levels of dusty starburst, and furthermore these different mechanisms have different predicted characteristic IR luminosity levels, as well as different time scales for sustaining these different IR luminosity levels. Given the known BO transformation rate (which can be determined from a statistical study of the galaxy morphologies in rich clusters at the intermediate redshifts, compared to a similar study for the local universe), we can infer the roles of different proposed mechanisms by observing the different frequencies of occurrence of the varying IR-luminous galaxies in the different regions of intermediate clusters. This quantitative approach has major advantages in constraining the importance of the candidate dynamical mechanisms in transforming cluster galaxy morphologies compared to the simple-minded counting of the IR sources of a given luminosity range in an observing epoch because the time scales of the operation of these dynamical mechanisms are vastly different, and a simple counting approach can result in serious misconceptions of the relative contributions of the different processes, as our example in the next section will show.

For a cluster region of total galaxy number N, if there are n galaxies observed to be in a luminosity range characteristic of a type of starburst event, we denote the observed frequency for that type of event f_{event} as

$$f_{event} = \frac{n}{N}. \tag{4.56}$$

Assume the duration of the starburst event is τ_{event} (usually believed to be on the order of 10^8 years for strong-interaction-type events such as merger and harassment, and much longer for gas stripping-type events and for post-interaction secular evolution), after a cluster evolutionary time scale of T (on the order of a few Gyr, the exact number depending on the redshift z of a cluster), we expect a fraction

$$F = f_{event} \cdot \frac{T}{\tau_{event}} \tag{4.57}$$

of the cluster galaxies to have gone through morphological change due to the particular type of star-forming event alone. The total fraction F for BO transformation between an intermediate redshift and the local universe, due to the yet-unsettled physical mechanism(s), is now fairly well determined for rich BO clusters (e. g., see van Dokkum & Franx 2001; Franx 2004), and its value is around 20%, for example, between $z = 0.4$ and $z = 0$. Since this 20% conversion fraction is for cluster as a whole, and in the later calculations we divide the cluster into three broad regions (post-virialization, active virialization, and outskirts), we estimate the morphological conversion fraction between $z = 0.4$ and $z = 0$ for the active virialization region to be 30% and for the cluster outskirts about $F = 10\%$. No estimates for the post-virialization region are given since the IR diagnostics are not available there, but it is expected that the morphological transformation process is going on at some level even in that region where the star formation has ceased.

We stress that not all of the mechanisms that can transform galaxies will include a dusty-starburst phase (e. g., one proposed pathway for morphological transformation not involving a dusty starburst is gas-free dry mergers). Even for those mechanisms which can potentially invoke a dusty starburst, this phase will only be prominent when there is sufficient gas reservoir in the galaxy at the observing epoch. This method thus predicts the *maximum possible abundance* of dusty starbursts that can be attributed to each mechanism, to be compared with that of the observed abundance. So for the proposed approach to work effectively, we make the implicit assumption that the BO precursor galaxies are gas-rich galaxies, which seemed to be a valid one since that was how Butcher and Oemler first discovered them (i. e., by observing their blue colors due to enhanced star formation in a gas-rich environment). For the later stage of the evolution, after the gas is consumed, we can still have the interaction and the associated morphological transformation events but no star-formation signature. However, the arguments for the validity of the fractions estimated by this approach require only that the relevant galaxies are gas rich at the observing epoch z, and that the interaction strengths of the relevant mechanisms stay roughly constant throughout z. They do not require the gas fraction to remain the same throughout z. This is because we are relating the star-forming events at z with the overall morphological transformation rate F throughout z, and the latter does not require a constant gas supply. So in this sense the star-formation events at the observation epoch z are only a diagnostic signature of the underlying morphological transformation process.

We therefore are ruling out the possibility that there are other processes which potentially could be transforming galaxies in a way that does not generate an IR bright phase, rather we are ruling out these mechanisms *as the explanation for the observed incidence of dusty starbursts* in those clusters where abundant starbursts are observed. Furthermore, using the same principle this kind of "frequency test" can in fact be used also for sources observed in other wave bands (such as optical) as long as there is a clearly identifiable spectral/photometric feature which has a known time scale of operation.

If the time scale for the individual interaction event τ_{event} is known, we can invert the above expression to calculate the appropriate value of f_{event} (i. e., the observed fraction of a particular type of IR-luminous galaxy at z) for each type of interaction event in the appropriate luminosity range and for the relevant cluster region. These calculated results are tabulated in Table 4.2 for a hypothetical BO cluster redshift of $z \sim 0.4$. In this table we list calculated values of f_{event} for each of the proposed physical mechanisms we consider, split into several radial cluster regions, and for transforming galaxies that fall into several different classes of IR luminosity. We will describe in more detail below on how the frequencies in Table 4.2 are obtained.

Note that if the predictions/observations were made at a different redshift z (say $z = 0.2$ instead of $z = 0.4$), then the total observed fraction of IR sources is expected to be lower due to the smaller amount of available gas (i. e., some of the galaxies have already completed the late-type to S0 conversion between $z = 0.2$ and $z = 0.4$). This effect is partly taken into account by the outward migration of the boundaries of the active virializing region with decreasing z, partly by the corresponding change of $F(z)$ at the new observing epoch z. Of course part of the $F(z)$ dependence accounts for the reduced elapsed time for a smaller z as well. Yet the argument should hold no matter what observation epoch one uses as long as one concentrates on the regions

Table 4.2: The expected fractions and numbers of galaxies to be observed in each luminosity range for $z \sim 0.4$ rich clusters in two different cluster environments, for different proposed physical processes acting alone, generated using eq. (4.57).

	f_{merger}	$f_{harassment}$	$f_{gas\ stripping}$	$f_{secular\ evolution}$
Post-virialization region	N/A	N/A	N/A	N/A
Active virializing region	~1.2% BIRGs (12)	~3% BIRGs (30)	~0.6–6% BIRGs (6-60)	~15–30% BIRGs (150–300)
	~0.3–0.9% VBIRGs (3–9)	~3% VBIRGs (30)	~0% VBIRGs (0)	~0.6% VBIRGs (6)
	~0.3–0.9 % ULIRGs (3–9)	~0% ULIRGs (0)	~0% ULIRGs (0)	~0% ULIRGs (0)
Outskirts	~0.4% BIRGs (4)	~0% BIRGs (0)	~0% BIRGs (0)	~3% BIRGs (30)
	~0.1–0.3% VBIRGs (1–3)	~0% VBIRGs (0)	~0% VBIRGs (0)	~0.2% VBIRGs (2)
	~0.1–0.3% ULIRGs (1–3)	~0% ULIRGs (0)	~0% ULIRGs (0)	~0% ULIRGs (0)

The expected number of detected galaxies for the different scenarios are given in parentheses, assuming 10 clusters, each having an average of 200 galaxies in the two outer regions (and therefore approximately 100 galaxies in each of the two regions described above). Here ULIRGs (UltraLuminous InfraRed Galaxies) have IR luminosities in the range of $L_{IR} = 10^{12} - 10^{13} L_\odot$, (Sanders & Mirabel 1996), VBIRGs (Very Bright InfraRed Galaxies) have $L_{IR} = 4 \times 10^{11} - 10^{12} L_\odot$, and BIRGs (Bright InfraRed Galaxies) have $L_{IR} = 6 \times 10^{10} - 4 \times 10^{11} L_\odot$ (adapted from Zhang 2008).

where virialization is going on. Since the BO effect refers to galaxy evolution between the intermediate redshift z and the local universe, such a virializing zone at z always exists. The gradually enlarging central part of the cluster then falls into the N/A no-prediction zone.

Note also that in Table 4.2 we did not break the IR luminosities of the moderately luminous galaxies at the usual LIRG (Luminous InfraRed Galaxies) level (which has the lowest luminosity value of $L_{IR} = 10^{11} L_\odot$), but rather at $4 \times 10^{11} L_\odot$. This choice of division is due to our preliminary examination of the results of Geach et al.'s (2006) Spitzer 24 μm observation of cluster CL 0024+16 at $z = 0.39$. In this cluster the majority of the IR sources detected have colors intermediate between star-forming and passive galaxies, indicating that this is possibly a transitional-type population, and these transition-type galaxies have luminosities extending to about $4 \times 10^{11} L_\odot$.

Furthermore, in Table 4.2 we have divided each cluster into three approximate zones (Goto et al. 2003a): (1) the very central post-virialization region, where the star-formation activity has ceased by the epoch of observation. For this region we do not give predictions of IR bright sources since we assume these are rare due to the exhaustion of the gas for galaxies in this region. However, some post-interaction morphological transformation could still be going on here after the extinguishing of star formation. (2) The active virialization region. For this region the virialization-triggered star-formation activity is in full swing, as will be the star-forming activities due to a number of proposed mechanisms. For clusters at different redshift ranges, this intermediate zone has been found to propagate from the inner cluster toward outer cluster region as redshift decreases. (3) The cluster outskirts. In this region the star-formation activity is elevated compared to the field, but not as vigorous as in the active virialization region.

The predictions for the frequencies of IR bright sources are thus given only for the latter two regions of clusters since for the first region all of the proposed mechanisms can be made compatible potentially with the observation of a lack of star-formation activity there. The estimates for the observed fractions for regions (2) and (3) are further divided into sub categories for each proposed dynamical mechanism (assuming it operates alone or is the dominating one in a particular region).

4.9.2.1 Galaxy Merger

In the different variants of the cold dark matter (CDM) paradigm, galaxy mergers are the preferred means of morphological evolution of galaxies in clusters (e. g., see Kauffmann 1995). Even though the likelihood of mergers in the dense regions of clusters has been questioned because of the encounter-speed arguments, the issue of to what extent mergers play a role in cluster galaxy evolution, in both the dense regions as well as in the outskirts, has never been firmly settled, especially due to the paucity of high angular resolution mid- and far-IR observations of distant clusters before the launch of Spitzer Space Telescope.

For major mergers between large, gas-rich disk galaxies, we expect to see UltraLuminous InfraRed Galaxies (ULIRGs) with peak IR luminosities L_{IR} in the range of 10^{12}–$10^{13} L_\odot$ based on the local ULIRGs observations, though lower luminosity-ranged mergers had been observed as well presumably between smaller galaxies or less gas-rich systems (Sanders & Mirabel 1996, Table 3). The assumptions we made to arrive at the numbers used in Table 4.2 for mergers are that the galaxies involved in these mergers are *large*, as well as *gas-rich* disks, a condition clearly satisfied by the BO progenitor galaxies (Dressler et al. 1997). For the virializing region of a dense cluster, if we expect major mergers to be responsible for producing the observed level of BO effect, or the transformation of $F = 30\%$ of the galaxy population from spirals to S0s and other early types in the past 5 Gyr, and if we assume an interaction time scale of 5×10^7–1.5×10^8 years for each of the two higher luminosity ranges listed in Table 4.2 (Sanders & Mirabel 1996, who further cited the merger simulations of Mihos & Hernquist 1994a, 1994b; also Murphy et al. 1996), eq. (4.57) gives an observed merger fraction of $f_{merger} = 0.3$–0.9% for each of the higher IR luminosity ranges. We have also assumed the BIRGs (Bright IR Galaxies, with L_{IR} in the range of $6 \times 10^{10} L_\odot$–$4 \times 10^{11} L_\odot$) phase to last about twice as long as the average VBIRGs (Very Bright IR Galaxies, with L_{IR} in the range of $4 \times 10^{11} L_\odot$ – $10^{12} L_\odot$) and ULIRGs phases (with $L_{IR} > 10^{12} L_\odot$, i. e., it has a time scale of 2×10^8 years, tracing precursors and aftermaths of merger events, which leads to the estimate of $f_{merger} = 1.2\%$ in the BIRG phase.

We note that different individual merger events have different interaction kinematic configurations, which will also result in the difference in how long the merging process will last. This uncertainty in the time scales of merging events should not hamper our ability to determine the possible roles of merger in a given cluster since the predictions of the merger scenario, both in terms of observed fractions and in terms of the luminosity of its most luminous members, are orders of magnitude different from other scenarios, and thus can tolerate this level of uncertainty. In the example we present in the following, we will see that the predominance of moderately luminous IR sources in that cluster from many different angles precludes the possibility that merger had played a significant role in transforming the galaxy morphologies in that cluster.

For the outskirts of the cluster, which we had assumed to have a morphological transformation rate of $F = 0.1$, it leads to an average $f_{merger} = 0.2\%$ expected for the outskirts if merger is the major cause of the BO transformation, slightly higher than for the field of $f_{merger} = 0.1\%$ which is the observed value (Sanders & Mirabel 1996).

4.9.2.2 Galaxy Harassment

Since galaxy harassment is a mechanism proposed based on numerical simulations (Moore et al. 1996, 1998), rather than on observations of physical processes, the precise time scales of this mechanism are not prescribed. We expect, from the general dynamics of high-speed swing-by between galaxies, that the time scale should be on the order of 10^8 years for each of the BIRG and VBIRG phases, similar to the less-violent phases of a major merger. Based on this assumption, as well as the proposed *multiple* nature of the harassing encounters, which we take to be 5 for an average galaxy to go through the late-to-early-type transition through the harassment process, we arrive at an overall time scale τ_{event} of 10^9 years (5 times of 10^8 years in each of the two luminosity ranges), which is reasonable considering that this is essentially the time scale of the cluster-core crossing for cluster galaxies. The high-density cluster core region is believed to be where the harassment mechanism is most effective (Moore 2004). These time scales gave the estimates of the observing frequencies quoted in Table 4.2.

The harassment process is expected to be much less effective away from the high-density cluster core regions since it depends on the strong tidal interaction between close-neighbor galaxies. We therefore set the expected observing frequency for IR bright galaxies due to harassment in the outskirts of clusters to be nearly zero. Any observed IR bright galaxies in those regions must be due to physical processes other than galaxy harassment. We would not expect to see ULIRGs either in the core or in the outskirts due to the action of galaxy harassment process alone.

4.9.2.3 Ram Pressure Stripping

Like the harassment mechanism, the ram pressure-stripping mechanism is also expected to operate mostly in the dense region of a cluster since that is where the intracluster medium (ICM) is mostly concentrated. If ram pressure is the dominant mechanism for the BO transformation, we would expect BIRGs to be preferentially located in the central virializing region of clusters since this is a gentle process which is not expected to produce ultra-bright IR sources. Furthermore, the time scale of operation of this mechanism is on the order of a cluster-core crossing time, or about 10^9 years (Boselli & Gavazzi 2006), though shorter time scales (down to 10^8 years) have been proposed as well (Fujita & Nagashima, 1999; Quilis, Moore, & Bower 2000). This leads to the frequencies of observations derived for the active virialization region and the outskirts of a cluster due to the operation of ram pressure-stripping alone, as shown in Table 4.2. We would not expect to see ULIRGs in any regions of the cluster due to the ram pressure-stripping process alone.

4.9.2.4 Post-Interaction Secular Evolution

The post-interaction secular evolution process refers to the enhanced star-formation and radial mass accretion process mediated by the large-amplitude, open density wave patterns such as spirals, bars, as well as skewed three-dimensional mass distributions (Zhang 1999; Zhang & Buta 2007). Secular evolution can operate in both the core as well as the outskirts of a cluster (or even in the group environment or in isolation), the only difference between these environments is the different evolution rates caused by the different strengths of the density waves, which in turn are caused by the different amount of tidal perturbation in these environments.

For the tidally enhanced secular evolution process we expect a large population of BIRGs with luminosities up to the lower range of VBIRGs. The IR manifestations of the secular evolution process depend not only on the strength of tidal perturbation, but also on the availability of gas supplies for star formation. In the post-virialization region, even though the mass accretion process may still be going on, the star-formation activity can be significantly quenched due to the paucity of gas in this region of the galaxy disk.

For the active virializing region, since the member galaxies are effectively always feeling the gentle tidal nudges of the near and as well as far neighboring galaxies (a condition characterizing the virialization itself), we therefore set the highest time scale of secular evolution in the BIRGs phase to be 5×10^9 years, that is, exactly equal to the amount of elapsed time between $z = 0.4$ and $z = 0$. We also set the lower limit to be half of the maximum. Note that these time scale estimates for secular evolution assumes spiral-induced star formation and mass accretion which are a disk-specific processes. There could exist more violent primordial, clumpy mass accretion at higher z due to dynamical friction in proto-galaxies, as might be what have terminated the star formation in large and early virialized clusters such as MS 0451, which we will discuss later.

For the outskirts we have assumed the star-formation time scale to be about 1.5×10^9 years, typical of post-starburst evolution time scale (Poggianti et al. 1999), and interaction is not continuous as in the active virialization region.

With these assumptions we derive the expected fractions of the BIRG sources as given in Table 4.2. The fraction of the VBIRGs is calculated assuming an interaction time scale of 10^8 years, describing the precursor starburst events leading to *some* of the subsequent secular evolution. We would not expect to see ULIRGs due to the secular evolution process alone.

Finally we want to comment here that in this entire derivation for the frequency table (Table 4.2), we have implicitly assumed: (1) the constancy of the rates of the interaction processes across the redshift range under concern, (2) the availability of the gas supply to feed the star-formation activity at the epoch of the observation z, and (3) a single process dominates the BO transformation. All of these assumptions may need to be modified, at least for some clusters and during some periods of the evolution. So corroborating data from other wavelengths for the clusters under study will help to arrive at a satisfactory picture of the major drivers for the morphological evolution in a given cluster, or in a population of clusters studied in a statistical sense. Apart from these implicit assumptions the main uncertainty of the table comes from the time scale value τ_{event} for each proposed physical process (since F and T in eq. 4.57 are well-determined and are common for all processes in a given cluster).

4.9.3 Application of the Infrared Diagnostic Approach

The 24 μm Spitzer observations of two distant clusters (Geach et al. 2006) have revealed elevated levels of star formation throughout the cluster, up to and slightly beyond the cluster turnaround radius, both compared with that determined using optical observations and compared to values typical of the field environment at the same redshift. It also revealed very different frequencies of IR-luminous sources in these two clusters: for MS 0451-03 at $z = 0.55$, very few 24 μm sources were detected; whereas for CL 0024 + 16 at $z = 0.39$, a large excess of the 24 μm sources were detected (~ 150 IR galaxies over a $25' \times 25'$ area, or 9×9 Mpc^2).

The difference in the rate of IR luminous sources may reflect a correlation of the IR sources with the virialization state of the cluster, that is, even though MS 0451-03 is at slightly higher z than CL 0024+16, it has larger mass and also appears to be at a more advanced stage of virialization than CL 0024+16 (Geach et al. 2006; Moran et al. 2007b), and therefore the most intense star-formation episodes may have already been over by the observation epoch (as evidenced by the presence of many passive spiral galaxies in this cluster, Moran et al. 2007b). In this case the entire cluster belongs to region 1 (post-virialization region) of the table, which can be consistent with essentially all of the proposed mechanisms (i. e., we will not be able to discriminate among the proposed mechanisms for this particular cluster). The star-formation condition of MS 0451 is not expected to be typical for an average BO cluster at intermediate redshifts, though, since BO clusters are usually selected to contain a large population of gas-rich star-forming galaxies. We will focus our attention of the application of the proposed approach thus only on CL 0024+16.

Most of the 24 μm galaxies in CL 0024+16 observed by Geach et al. (2006) show enhanced luminosity in the range of BIRGs, or $6 \times 10^{10} L_\odot$ to $4 \times 10^{11} L_\odot$, compared to their counterparts in the field ($\sim 2 - 3 \times 10^{10}$ $L_\odot$ for normal disk galaxies, see Sanders & Mirabel 1996). The majority of the 24 μm sources detected by Geach et al. (2006) displays optical/NIR colors that are intermediate between star-burst galaxies and passive galaxies, which indicates that this is likely a population of galaxies in the post-interaction stage, and are going through the slow and prolonged evolution under the influence of the interaction-enhanced density waves. We now apply the frequency approach developed previously to further confirm the nature of these 24 μm sources.

Geach et al. (2006) found that there are about 150 IR luminous sources in an area of $25' \times 25'$, the majority of them are in the IR luminosity range of $6 \times 10^{10} - 2 \times 10^{11} L_\odot$, which, compared with the total galaxy number of about 500 determined by Moran et al. (2005) for the same region, and taking into account of completeness estimates of Moran et al. ($\sim$65% completeness for galaxies in the range of $17.75 < I < 21.1$ and $\sim$40% for galaxies in the range of $17.75 < I < 22.5$), leads to an estimation of a fraction of about 15% of IR bright sources among the entire optical sample[†]. Note that Geach et al. (2006) did not carry out the luminosity statistics of their 24 μm sources according to the different radial ranges in the cluster, but rather estimated the luminosity statistics for the cluster as a whole. Therefore we do not have more detailed information at the present time to do a more detailed analysis in terms of the *radial distributions* of the frequency of the different kinds of IR bright sources.

The derived fraction of IR bright sources is in the correct range as predicted by Table 4.2 for secular evolution to account for the majority of the BOs transformation between the intermediate redshift and the present for this cluster: the fraction predicted in Table 4.2, when averaged over the entire cluster region, would lead to a predicted BIRG fraction of around 15%. From Geach et al.'s (2006) result, it is also seen that the brighter IR sources (those in the VBIRG range of 4×10^{11}–$6 \times 10^{11} L_\odot$) consist of about four sources, or 0.4%, which is also consistent with the secular evolution scenario. Most significantly, there were no sources detected above the flux level of $6 \times 10^{11} L_\odot$, therefore no ULIRGs or strong interactions are present in this cluster at this observing epoch. Improved statistics from a much larger sample would help to quantitatively assess the roles of other proposed mechanisms across the entire population of BO clusters.

Dressler et al. (1999) showed that the post-starburst galaxy fraction in their sample of 10 intermediate redshift clusters observed using the HST in the MORPHS project is about 20%, similar to the fraction of IR bright sources in CL 0024+16. Poggianti et al. (1999) confirmed that the total a+k/k+a type galaxy (the type of emission-line galaxies signifying a post-starburst population) fraction is also about 20%.

In the MORPHS sample analyzed by Poggianti et al. (1999) and Dressler et al. (1999), 10% of cluster galaxies were classified as e(a) (which is a dusty-starburst spectral class), and about 40% of these (or 4% of the total) showed morphological signs of tidal interaction or were observed to be directly involved in a merger/close interaction. This is much larger than the maximum allowed fraction of mergers we had estimated in Table 4.2 (i. e., 0.9%), which was calculated based on the observed BO transformation rate between $z = 0.4$ and the local universe. Therefore what Poggianti et al. and Dressler et al. observed could not mainly be merger events but rather interactions that were confused with mergers due to limited spatial resolution. This conclusion is also supported by the fact that in both CL 0024 and MS 0451, the

---

[†] In arriving at these fractional estimates we have considered the fact that we might need to extend the lower bound of the IR flux range from $6 \times 10^{10} L_\odot$, which is the 5 σ detection limit of the Geach et al. observations, to closer to $3 \times 10^{10} L_\odot$ which is the typical flux level of the local dusty starburst galaxy, such as NGC 253. However, we note that not all dusty starburst at the lower IR luminosity range will lead to the conversion of a late-type to an early-type galaxy in 5 Gyr, therefore the more numerous IR sources at the lower luminosity range probably should not enter into the current statistical estimates. This consideration is also supported by the result of Elbaz et al. (2007) who found higher redshift starburst galaxies to be much more IR luminous on average than the local starburst galaxies of similar morphology.

Spitzer 24 μm observations by Geach et al. (2006) had revealed essentially no ULIRGs population at all.

There exists the question of whether the few highest luminosity sources (currently estimated to be in the mid-range of $10^{11} L_\odot$ in the Geach et al. observations) in CL 0024 could have been undergoing mergers, because there is uncertainty in the conversion of 24 μm, or the rest frame 15 μm flux to the IR luminosity. We argue that this could not be true for this cluster, based on the entire distribution of the IR sources in this cluster. In Table 4.2 the predicted rates for mergers are not only for the ULIRGs, but also for the BIRGs and VBIRGs. For a given dynamical scenario the IR source statistics in all three ranges of the IR luminosity need to be accounted for, not just the highest luminosity range. The overabundance of moderately IR bright sources in this cluster is not consistent with a scenario in which merger is the dominant process for BO transformation in this cluster, because if so a roughly 0.3–0.9% ULIRG fraction would by itself be able to account for the $F = 30\%$ BO conversion, which leave the rest of the 15% moderately luminous IR sources unaccounted for. Or if these moderately bright IR source also lead to transformation by some other dynamical mechanisms such as secular evolution in addition to the role played by merger, together they would be overproducing the observed 30% BO conversion rate between $z = 0.4$ and present.

Furthermore, the observed IR sources could also not have been consistent with their being the product of either harassment or ram pressure stripping since these should have produced fewer IR sources based on the values given in Table 4.2. Plus, harassment should have erased the spiral structures whereas most of these post-interacting galaxies were found to be spiral disk galaxies (Poggianti et al. 1999; Moran et al. 2007b). Ram pressure stripping would also have difficulty producing the brighter portion of these IR sources in CL 0024. Therefore, tidally enhanced secular evolution process becomes the only possible remaining candidate to account for the origin of the population of IR sources in CL 0024.

Other ground-based optical observations on CL 0024 appear to corroborate our conclusions about the role of interaction-enhanced secular evolution in accounting for these excess IR sources. Moran et al. (2007a) found from the analysis of resolved optical spectroscopy of CL 0024 galaxies and a control sample of field galaxies at similar redshifts that the cluster Tully–Fisher relation exhibits higher scatter than its field counterpart. They also found that the central mass densities of the spiral galaxy population they examined were higher within the cluster virial radius than outside, with a sharp break exactly at the cluster virial radius. The cluster environment thus appears to be responsible for the creation of the increase in scatter of the Tully–Fisher relation (which is one manifestation of the Hubble-type transformation of cluster galaxies), as well as for the increase in central density of cluster spirals within the cluster virial radius.

Moran et al.'s (2007a) tentative explanation of these observations is that galaxy harassment process has eliminated the less-dense spirals in the cluster central regions. This explanation however could not account for the lack of high-central-density spirals in the cluster outskirts. So they proposed that these cluster-outskirts high-density spirals might have been eliminated by mergers. However, an obvious problem with this proposed explanation is why mergers in cluster outskirts would selectively eliminate only the high-density spirals but not the low-density spirals. Furthermore, 24 μm observations of this cluster have not revealed a significant merging population in this cluster.

The observations of Moran et al. (2007a) were also difficult to explain through any kind of ram pressure-stripping mechanisms as these usually lead to unusual star formation gradient, whereas the enhanced scatter observed in Tully–Fisher relation was present in both the V and the K_s band, and thus is not due to the influence of star formation or dust obscuring, but rather has to be related to the kinematic and structural changes of the cluster galaxies, as Moran et al. (2007a) had concluded.

The observed trend would be most consistent with a mild level of tidal perturbation for cluster galaxies, as well as the post-interaction secular evolution of the galaxy morphology and kinematics. These tidal perturbations are obviously more pronounced for galaxies within virializing region of the cluster because the condition of virialization essentially guarantees a continuous interacting state of the member galaxies in the cluster virial radius. The secular evolution scenario could naturally produce the observed difference in the central mass density of disk galaxies inside and outside the cluster virial radius, as well as the increase in structural and kinematic scatter of these disks (Moran et al. 2007a).

Note however that the increase in scatter is only observed for the Tully–Fisher relation of the late-type cluster galaxies within the cluster virial radius. For S0s, for example, the fundamental scaling relation in fact becomes tighter within the virial radius (Illingworth et al. 2000). This phenomenon can be understood in the secular evolution scenario as the settling onto an "attractor" of the dynamical evolution. Thus the tidal agitation in the cluster environment is a way to disturb one dynamical equilibrium (that of late-type disks) to facilitate the settling onto a new dynamical equilibrium (that of early-type morphology, which is of higher gravitational entropy).

Once again, we emphasize that the conclusions we have reached based on the analysis of CL 0024+16 serves only to illustrate the application of the approach we proposed, and the conclusion on the dominating mechanism to account for the BO effect for the majority of the intermediate redshift clusters can only be reached once we have analyzed enough samples of rich clusters by the same approach. As we have seen in this preliminary study, even the information obtained in the core region of a single cluster already provides hints at the underlying dynamical mechanism responsible for cluster galaxy transformation, though the addition of the cluster outskirts and group observations will help to complete the picture.

4.9.4 Other Cluster Observations in Support of Secular Evolution

In addition to the direct support from applying our infrared diagnostic approach to cluster CL0024+16, we present below supporting evidence for the importance of secular evolution process in transforming the morphology of cluster galaxies from two other well-known studies.

4.9.4.1 Results from Sloan Digital Sky Survey

Goto (2005) analyzed the velocity dispersion of 355 galaxy clusters (with 14,548 member galaxies) from the Slone Digital Sky Survey (SDSS) and found that bright cluster galaxies have significantly smaller velocity dispersion than faint galaxies, consistent with a picture of dynamical friction (galaxy–galaxy interaction) operating during the process of cluster virialization which reduces the velocity dispersion of massive cluster galaxies at the expense of the increase of velocity dispersion of the less massive cluster galaxies. He also found that star-forming late-type galaxies in his sample have a larger velocity dispersion than the passive late-types (i. e., those having spiral morphology, but do not show any ongoing star-formation activity, see Goto et al. 2003b), which is once again consistent with dynamical friction reducing the velocity dispersion of the more evolved (passive) population of galaxies.

The result of Goto (2005) is consistent with our conclusion reached from CL 0024+16 that tidally enhanced secular evolution appears to be the driver for transforming the cluster galaxy morphology, since galaxy–galaxy interaction and dynamical friction in clusters are known to excite large-amplitude density waves which leads to subsequent radial accretion of mass and the evolution of galaxy morphological types (Zhang & Buta 2007, 2015). The Geach et al. (2006), Goto et al. (2003b), and Goto (2005) results are also mutually consistent in the sense that the 24 μm sources which show intermediate colors between star-burst and passive populations are likely to be the population in transition from star-burst to passive galaxies, and these passive galaxies, through further secular evolution, will become the earlier Hubble-type cluster galaxies observed in nearby clusters.

The role of dynamical friction during the cluster relaxation process is reflected in the often sharp transition in galaxy properties across the cluster virial radius or at a critical local surface density (Couch, Colless, & Propris 2004; Tanaka et al. 2004; Moran et al. 2005, 2007a. Figure 5 of Illingworth et al. 2000 shows another example of the change of the tightness of scaling relation for the S0 population across the cluster virial radii). These characteristics, as we commented before, can be naturally explained by the enhanced secular evolution due to the tidally induced interaction processes.

Goto's (2005) result is on the other hand inconsistent with ram pressure having played a significant role in causing cluster galaxy evolution. The effectiveness of the ram pressure is proportional to σv^2 (where σ is the gas density and v is the velocity dispersion), which should be more effective for galaxies with larger velocity dispersion, that is, it should have predicted that the larger-velocity-dispersion population is the more evolved one (i. e., the passive population should have larger velocity dispersion if ram pressure is the main cause), contrary to what is observed.

4.9.4.2 Results from the MORPHS Collaboration

The MORPHS team conducted imaging and spectroscopic studies of 10 distant clusters in the redshift range of 0.37–0.56. Poggianti et al. (1999) found that among the MORPHS cluster galaxies a significant overabundance of the so-called "post-starburst" galaxies with the characteristic k+a/a+k spectral features, which exhibit spatial and kinematic distributions intermediate between the passive and active populations. The number density of these post-starburst population is significantly higher in their sample clusters than in the field, so there the star-burst activities prior to the k+a/a+k phases were obviously triggered by the cluster environment, so are the subsequent termination of the star-forming events.

One interesting thing noticed by Poggianti et al. is the mostly spiral-like disky morphology of these post-starburst galaxies from the HST images, showing that whatever interaction processes which triggered the starburst had not changed the galaxy morphology immediately. This fact is contrary to both the merger and the harassment predictions, which require immediate change to galaxy morphologies after the interaction. The evidence for the milder tidal perturbation, on the other hand, is entirely consistent with an interaction-enhanced secular evolution scenario.

Poggianti et al. also concluded that either two time scales or two different physical processes have caused the two observed transformation processes (the halting of star formation and the spiral-to-S0 transformation). This result is supported by the later investigation of Goto et al. (2004) using the SDSS data, who found that the color evolution of the BO galaxies in their sample is much faster than the morphological evolution. In the secular evolution scenario, these two different time scales acquire a most natural explanation: the interaction events which serve to excite the large-amplitude spiral density waves and bars may only last on the order of 0.1–1 Gyr, but the subsequent secular mass accretion process could continue for much longer, that is, could be a significant fraction of a Hubble time. Furthermore, the same process could operate on the passive phase of the galaxies involved as well, even when most of the star-formation activities have extinguished.

4.9.5 Further Comments on the Different Proposed Mechanisms for Cluster Galaxy Evolution

The results of the afore-mentioned cluster studies as well as several additional ones shed light not only on the relevance of the secular evolution process, but also on the implausibility of a few other proposed mechanisms for cluster galaxy transformation, which we now substantiate.

4.9.5.1 Infall, Stripping, Mixing, and the LCDM Paradigm

Poggianti et al. (1999) found no evidence among the MORPHS cluster galaxies of a difference in the radial distributions of passive and emission-line spiral galaxies. A difference of distribution in the distribution of these two populations is to be expected if significant infall had occurred. Cooper et al. (2006, 2007) described the results from the DEEP2 survey and found that the steepening of the morphology–density relation works in the group environment just as effectively as in the cluster environment. The results from SDSS and 2dFGRS also support this conclusion. Therefore, cluster-specific mechanism, such as ram pressure stripping, will not be a justified cause for the group galaxy evolution. Dressler et al. (1997) concluded that the morphology–density relation indicated that more local processes are at work. Couch Matthew, and De Propris (2004) reached similar conclusion from the 2dF survey results.

Elbaz et al. (2007) showed that at the high redshift the mechanism which relates galaxies to their environment is not simply a quencher but also a trigger of star formation: the star-formation/density relation at $z = 1$ in fact reverses the morphology–density relation trend. These authors concluded that their results are most consistent with a scenario where star formation is accelerated by the dense environment of galaxies, resulting in the faster exhaustion of their gas reservoir. This new result of the star-formation and galaxy environment dependence is inconsistent with the ram pressure-stripping picture.

LCDM paradigm predicted the late assembly of early-type galaxies, including cluster galaxies, which superficially is consistent with the morphological BO effect and with the increase in number density of early-type galaxies with time, as observed in both the cluster and the field environment (Treu 2004 and the references therein). However, unequal-mass mergers completely destroy the color–magnitude relation (Bower, Kodama, & Terlevich 1998), whereas equal-mass merging among massive galaxies reduces the early-type number density, contrary to the observed early-type-galaxy number density evolution. Furthermore, collisionless mergers simply cannot reproduce the high phase space density observed for the central regions of early-type galaxies (Ostriker 1980). The rotationally supported S0 disk, which is the end product of the majority of BO transformation in clusters, is also problematic for a merger-based morphological transformation process. These observational constraints, coupled with the high-velocity dispersion of cluster galaxies, means that mergers, including dry mergers, could not have played an important role in the BO transformation.

As we have commented earlier, the existence of the morphology–density relation (Dressler 1980; Dressler et al. 1997) means that there could not have been a significant amount of persistent mixing of late-type galaxies from the field environment into the cluster environment during the cluster evolution process. Similar minimal-mixing conclusions have also been reached by Abraham et al. (1996) and

Morris et al. (1998). The hierarchical assembly scenario, on the other hand, advocates the gradual and continuous assembly of higher density clusters from lower density subclusters. Through numerical simulations, Dressler (2004) and collaborators found that under the LCDM paradigm there is actually significant mixing on cluster dynamical time scales.

One of the central assumptions of the LCDM paradigm is that the cluster formation is a purely gravitational process (apart from the influence of dark energy). There are in fact very well established observational evidence contradicting this assumption. Evrard (2004) reviewed many existing theoretical and observational tests of cosmology using galaxy clusters as probes, and highlighted the deficiency of the cluster physics according to the LCDM paradigm as "a cluster energetics problem." Under the assumption that the principal mechanism of ICM heating is through gravitationally induced shocks, there is about a 70% excess heat that could not be accounted for between what is required for virial equilibrium and what is observed. Furthermore, galaxy velocity dispersion in clusters shows a comparable level of excess "heat" compared to the dark matter velocity dispersion normalized to the WMAP (Wilkinson Microwave Anisotropy Probe) and large-scale structure distributions. Therefore, the cluster environment is in general not in a gravitationally unstable collapsing configuration as predicted by the LCDM theories.

Many of the observed clusters form through their inherent hydrodynamical large-scale systematic velocities and the resultant high-speed subcluster clump mergers (such as what had happened for CL 0024, as well as for many other well-known clusters in the intermediate redshifts as studied by the MORPHS collaboration and by Couch et al.), but the cause of these mergers is not gravitational collapse, but rather hydrodynamical collision due to their inherent relative velocities (see further the discussion on this issue in the next chapter). A portion of the observed clusters such as MS 0451 does indeed appear to form out of gravitational collapse, however. These clusters are likely to be situated in a high-density region in the primordial mass fluctuation spectrum, and thus are much better virialized as a whole because they satisfy better the gravitational instability condition. This latter class will appear to be more X-ray luminous for a given amount of baryonic mass. They also tend to evolve faster in terms of member galaxy morphologies as the example of MS 0451 had shown.

The excess systematic velocities of many of the proto-cluster clumps could have originated from the so-called "primordial turbulence" at the time of decoupling (von Weizsacker 1951a; Gamow 1952; Ozernoy 1974a, 1974b; Bershadskii & Sreenivasan 2002, 2003; Bershadskii 2004). If the primordial turbulence underlies the formation of the large-scale structure and the clusters, then the paradoxical excess heat of the ICM and excess velocity dispersion of the cluster galaxies (Evrard 2004 and the references therein) would all acquire a very natural explanation, which the current LCDM paradigm and the gravitational collapse picture failed to provide an explanation. In the primordial turbulence picture mergers due to inherent systematic velocities between proto-cluster clumps will play a much more important role, and infall and mixing which are associated with the LCDM gravitational collapse will not be an important physical process during the cluster formation and evolution process.

4.9.5.2 Galaxy Harassment

Contrary to the later usage by many investigators, the early proposers of harassment intended it mainly as a mechanism for the transformation of dwarf disk galaxies to dE galaxies (Moore et al. 1996). The harassment mechanism has been shown to be ineffective during the simulations for transforming the morphologies of large disks, especially in increasing the bulge-to-disk ratio (Gnedin 1999; Moore et al. 1999). As is well known, a large fraction of the disk galaxies undergoing BO transformation are normal spiral galaxies which have large disks (Dressler et al. 1997).

Furthermore, harassment is expected to be effective only in the dense regions of clusters (i. e. virialized cores) since it depends on multiple high-speed encounters, whereas the morphology–density relation is shown to hold all the way from the dense core, to cluster outskirts, and to the group environment (Dressler 1980; Dressler et al. 1997; Cooper et al. 2007). So harassment cannot be a general mechanism for transforming the morphology of the majority of cluster galaxies.

Moreover, harassment mechanism is expected to operate by tearing apart the outer disks of a galaxy, or at least destroying the spiral structure on the disk, and yet most of the cluster post-burst population have been found to have well-preserved disks which contain spiral structures (Couch et al. 1998; Poggianti et al. 1999; Goto et al. 2003b; Moran et al. 2007b).

We also emphasize that despite the superficial similarities of the harassment mechanism and tidally enhanced secular evolution, the harassment simulations in fact produced very different remnants as a result of interaction compared to interaction-enhanced secular evolution. For example, Moore et al. (1999) state that in their simulation of galaxy harassment, for LSB galaxies "the bound stellar remnants closely resemble the dwarf spheroidals (dEs) that populate nearby clusters," whereas for HSB galaxies "although very few stars are stripped, the scaleheight of the discs increases substantially and no spiral features remain." It is clear that the dEs are not related to cluster S0s which are the end product of BO-type evolution. But even for the HSB galaxies, we know from the work of Poggianti et al. (1999) that the spiral structure and thin disks survive the star-formation events induced by interactions in clusters. So these remnants cannot be the same as those from the harassment simulations, which had their spiral structure completely destroyed by the strong tidal shocks. Furthermore, since secular evolution depends precisely on the presence of density wave features (i. e., spiral arms, bars, etc.) in the disks of galaxies to induce radial mass accretion, the operation of tidally enhanced secular evolution requires a completely different galaxy-disk physical condition (cold disk with prominent density wave structure) than what the harassment mechanism can supply (the hot and puffed up disk devoid of density waves), despite the superficial similarities of these two proposed mechanisms.

4.10 Secular Evolution and the Origin of Color–Magnitude Relation

The empirical color–magnitude relation (CM, e. g., see Bower, Lucey, & Ellis 1992 and the references therein) has long puzzled astronomers as to its origin, especially in the face of obvious morphological transformation of galaxies as reflected in the BO effect, as well as the cluster early-type galaxy number density evolution (Franx 2004). The conventional wisdom had always been that the constant slope of CM relation across redshift z means that this is not an age–mass relation, but rather a metalicity–luminosity relation (Ellis et al. 1997; Kodama et al. 1998), and the fact that most intermediate and nearby cluster early-type galaxies fall on a tight color–magnitude relation means that their stars formed at high redshifts (Franx 2004). This, however, implicitly assumed that the total luminosity L of a galaxy does not increase with time.

The existence of this relation in the face of morphological transformation, however, could also naturally follow from the secular evolution scenario. When the density-wave-induced secular evolution process is considered, which results in a continued ignition of more and more baryonic star-forming material at the density-wave crest (see also Section 5.2.2.2), as well as a continuous radial inflow of baryons to increase the central concentration of luminous mass of early type galaxies, the total luminosity L of a galaxy can increase with time. Although with the aging and reddening of the stellar population, the M/L ratio of the early-type galaxy increases (i. e. the average stellar population dims with age), the total L can nonetheless increase because the increase of M due to secular evolution is much more substantial. The aging and reddening of the stellar population, together with the gradual turning on of the star-forming mass reservoir, thus keep the galaxies on the CM relation as they redden and get more luminous.

Observationally, it was found that many elliptical galaxies contain substantial outer disks (Franx 2004 and the references therein). The L^* disky ellipticals are found to form a continuous sequence with the S0s (Cappellari et al. 2013). Thus the continuous secular evolution through disk-related mass accretion process can conceivably lead to the increase of the early-type disk galaxies', as well as elliptical galaxies' mass throughout the Hubble sequence. For the cluster early-type population, it was found by Fasano et al. (2000) that for clusters which have a large elliptical population, their S0 population is correspondingly reduced, and vice versa. This gives indirect support to the idea that the secular evolution process can in fact carry an early-type galaxy across the S0 Hubble type into disky ellipticals, through the 3D skewed mass distributions often found in the central regions of early type galaxies.

So far, high-z surveys have not uncovered many red cluster-galaxy progenitors, presumably this means they are yet to be transformed into such. The thought-to-be high-z progenitors of the nearby early-type galaxies, the Lyman break galaxies (Steidel et al. 1996), were found to be much less massive than the average early-type galaxies in the nearby universe (Papovich, Dickinson, & Ferguson 2001; Shapley et al. 2001).

It is likely that the mass of these high-z early-type galaxies have grown significantly since these earlier epochs through the secular mass accretion process. This explanation is one way to account for the "progenitor bias" as well as the increase in number density of early-type galaxies with decreasing redshift (Treu 2004 and the references therein).

The explanation of the nature of the color–magnitude relation as effectively an age–mass relation resolves the dilemma faced by the conventional explanation, that while on the one hand the observed morphological and spectroscopic BO effect for cluster galaxies means that not all cluster galaxy stars are old, on the other hand the CM relation in the nearby universe is so tight which would have implied a uniform high formation redshift of the major population of stars in these early-type cluster galaxies according to the conventional wisdom. This explanation also naturally accounts for the fact that for every cluster observed (even those in the nearby universe, if one goes far enough into the cluster outskirt), there are always some blue outliers which were off the main track of the cluster CM relation. This shows that galaxies only settle onto the CM relation as they age since when starting off they are generally not already on the CM relation. The CM relation is thus another "attractor" of the secular evolution process.

This new explanation of the meaning of color–magnitude relation is also consistent with the downsizing trend (e. g., see Kodama et al. 2004 and the references therein; Cimatti et al. 2006) for structure formation and mass assembly, which requires that the oldest object to be the most massive ones: this, coupled with the empirical CM relation, means that the red objects are indeed older, in addition to being metal rich. Recent observations have by now given direct confirmation that the colors of the early-type galaxies in the local universe are indeed correlated with the mean age of the stellar population (Barr et al. 2007).

4.11 An Example of Secular Evolution in Interacting Galaxies[†]

In this section we will present a close-up view of the effect of galaxy interaction in triggering a sequence of events which coordinate with one another to accomplish the task of accelerating the rate of radial mass redistribution of disk galaxies involved in the interaction, which in turn leads to enhanced nuclear activity in these post-interacting galaxies.

A nearby normal spiral galaxy, NGC 3627, which is a member of the interacting group the Leo Triplet, was mapped in the CO 1-0 transition by Zhang, Wright, & Alexander (1993) for its central 4.5 kpc region using the Berkeley-Illinois-Maryland

---

[†] Portions of this section used material first published in Zhang et al. (1993), reproduced with modifications from The Astrophysical Journal @ AAS. Reproduced with permission.

Association (BIMA) interferometer array with $7''$ resolution, and in the 21 cm emission of neutral hydrogen (HI) for the entire galaxy using the very large array (VLA) with $30''$ resolution. The combined CO and HI data provide new information, both on the history of the past encounter of NGC 3627 with its companion galaxy NGC 3628, and on the subsequent dynamical evolution of NGC 3627 as a result of this tidal interaction. In particular, the morphological and kinematic information indicates that the gravitational torque experienced by NGC 3627 during the close encounter triggered a sequence of dynamical processes, including the formation of prominent spiral structures, the central concentration of both the stellar and gas mass, the formation of two widely separated and outwardly located inner Lindblad resonances (ILRs), and the formation of a gaseous bar inside the inner resonance. These processes in coordination allow the continuous and efficient radial mass accretion across the entire galactic disk. These results show that NGC 3627 is likely in the process of evolving into a nuclear active galaxy. It also suggests one of the possible mechanisms for the formation of successive instabilities in post-interaction galaxies, which could very efficiently channel the ISM into the center of these post-interacting to fuel nuclear star-burst and Seyfert activities.

4.11.1 Background

Interactions among galaxies are known to play important roles in the evolution of galaxies in a diverse range of environments. Statistical studies of the distribution of interacting galaxies at different redshifts indicate that most galaxies have suffered various degrees of gravitational interactions with neighboring galaxies in their lifetime (Zepf & Koo 1989; Barnes & Hernquist 1992). It is also found that nuclear activity and starburst phenomena observed in many galaxies seem to correlate with these galaxies being in compact groups or clusters, or else have companion galaxies (e. g., see Adams 1977; Simkin, Su, & Schwarz 1980; Stockton 1982; Hutchings & Campbell 1983; Dahari 1985; see also the many contributions included in Wielen 1990; Sulentic, Keel, & Telesco 1990).

The effect of galaxy interaction on the galactic mass distribution was first investigated by Toomre & Toomre (1972) using restricted three-body calculations. Their models showed furious response of the disk mass for prograde encounters in the form of bridges, tails, and spiral patterns. Retrograde encounters showed much weaker response, which is partly physical and also partly because self-gravity of the particles constituting the disk was not taken into account in these kinematic models. Subsequent theoretical and numerical studies have been carried out to investigate the effect of tidal interactions on the excitation of disk instabilities, and on the removal of angular momentum from the ISM, so as to channel the gas into the center of a galaxy to fuel nuclear activity.

Byrd et al. (1986) studied the effect of the gravitational tides produced by companion galaxies in triggering large-scale gravitational instabilities in the galactic disk.

They found that the tidal interactions decreased the disk instability threshold, and the required level of tidal perturbation for inducing significant nuclear inflow matched the strength of the observed galaxy interactions. Lin, Pringle, & Rees (1988) supplemented the analysis of Byrd by proposing that the propagation of the effect of tidal perturbation to the inner disk is enhanced if the "swing-amplification" mechanism for the global spiral pattern is taken into account. These large-scale mechanisms could account for radial mass accretion down to the central 1–2 kpc radius, where the large-scale spiral structure weakens. In the same paper, Lin et al. (1988) also proposed a mechanism for the continuous central channeling of gas, needed to bridge the gap between the radius of 1 kpc, where the swing amplifier runs out of steam, and the radius of 0.1–1 pc, where the accretion disk around the central black hole takes over. Lin et al. (1988) proposed that the gas in the inner kiloparsec of a galaxy is in nearly circular motion in a gaseous disk. The effect of the galactic interaction propagates inward and perturbs the cloud orbits. This results in increased cloud–cloud collision and the subsequent rapid cooling of the gas material. The increase in the importance of gas self-gravity as a result of cooling leads to an increase in gas viscosity and an enhanced mass accretion rate. Another model by Shlosman, Frank, & Begelman (1989) proposed a so-called bars-within-bars scenario. In this model, within the background of a large-scale stellar bar, a central gaseous bar forms as a result of the gravitational instability, induced by the increased mass accretion from the outer disk onto the central self-gravitating gaseous disk. This process of the formation of successive gravitational instabilities at different radii provides a possible mechanism for the continuous central channeling of gas across the galactic disk.

In order to obtain detailed information for the central region of a post-interacting galaxy, where the global spiral pattern and the nuclear instability structure meet, Zhang et al. (1993) mapped a nearby interacting spiral galaxy NGC 3627 in the CO 1-0 transition and in the 21 cm emission of neutral hydrogen (HI) using the aperture synthesis technique. NGC 3627 is a member of an interacting group, the Leo Triplet, whose other two members are NGC 3628 and NGC 3623. It is relatively nearby, at a distance of 6.7 Mpc according to de Vaucouleurs (1975). It has an inclination angle of about 60°, which allows its velocity field to be obtained and at the same time the components which generate these velocities to be identified.

Although classified as a normal spiral [RC2 type SAB(s)b], NGC 3627 has certain unusual properties, such as a high H_2 to HI mass ratio, close to that appropriate for Seyfert galaxies (Young, Taccono, & Scoville 1983), and a reasonably strong nuclear H_α emission line indicating that it is currently undergoing at least a weak form of nuclear starburst (Filippenko & Sargent 1985). It also has a weak optical bar, two prominent spiral arms, and prominent dust lanes (Arp 1966). These characteristics indicate that NGC 3627 is possibly a borderline galaxy, with properties between those of normal and active galaxies, and thus is an ideal candidate for studying the post-interaction evolution and the process of the transition from normal galaxies to nuclear active galaxies as a result of interaction-triggered instabilities.

4.11.2 Previous Observations and Simulations of the Leo Triplet

NGC 3627, NGC 3628, and NGC 3623 form the well-known interacting group, the Leo Triplet. The past encounters between NGC 3627 and NGC 3628 left traces of distortion in both galaxies. The most noticeable one is a plume in the galaxy NGC 3628 which shows up in the optical (Zwicky 1956; Kormendy & Bahcall 1974), in HI (Rots 1978; Haynes et al. 1979), and in 100 μm (Appleton & Hughes 1988), emerging from the eastern tip and extending about $50'$ from the center of the galaxy. Distortions in NGC 3627 have also been found, including large-scale asymmetry and a counter-rotating HI envelope near the southeast part of the galactic disk. Young et al. (1983) observed the Leo Triplet galaxies in the CO 1-0 transition using the FCRAO 14 m (HPBW = $50''$) telescope. The total $M(H_2)/M(HI)$ ratio is 3.7 for NGC 3627, similar to that observed in the Seyfert galaxy NGC 1068 (Scoville, Young, & Lucy 1883). This situation is unique among normal Sb and Sc galaxies (Young & Scoville 1982a, 1982b).

Toomre (1977) and Rots (1978) have separately modeled the encounter of NGC 3627 with NGC 3628 using similar model parameters, treating it as a restricted three-body problem. In their models, NGC 3627, modeled as a point mass, passed NGC 3628 in a parabolic orbit, and the orbit of the encounter is in the direct sense of galactic rotation for NGC 3628 (but in the retrograde sense for NGC 3627 in reality). These kinematic models accurately reproduced the current morphology of NGC 3628 in both the spatial and the velocity space. For the best-fit model parameters, the perigalacticon happened 8×10^8 years ago, and the perigalacticon distance is about 25 kpc.

Although experiencing a retrograde encounter, previous observations have shown that NGC 3627 also displayed signs of distortion as a result of the interaction, such as the optical "hook" extending far to the northwest of this galaxy (Arp 1966) and the distorted neutral hydrogen distribution in the outer disk of NGC 3627 (Haynes et al. 1979). The HI velocity in the southeastern part of the envelope is almost in the opposite sense of the galactic rotation for that region. Nevertheless, NGC 3627 shows a lesser degree of damage compared with NGC 3628. This is mainly because the orbit of encounter for NGC 3627 is almost retrograde with respect to the rotation of this galaxy while for NGC 3628 the encounter is in the prograde sense. Retrograde encounters usually cause less distortion to galaxy mass distribution than direct encounters (Toomre 1977).

4.11.3 CO 1-0 and HI Aperture Synthesis Observations of NGC 3627

The observational results described below were first reported in Zhang, Wright, & Alexander (1993). The CO 1-0 aperture synthesis observations of the galaxy NGC 3627 were obtained between the spring of 1990 and the spring of 1992, using the Berkeley-Illinois-Maryland Association (BIMA) array of (at the time) three aperture interferometer. The velocity coverage is about 832 km/s and a velocity resolution of 3.25 km/s/channel is achieved. The FWHM of the main beam at this frequency is a little less than $2'$. The synthesized beam size is about $9'' \times 12''$. The flux calibration is believed to be accurate to within 20%. Since the observation was make with both the compact and extended array configurations, the flux loss was not significant. Maps were made by averaging 10–15 channels of visibility data to enhance the signal-to-noise ratio. The maps were deconvolved using the CLEAN algorithm to remove the synthesized beam sidelobes. Primary beam taper correction was not made.

The HI observations of NGC 3627 were obtained with the VLA in C- and D-array configurations during 1988. In the analyses of the HI data we used the same central position as in our CO aperture synthesis observations, which is the same as that used in the previous single-dish observations in HI by Haynes et al. (1979) and in CO by Young et al. (1983). The angular resolution is about $30''$. The velocity resolution is about 20.6 km/s, and the total bandwidth is about 6.25 MHz (1,320 km/s).

Figure 4.23 shows an overlay of the CO (1-0) and HI integrated intensities for NGC 3627. Note that the HI image covers the entire optical galaxy, whereas the CO observations cover only the central $140'' \times 140''$. The CO bar-like feature is seen to reside inside the central HI deficiency. Two CO clumps farther out coincide with the HI clumps at the same locations, which trace the beginning of the spiral arms. The HI and CO distributions show evidence of asymmetry and distortions. A separate clump at the lower left (southeast) side of the figure was determined from the detailed analyses of velocity channel maps (Zhang et al. 1993) to be a piece of material torn apart from the right-hand spiral arm during the galaxy encounter by the ram pressure of the companion.

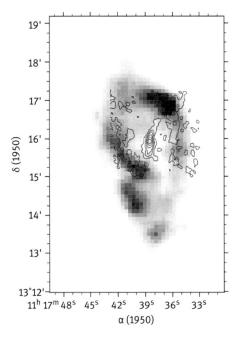

Figure 4.23: Overlay of the CO (contours) and HI (halftone) aperture synthesis images. The CO observation covers only the central $140' \times 140'$ (adapted from Zhang et al. 1993).

4.11.4 Analysis and Discussion

From the overlay map of CO and HI, we note that the gas distribution has large-scale asymmetry both in the central region and at the outer envelope, with signs of distortion especially near the southeastern part of the galaxy. The distribution of H_2 (inferred from the CO distribution using the standard conversion ratio) and HI also display the characteristic "twin peaks" near the beginning of the spiral arms, suggesting that this is near the radius of an ILR where the crowding of the gas streamlines causes the accumulation of the gas mass similar in some ways to those discussed by Kenney et al. (1992) for a sample of four galaxies, where the "twin peaks" were located much closer to the nuclear region. The locations of these ILRs have significant implications for the dynamical evolution history of NGC 3627, and these will be discussed in the following. At a radius of 2 kpc (or 1 arcminute at the distance of this galaxy), there is a deficiency of H_2. A bar-like molecular concentration is located in the innermost 1 kpc.

Figure 4.24 shows the CO-HI composite position-velocity map along the major axis of the galaxy, together with a fitted rotation curve. We observe that the rotation curve is asymmetric between the southern and northern part of the galaxy, as had been found before by Young et al. (1983). Since the southern part of the major axis (top part of the rotation curve on this plot, for a position angle of 173°) cuts through the larger portion of the galaxy, we will use the southern rotation curve in most of the following

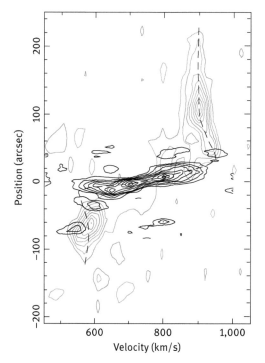

Figure 4.24: Position-velocity map derived from the CO (dark contours) and the HI (lighter contours) data along the major axis of the galaxy. The dashed line shows the fitted rotation curve (adapted from Zhang et al. 1993).

discussion. Note that the derived rotation curve is not corrected for inclination; in order to obtain the true circular velocity, the projected velocity at each radius should be divided by $\sin\theta_i$, where θ_i is the inclination angle of the galaxy. In the following discussions, all the velocities and angular velocities refer to the projected velocities in the plane of the sky.

Using the derived rotation curve, we can further calculate the angular speed, the pattern speed, and the location of the ILRs. The angular speed Ω is determined at each radius through $\Omega(R) = V(R)/R$ where $V(R)$ is the circular speed at that location. The pattern angular speed Ω_p is the same as the circular speed at the radius of corotation of the orbiting stars and spiral pattern. The corotation radius in this case is determined as the radius where the optical spiral arm terminates (Roberts 1975; Rohlfs 1977; Kenney et al. 1992), which is about 7 kpc for this galaxy. Later PDPS determination of the corotation by Zhang & Buta (2015) for this galaxy also supported this conclusion. The locations of the ILRs are defined as the positions where the $\Omega = \Omega_p$ line intersects the $\Omega(R) - \kappa(R)/2$ curve. where κ is the epicyclic frequency.

The derived rotation characteristics for the southern part of the galaxy are shown in Figure 4.25. Note that the density of the error bars does not indicate the density of data points observed. These are merely the arbitrary locations where the angular velocity and their derivatives are calculated. Also note that since there is a short piece of linear or solid-body rotation curve near the center of the galaxy, as seen in Figure 4.24, the $\Omega(R)$ curve at the center of the galaxy is expected to have zero derivative. This is not represented very well here due to the sparse data points we chose to calculate velocities and their derivatives. Again the inclination angle correction is not made; thus the true angular velocities can be obtained from the angular velocities drawn here by dividing by $\sin\theta_i$. The locations of the ILRs will not be influenced by this rescaling of all of the velocities. We see that there are two intersections of the $\Omega = \Omega_p$ line with the $\Omega - \kappa/2$ curve. These are designated the IILR, which is the one closer to the galactic

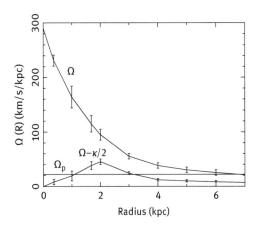

Figure 4.25: Projected angular speed, pattern speed, and the locations of the inner Lindblad resonances for NGC 3627 (adapted from Zhang et al. 1993).

center, and the OILR, which is the one further out. This configuration is similar to that which we had obtained for NGC 1530 previously (see Figure 4.4).

Comparing with the CO-HI overlay map of Figure 4.23, we see that the radius of the IILR (about 1 kpc, or 30 arcsec) is close to the end of the molecular bar, and the radius of the OILR (about 3 kpc) is close to the location of the twin peaks. Note that since the twin peaks are not on the major axis of the galaxy, the radii of the twin peaks from the galactic center are projected through a position angle of ∼30°.

The dynamical basis for the morphology and kinematics of the central gas distribution can be understood from an analysis of orbits in an asymmetric background stellar potential[†]. Contopoulos & Mertzanides (1977) showed that in a weakly barred potential, if the mass distribution and the pattern speed are such that two ILRs are present, then outside the OILR and inside the IILR the dominant periodic stellar orbits are those elongated in a direction parallel to the bar potential (the so-called x2 orbits), between the two ILRs the dominant periodic orbits are those perpendicular to the bar potential (x1 orbits). An immediate consequence is that a self-consistent bar model cannot be constructed by orbits mainly of the x1 type, in which the orbiting mass and the gravitational potential are out of phase. Or, as the numerical models of Schwarz (1984) show, the effect of the appearance of the two ILRs is to clear an annulus between the two ILRs by sweeping the mass toward the center. which may lead to the formation of a bar inside the IILR. On the other hand, near the OILR, the crossing of the two families of periodic orbits, when combined with the pressure of the gas, leads to the crowding of the gas streamlines and thus the formation of the "twin-peaks" gas concentration near the OILR (Schwarz 1984; Kenney et al. 1992).

The spiral pattern formed near the OILR (the "twin peaks") should lead in phase with respect to the background stellar bar potential (the orientation of the large-scale stellar bar cannot be easily discerned from the optical image of this galaxy, but can nevertheless be inferred from an angular momentum argument). This allows torque exerted by the bar potential on the orbiting mass to remove angular momentum from the mass near the OILR, causing it to accrete inward. Once the inward accreting gas is swept across the space between the OILR and the IILR, it will reach the central gaseous bar structure inside the IILR. The linear offset shocks which usually form at the leading edge of the central bar can continue taking away energy and angular momentum, thus facilitating further mass accretion toward the center. The inferred orientation of the large-scale stellar bar potential, the orientation of the "twin-peaks," and the orientation of the central gaseous bar (which also seems to coincide with a central small stellar bar shown on the optical image) indicate a continuous shift in phase, reminiscent of the isophote twists observed in Shaw et al. (1993).

[†] Since we are here analyzing the response of gas under a stellar potential, the passive orbit approach appears to be reasonably successful, as long as the gas contributes only a small fraction of the total potential.

It should be emphasized here that the presence of the ILRs, especially two widely separated ILRs which are located relatively outward in radius, is an indication that the galaxy mass distribution is very centrally condensed. This concentration includes both the stellar mass and the ISM mass. The means for the *stellar* mass distribution to reach higher central condensation is the focus of our previous discussions, and we have found in Section 4.3 that it takes about 5×10^9 years for mass to accrete inward for about 1 kpc for spiral structure of the Galactic strength. This explains that for the case of NGC 3627, 10^9 years after interaction the high central mass concentration has just started to form (and NGC 3627 has a more prominent spiral than our Galaxy, thus will have larger mass accretion rate than the above value estimated for our Galaxy), and only weak nuclear activity can be detected, which appears as a moderately strong nuclear Hα emission (Filippenko & Sargent 1985). The similar morphology and kinematics of HI and H$_2$ near the radius of OILR indicate that the central gaseous disk as a whole responds coherently with the instabilities in the stellar disk. Away from this radius the difference in the distributions of HI and H$_2$ reflects partly the difference in the interconversion of HI and H$_2$.

The morphology and kinematics of the molecular and atomic gas in the central region of galaxy NGC 3627 imply certain characteristics of the post-interaction galactic evolution, which might turn out to be common for galaxies which had suffered intermediate to long-range tidal disturbance, as is the case for NGC 3627. The most significant is that post-interaction evolution is a well-coordinated process in the combined stellar and gaseous disk. This characteristic is demonstrated through the type of instability structures present in this galaxy: the two prominent spiral arms are the largest scale dynamic instabilities in the galactic disk, and most certainly also the first that were formed or enhanced as a result of the nonaxisymmetric gravitational perturbations to the galactic disk during the tidal encounter (Toomre 1981; Noguchi 1987; Barnes & Hernquist 1991). The large-scale spiral structure serves to transfer angular momentum outward and thus allows the central concentration of both the stellar and ISM mass. This initial stage of radial mass accretion in the outer disk allows two inner Lindblad resonances to form and, as in the case of NGC 3627, move to larger radii as the degree of central mass concentration becomes higher. Our analysis shows that these ILRs, as well as the bar structure formed inside the IILR, take over the task of removing angular momentum from the disk mass and continuing the central mass accretion process, right from the location where the spiral pattern diminishes.

The structures mapped in NGC 3627 strongly suggest that the formation of successive instabilities and the enhancement of the radial mass accretion process after galaxy interaction happen as a well-coordinated sequence of events on the combined stellar and gaseous disk, down to even the central sub-kiloparsec scale. It is clear that the kind of instabilities seen in the central region of NGC 3627 are not instabilities in the isolated self-gravitating gaseous disk (Shlosman et al. 1989), nor is the increase in the rate of central mass accretion caused chiefly by the increase in effective viscosity as a result of increased cloud–cloud collision (Lin et al. 1988). These are due to the following observations: (1) the central bar terminates right at the boundary of the IILR, and there is a cleared-up annulus in between the two ILRs, and (2) the gas distributions are in the form of "twin peaks" at the beginning of the spiral arms which coincide with the radius of the OILR. Neither of these morphologies can be explained by a self-gravitating viscous gaseous disk model.

Although the particular type of coherent instabilities has here been discussed only in the case of galaxy NGC 3627, the formation of a bar inside the IILR under a general nonaxisymmetric galactic potential, for somewhat evolved galaxy after interaction, appears quite common (NGC 1530, and our own Galaxy, also appear to fall into this category). Also, in the sample of four galaxies in the Kenney et al. (1992) study, the only galaxy out of that sample which has a central molecular mass peak, which resides inside the IILR, turns out to be an interacting galaxy. The observations of Thompson (1981) showed that a larger fraction of galaxies were barred in the core of the Coma Cluster than outside, which suggests a possible correlation between galaxy interaction and the formation of bars in the center of a galaxy. Numerical simulations by Byrd et al. (1986) on the effect of the interactions on the subsequent galactic evolution indicate that a bar usually forms in a model galaxy under strong perturbations. The bar-forming phenomena in the perturbed disks are also observed in the numerical models of Noguchi (1987). It should be emphasized, however, that oftentimes the tidal torque during a galactic encounter does not influence the mass distribution of the central region of the disk directly and immediately, but instead is mediated by the disk spiral density wave, which is excited or amplified by the tidal torque of the encounter. This is true especially for high-speed encounters. This shows up in the numerical simulations of Byrd and Hernquist (1990) and Hernquist (1990). Our current work provides the detailed dynamical mechanisms that this process can happen. The prevalence of the nested resonance phenomena in nearby galaxies (Zhang & Buta 2007, 2012, 2015; Buta & Zhang 2009), and the fact that nested resonances tend to happen in earlier-type galaxies with deeper central potential wells, shows that the secular evolution of the outer stellar disk is an integral part of the process of the channeling of material into the central region of galaxies to fuel nuclear activities, a topic we will turn to in the next section.

4.12 Black Hole Mass and Bulge Mass Correlation

We have shown that the nested resonances in the central regions of intermediate and early-type galaxies, such as those in NGC 1530 and NGC 3627, are important to the channeling of matter to fuel nuclear activity. Since these nested resonances only form when the potential near the central region of a galaxy is deep enough, that is, its central angular velocity needs to rise sufficiently steeply to allow secondary density wave modes to form, the outer disk mass accretion, which transforms a galaxy's morphology from late to early Hubble type, appears to serve as a necessary preparatory stage for the fueling of the active galactic nuclei. The same mass accretion process across a hierarchy of modal patterns throughout the outer and inner regions of the galaxy disk could in fact be the dominant process in the formation of the tight black hole mass/bulge mass correlation relation observed for many early-type galaxies, as the recent results of the COSMOS group have demonstrated (Cisternas et al. 2011a, 2011b).

The well-known black hole mass and bulge mass correlation was found to exists across many orders of magnitudes of the bulge and black hole masses, and this correlation is especially tight for earlier type disk galaxies (Gebhardt et al. 2000 and the references therein; Merritt & Ferrarese 2001 and the references therein). On the other hand, Kormendy et al. (2011) found that the so-called pseudo-bulges in late-type galaxies do not show as strong a correlation with the black hole mass compared to the correlation observed for early-type bulges. This decreasing tightness of the black hole mass/bulge mass correlation with later galaxy types, when carried to one extreme, corresponds to the bulge-less AGNs observed in certain very late-type galaxies (Satyapal et al. 2009 and the references therein).

In the secular evolution scenario, the existence of bulge-less AGNs as well as the weak correlation trend of the black hole mass/late-type-bulge mass (the data points that lie above the linear correlation derived for early-type galaxies, assuming the black hole mass is plotted on the vertical axis and bulge mass on the horizontal axis) can be explained by the different accretion time scales of the black hole accretion disk and the galactic (accretion) disk. Black holes have a much smaller accretion disk than galactic bulges whose accretion disk is essentially the stellar and gaseous galactic disks. Thus the black hole mass can be built up much sooner in time compared to bulge mass. Even the onset of the black hole accretion event could be much earlier, that is, it can happen before much of the galactic disk of the host galaxy even forms, because a primordial central mass concentration adequate to the formation of the central AGN accretion disk can be formed in the very early stages of a galaxy's life. This provides the most natural account for the observed bulge-less AGNs. Of course, the noisy data can lie also below the linear correlation curve due to the fact that bulges can also accumulate (stellar as well as gaseous) mass in a shallow-potential-well configuration from the outer galactic disk, without first forming an inner AGN accretion disk.

The gradual settling onto the black hole mass/bulge mass correlation as a galaxy evolves from the late to the earlier Hubble types is an indication that the formation of the central black hole and the evolution along the Hubble sequence of an average disk galaxy are both attractors in the galactic configuration.

The establishment of the secular evolution origin of the black hole mass and bulge mass correlation shows from a different side the importance of *stellar mass accretion* on the galactic disk since, as the COSMOS team had shown, merger events are not significant for continued black hole mass growth, and the galactic disks (in particular the stellar disks) were already in place by $z = 1$ but bulges had not fully formed at that time.

Thus we see from the evidence presented in this (and the previous) section that the formation and evolution of the black-hole-mass and bulge-mass correlation have to have happened mostly beyond the pseudo-bulge stage, and the process has to have involved significant stellar mass accretion on a preexisting galactic disk that had little or no bulge. Nested resonances in the central regions of early-type galaxies are important manifestations of the effect of outer disk mass accretion, and they are also important instrument for channeling the gas into the nuclear region of a galaxy to grow the black hole mass and to fuel AGN activity.

5 Putting It All Together

Having presented extensive evidence in support of secular evolution from the analytical, numerical, and observational aspects, we now put together what we have learned so far to examine the foundations and implications of the new secular evolution paradigm in more depth. Part of the content in this chapter, especially Section 5.1, has previously appeared in Zhang (2016).[†]

5.1 Reexamine the Foundations

In this section, we address a few issues that had been at the foundations of the density wave theory since its inception, and had also been the greatest source of controversy and debate. The lessons we learned in trying to understand the role of collective effects in, and self-consistent evolution of, disk galaxies possessing density wave modes can finally put to rest some of these controversies.

5.1.1 On the Modal and Quasi-Steady State Hypotheses of Density Waves in Physical and Simulated Disk Galaxies

The mass flow rate equation we have obtained in this work is exact when the density wave pattern involved is a *spontaneously formed mode* that has achieved *quasi-steady state* (QSS). When this state is achieved, the rate of global wave amplification through over-reflection at corotation and feedback near the galactic central region is balanced by local dissipation at the density-wave collisionless shock. The wave amplitude as well as its shape is unchanged (at least on the timescale of a local dynamical time, or the galaxy rotation period), and the only measurable secular change is the slow redistribution of disk matter both inside and outside corotation, as well as the heating of disk stars and the energy injection into interstellar medium (ISM). In this subsection, we examine the validity of the modal and QSS hypotheses.

5.1.1.1 Previous Work

The quasi-stationary spiral structure (QSSS) hypothesis had been used in the past few decades by many workers to explain the grand design spiral structures in galaxies

---

[†] In fact, most of the contents of Zhang (2016) paper were first written for an earlier draft of the current book. During its initial review handled by a different publisher, a referee objected to the fact that much of these new contents had not yet appeared in peer-reviewed journals. As a result, the author halted the revision of the book, and revised and reorganized a significant fraction of the new contents into article forms, which were then published in peer-reviewed journals prior to the book being accepted by De Gruyter for publication.

(Bertin et al. 1989a, 1989b and the references therein), though in these previous analyses the mechanisms for wave damping and for secular evolution of the basic state were not properly identified.

How good are the modal and QSSS hypotheses? Are there any objective ways we can judge these? Earlier workers (e. g., Elmegreen & Elmegreen 1983, 1989) employed statistical arguments to support the QSSS hypothesis, and N-body simulations provided support to the modal and QSSS hypotheses to varying degrees (Donner & Thomasson 1994; Zhang 1998, 2016).

Zhang & Buta (2007, 2015), Buta & Zhang (2009) used near- and mid-infrared images of galaxies to show that for nearby grand design galaxies, the level of QSS for the density wave patterns involved can be judged from the *coherence* of the potential-density phase shift (PDPS) curves, and from the *level of agreement* between the phase-shift positive-to-negative zero-crossing predictions for the corotation resonance (CR) radii and the actual resonance features in galaxies. Independent studies have confirmed the accuracy of the PDPS method for determining the CRs in grand design galaxies (i. e., Haan et al. 2009, who compared several methods for CR determination previously published in the literature, and stated "For our galaxies the phase-shift method appears to be the most precise method with uncertainties of (5–10)% ..."; as well as Martínez-García et al. 2009, 2011 who used a color gradient method to determine the CRs, and among their sample galaxies which overlapped with that of Zhang & Buta 2007 or Buta & Zhang 2009, there is good agreement in the CR locations determined by these two independent approaches).

As we have commented earlier in this text, the very validity of the PDPS approach requires that the density wave patterns involved in these grand design spiral and barred galaxies are global *modes* rather than transient waves. In the PDPS approach, we used only the surface density maps, obtained from the observed surface brightness maps, to derive a partially kinematic property such as the corotation radius. Without the pattern being a quasi-steady mode that has achieved global self-consistency, we would not expect such correspondence between surface density kinematics.

5.1.1.2 Analogy with Fully Developed Turbulence

In what follows, we demonstrate another effective approach for judging the QSS of the modes[†], in addition to the previous approach of Zhang & Buta (2007) and Buta & Zhang (2009), of comparing the phase shift curve positive-to-negative zero crossings with the resonant features on galaxy images. This time, instead of using the radial distribution of the PDPS, we will actually make use of the mass-flow-rate curve, even though the two curves have the same zero crossings, because as it turns out the magnitude of the mass flow rate curve provides crucial information about the quasi-steady modal status as well.

[†] We use QSS in the following instead of the original QSSS (meaning quasi-stationary spiral structure) abbreviation because we are considering the quasi-steady state of both the *spiral* and *bar* modes.

In Figure 5.1, we present the comparison of radial mass flow rates of six galaxies (previously analyzed in Chapter 4) calculated using the torque equation (2.124), first using only the $m = 2$ Fourier components (dashed curves) for the perturbation density and potential, and subsequently using the full set of Fourier components (solid curves). We see immediately that apart from minor differences, the agreement between the $m = 2$ (two-armed Fourier components) calculation and $m =$ full (short-hand for using all Fourier components) calculation agree to a high degree. In particular, slightly larger disagreements were obtained for the two galaxies (NGC 5194, or M51; and NGC 3627) that are known to be undergoing tidal interactions. We also notice that along the trend of Hubble-type evolution from late to early (left to right, top to bottom for these six galaxies in Figure 5.1), the agreement between the $m = 2$ and $m =$ full mass flow rates gets progressively better.

In Figure 5.2, we present similar comparison of $m = 2$ and $m =$ full mass flow rates for galaxy NGC 1530. This galaxy has exceptionally high mass flow rate even though it is in a relatively isolated environment. We observe that the $m =$ full mass flow rate was significantly higher than the $m = 2$ mass flow rate (though the zero crossings of the two curves are similar), indicating that the amplitude of the mode is yet to settle into its quasi-steady value even though the modal density profile appears to be stabilized: the amplitude of the mode is obviously on the wane.

Previously in Chapter 2, we have demonstrated that the growth rate of the mode γ_g due to global amplification and feedback cycle is positively correlated with the *fraction* of $m = 2$ sinusoidal modal component in the density wave composition (since only the $m = 2$ modal component has the correct phase relationship after each round trip of wave propagation to form growing mode; whereas the higher harmonic components cancel out due to their lack of correct phase relations). On the other hand, the effective dissipation rate γ_d of the mode, once the wave has acquired sufficient nonlinear amplitude, is approximately independent of the exact nonlinear distortions of the wave profile (since γ_d is determined mainly by the PDPS pattern for a wave with a given amplitude, and the phase shift was previously shown, in Figure 2.5, to be relatively independent of the degree of nonlinearity in the azimuthal density profile of the mode). Therefore, for high-amplitude, very nonlinear density wave mode such as present in NGC 1530, the mode is able to manipulate the degree of nonlinearity (i. e., non-sinusoidal-ness) of its azimuthal profile to boost up the ratio of the damping rate versus the growth rate, in order to evolve toward a quasi-steady equilibrium amplitude.

The results shown above, that at QSS of the wave mode the $m = 2$ and $m =$ full contribution to radial mass flow are approximately equal, might appear at first sight to be surprising, that is, how could something that is supposed to be only "a part" of the whole be equal to the whole? Here we need to realize the specialness of a self-organized density wave mode that has achieved QSS. The mode had gone through long period of fine adjustments among its various physical processes so that the global self-consistency condition is achieved. Among these processes, one of them is the randomization of the orbital energy of the basic state matter during the energy and

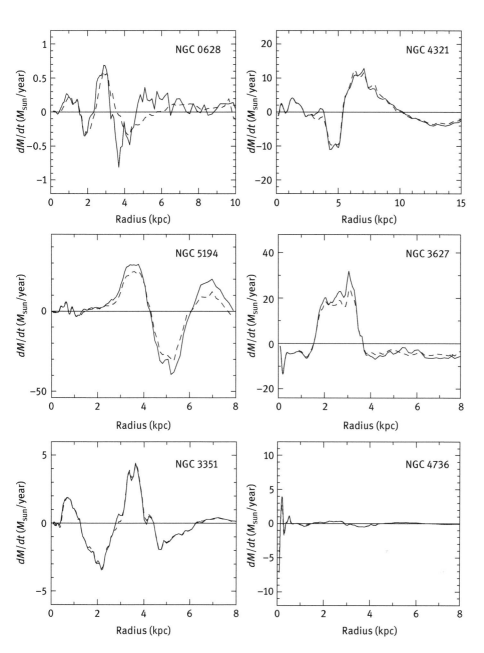

Figure 5.1: Radial mass flow rates for the six sample galaxies of Zhang & Buta (2015). The solid lines are calculated using the full Fourier spectra of density and potential, and the dashed lines using only the $m = 2$ components (adapted from Zhang 2016).

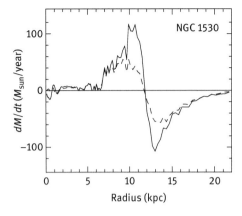

Figure 5.2: Radial mass flow rate for NGC 1530 first analyzed in Zhang & Buta (2007). The solid line is calculated using the full Fourier spectra of density and potential, and the dashed line using only the $m = 2$ components (adapted from Zhang 2016).

angular momentum exchange between the wave mode and the basic state, through the mediation of the collisionless shock at spiral arms. This randomization process happens both for the stellar and for the ISM components of the galaxy disk mass, and there are in fact two aspect to it: one aspect is the randomization of the orbital energy corresponding to $\Omega_p \times L_{\text{wave}}$, where Ω_p is the pattern speed of the wave and L_{wave} is the angular momentum density of the wave, this contributes to the damping of the wave and the secular decay of the mean stellar orbit; the second aspect is the randomization of the residual orbital energy corresponding to $(\Omega - \Omega_p) \times L_{\text{wave}}$ (for matter inside corotation. There is a corresponding sign change for matter outside corotation). This second aspect results in the secular heating of the disk stars (which produces the age–velocity dispersion relation of the solar neighborhood stars Zhang 1999), as well as the energy injection into the ISM to fuel a cascade process akin to the driven turbulence in fluid dynamics (which produces the Larson–Law size–line-width correlation of the interstellar clouds. See Larson 1981; Zhang et al. 2001; Zhang 2002; Falceta-Gonçalves et al. 2015; as well as our earlier discussions in Section 4.5).

Stars appear to accomplish the randomization of the orbital energy also through a kind of cascade/randomization process from the large down to the progressively smaller scales. The smaller-scale nonlinear exchange is nothing more than the cascaded down version of the large-scale exchange and does not create a new term in the energy balance equation. Another way to look at this matter is that at the QSS, the growth rate of the $m = 2$ mode is equal to the growth rate of the full nonlinear mode since the nonlinear mode does not have an independent global amplification mechanism other than the cascading of energy from the $m = 2$ component. Therefore, at the QSS the $m = 2$ mode dissipation rate (which is proportional to the $m = 2$ torque integral) is *formally* equal to the $m = 2$ growth rate[†], with the $m \neq 2$ components formally

---

[†] We say formally here because the torque integral acquires its meaning as corresponding to the dissipative angular momentum exchange between the wave and basic state matter *only for the overall full nonlinear set of components at the QSS*. For each m component individually or for waves not at the QSS the torque integral does not uniquely predict anything, since it is a spatial average of time-dependent

have zero growth rate as well as zero dissipation rate, even through in reality the $m = 2$ mode cannot accomplish energy dissipation without going through the cascade into nonlinear components.

This aspect of local interactions in density waves is in essence the same as what Kolmogorov had hypothesized for the energy cascade process of fully developed fluid turbulence in the inertial range (Frisch 1995). The fact that galaxies can randomize the originally coherent orbital velocity/energy is due to the global-self-consistency constraint of the spontaneously-formed density wave mode at the QSS, which produced both the local instability/collisionless shock condition at the spiral arms, as well as the detailed correlation between kinematic and positional distributions of disk matter that enable both wave growth and damping, accompanied by dissipative basic state evolution.

In what follows we present similar $m = 2$ versus $m =$ full mass flow rate comparison calculated using the surface density and potential obtained in the set-of-four N-body simulations we presented in Chapter 3. In Figures 5.3–5.6, we show the respectively calculated $m = 2$ and $m =$ full mass flow rates for the four runs, at six different time steps, and for a duration of 4,000 steps which correspond to about three and a half galactic rotation periods at radius 20, during the emergence and stabilization of the first dominant mode in each case.

From these four figures, we see that in simulations as in physical galaxies we can use the agreement between the $m = 2$ and $m =$ full mass flow rates to judge the QSS of the wave mode. Note that the QSS can be short-lived even after its initial acquirement because the disk surface density and modal shape continue to evolve throughout the simulation run for each softening case. Still, exploration into further time period of the modal evolution shows that a new QSS phase can once again be achieved (signified by good agreement between $m = 2$ and $m =$ full mass flow curves) after its temporary loss as a result of the basic state evolution.

Note that for the $a_{soft} = 0.1$ run in the last two frames there is the signature of a local instability clump, which produced poor agreement between the $m = 2$ and $m =$ full torque curves in the outer disk region, as is expected of such a transient nonmodal feature.

5.1.1.3 A Close-Up Look at the Modal Characteristics Evolution

In Figures 5.7–5.12 we present the morphological and power spectra evolution for the N-body spiral/bar mode using $a_{soft} = 0.25$, here with better spatial-temporal resolution in the plotting than we had presented in Chapter 3, and covering the entire duration of the simulation run.

---

variables. In other words, the torque integral, when used to predict the wave/basic state angular momentum exchange, is a form of *closure relation* enforced through the consideration of global energy balance, at the QSS of the wave mode.

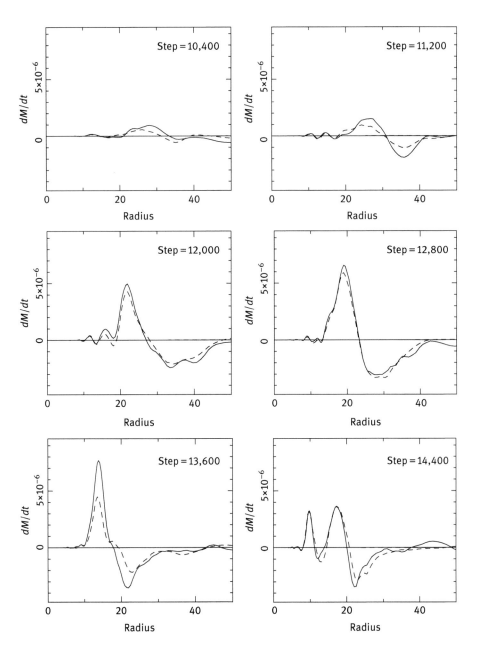

Figure 5.3: Radial mass flow rates for the N-body spiral mode with $a_{\text{soft}} = 1.5$, at six different time steps, using the full (solid lines) or $m = 2$ (dashed lines) Fourier components (adapted from Zhang 2016).

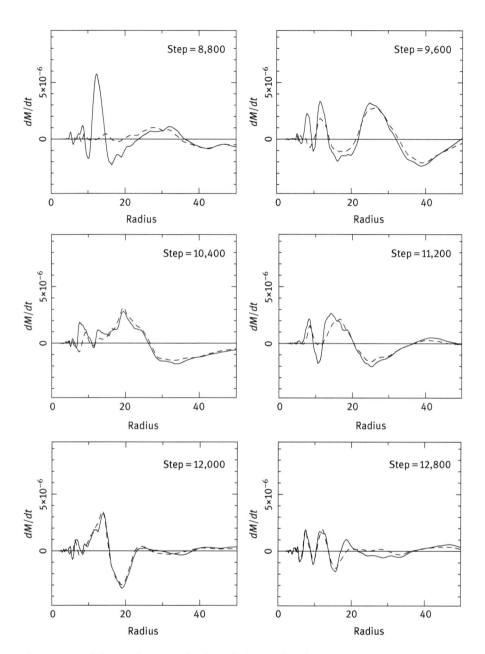

Figure 5.4: Radial mass flow rates for the N-body spiral mode with $a_{\text{soft}} = 0.75$, at six different time steps, using the full (solid lines) or $m = 2$ (dashed lines) Fourier components (adapted from Zhang 2016).

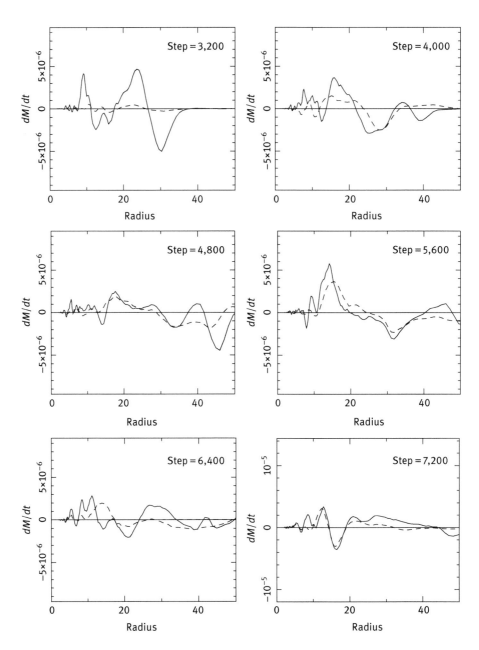

Figure 5.5: Radial mass flow rates for the N-body spiral mode with $a_{soft} = 0.25$, at six different time steps, using the full (solid lines) or $m = 2$ (dashed lines) Fourier components (adapted from Zhang 2016).

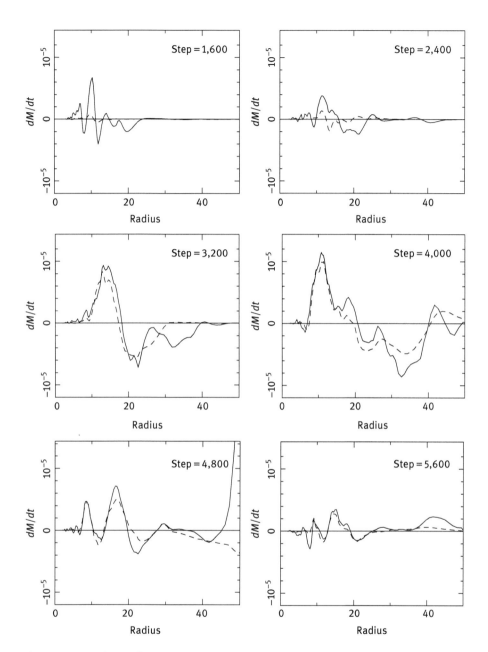

Figure 5.6: Radial mass flow rates for the N-body spiral mode with $a_{\text{soft}} = 0.1$, at six different time steps, using the full (solid lines) or $m = 2$ (dashed lines) Fourier components. The poorer $m = 2$ and m = full agreement in the outer disk region of the last two frames is due to a transient local instability clump (adapted from Zhang 2016).

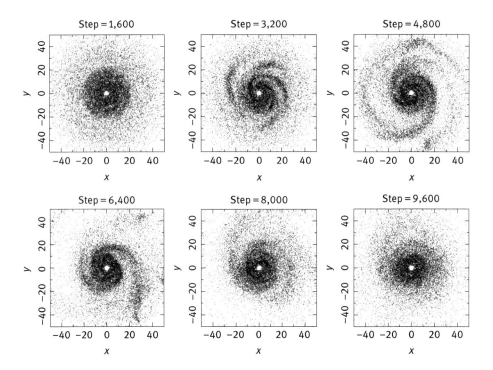

Figure 5.7: Morphological evolution of an N-body spiral/bar mode. The rotation period at $r = 20$ is about 1,256 steps. The softening parameter is $a_{\text{soft}} = 0.25$ (adapted from Zhang 2016).

We see from these figures that even though the spiral pattern involved is not perfectly steady, it is a mode (or a modal set, since during some time intervals there exist nested modes of different pattern speeds, which are present in physical galaxies as well) nonetheless, rather than a combination of transient wave trains (these wave trains would have filled up the spectral space with disorganized noisy components, rather than coherent and concentrated spectral clumps). During certain intervals, the mode transitions to a different shape in response to the changing basic state, and the power spectra may reveal additional features. But after a while the dominant mode (or sometimes the modal set) always settles back onto its quasi-stable location on the power-spectra plot. This is the kind of "attractor" or "asymptotic stability" behavior we would expect of a mode.

In Figures 5.13 and 5.14, we plot the $m = 2$ density and potential contours, covering also the duration of the simulation but with courser time spacing. We see that until the very end of the simulation, which is roughly 25 rotation periods at reference radius of 20, the underlying $m = 2$ mode is still alive and well, even though the heating of the basic state of the disk by the modal instability has submerged the nonlinear spiral pattern (the heating is a result of using smaller number of particles than in physical galaxies; of the near-collision condition at the modal collisionless shock; and of using a small softening length choice).

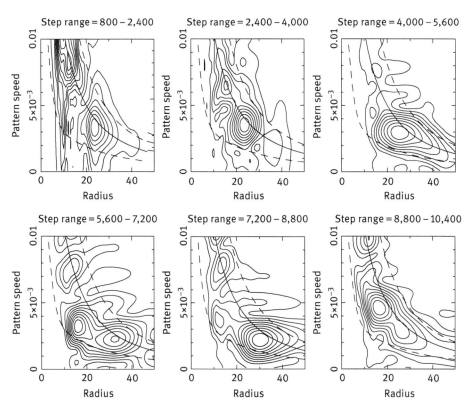

Figure 5.8: Power spectra of the above mode. Each frame is calculated with a central time step identical to that in the corresponding morphology frame and with a time step range of 1,600 (adapted from Zhang 2016).

In Figure 5.15, we plot the evolution of the radial mass flow rates due to the spiral mode, using once again the full density and potential perturbations (solid lines), as well as only the $m = 2$ density and potential perturbation components (dashed lines).

A comparison between this last group of figures and that of the power spectra evolution figures (Figures 5.8, 5.10, and 5.12) show several interesting features:

(i) As the evolution advances, the matter as well as the spiral activity become concentrated into the central region of the galaxy.

(ii) A better $m = 2/m =$ full agreement (especially for the inner disk, inside or around corotation radius of 20) is associated with a more concentrated power-spectrum peak location, indicating that the former is a good criterion for judging the QSS of the mode, as we have argued previously in Section 5.1.1.2.

(iii) The amplitude of the radial mass flow rate curves decreases with time (note the reduced maximum positive vertical scale with advancing time, among the three double-row-sets of figures, i. e., 10^{-5} for the first two rows, 2×10^{-6} for the second two rows, and 5×10^{-7} for the third two rows, respectively). Associated

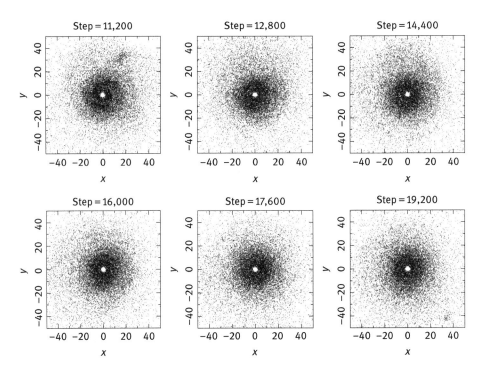

Figure 5.9: Morphological evolution of an N-body spiral/bar mode (continued from Figure 5.7). The rotation period at $r = 20$ is about 1,256 steps. The softening parameter is $a_{soft} = 0.25$ (adapted from Zhang 2016).

with this decreasing radial mass flow rate toward the end of the simulation, the agreement between the $m = 2$ and $m =$ full mass flow rates are better, especially for the central region where the mass and modal activity are concentrated. This shows that the slower secular evolution speed toward the end of the simulation (due to the reduction of nonlinear spiral amplitude) leads to slower modal evolution, and thus a better QSS or global self-consistency.

We conclude that the variation of the spiral pattern (both its morphology and pattern speed) is to a large extent a result of the secular evolution of the basic state induced by the very same density wave mode. We here recall the saying that "one cannot have the cake and eat it too." If we seek to obtain QSS of the mode, we cannot have fast secular evolution of the basic state at the same time. If the secular evolution is fast, the modal characteristics evolve rapidly with it, so as to be compatible with the evolving boundary condition. During these transitional times, the power spectra may indicate additional peaks other than that corresponding to the dominant modes, or possess slopes. This does not mean that the mode (or modal set) has become a transient wave. The mode is still a mode, only it is changing and evolving, just like when a person ages, he or she is still the same person (although some people would refer this process as the old self dying and a new self being born – which is OK as long as we all know what we are talking about).

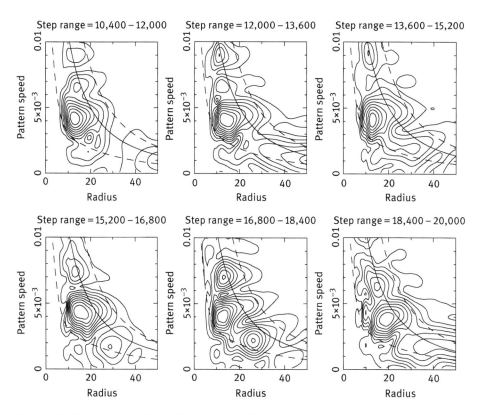

Figure 5.10: Power spectra of the above mode. Each frame is calculated with a central time step identical to that in the corresponding morphology frame and with a time step range of 1,600 (adapted from Zhang 2016).

Therefore we should relax our condition for calling a mode quasi-steady: As long as it satisfies the global self-consistency condition (growth rate = damping rate), we can regard it as quasi-steady even if it constantly evolves to accommodate the changing basic state boundary condition as a result of the secular evolution of the basic state induced by the very same mode. For the theoretical results (that of the torque integral and mass flow rate equations) to hold, the mode only needs to be quasi-steady on the local dynamical timescale (in this simulation the local dynamical time at radius 20 is 1,296 time steps, slightly shorter than the time duration between adjacent frames in Figures 5.7–5.12).

We would also like to point out once again that the observed galaxies, especially the nearby, grand design, noninteracting, intermediate to early type galaxies, appear to have even better agreement between the $m = 2$ and $m = $ full estimations of torque, than what we have been able to reproduce in the current generation of N-body simulations. The agreement between $m = 2$ and $m = $ full torques appears to span several sets of nested modes, as in the case of NGC 4321 (M100), NGC 4736, and NGC 3351. This agreement between the $m = 2$ and $m = $ full torques indicates a high degree of global

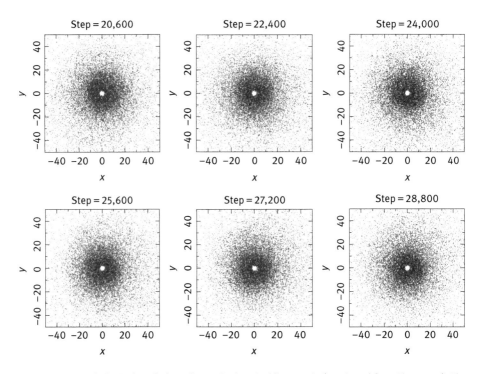

Figure 5.11: Morphological evolution of an N-body spiral/bar mode (continued from Figure 5.9). The rotation period at $r = 20$ is about 1,256 steps. The softening parameter is $a_{\text{soft}} = 0.25$ (adapted from Zhang 2016).

self-consistency for the modes in observed grand design galaxies, a result obviously partly due to a Hubble-time's worth of self-consistent evolution, including the active participation also of the bulge and halo components (which are held fixed in our N-body simulations), partly due to the much higher (by several orders of magnitude) numbers of degrees of freedom that physical galaxies have, compared to that available in simulated galaxies, to be used for adjusting the multitude of correlations to achieve a fine balance between modal growth and dissipation, between the secular evolution of the basic state and the transformation of density wave morphology to be compatible with the renewed basic state.

5.1.2 Role of Basic State Specification

As we have commented before, the correct choice of basic state in secular evolution studies involves the use of galaxy disk parameters that allows the global density wave *modes* to spontaneously emerge and be stabilized. Why is it so important to choose a basic state specification that allows unstable modes? Couldn't the transient waves on overstable basic state also accomplish the objective of secular evolution, as seems

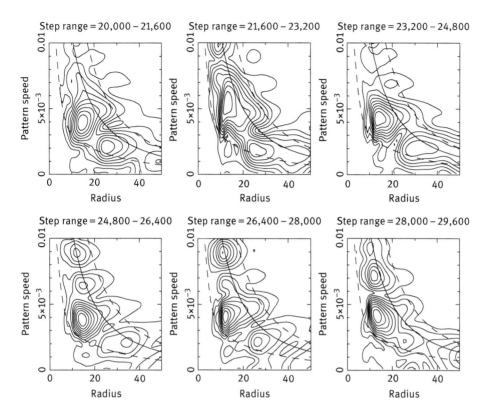

Figure 5.12: Power spectra of the above mode. Each frame is calculated with a central time step identical to that in the corresponding morphology frame and with a time step range of 1,600 (adapted from Zhang 2016).

to be advocated in a recent review article of Sellwood (2014), which is titled "Secular Evolution in Galaxies"? We have commented on this issue already in previous chapters, here we offer some further explanations.

First of all, we point out that in none of the transient wave simulations, carried out by Sellwood and his collaborators over the past three decades, as well as by more recent researchers such as Baba, Saitoh, & Wada (2013); D'Onghia, Vogelsberger, & Hernquist (2013); D'Onghia (2015), etc., was there a systematic trend of mass inflow inside corotation, and outflow outside, such as what we have demonstrated in the current work. The fundamental reason for this is that the transient waves rely on resonant interactions between the stars making up the wave trains and the basic state to exchange angular momentum. In transient waves, the resonances are broadened, but the interaction is not coordinated as in the modal case. For example, Carlberg & Sellwood (1985) state that the sense of angular momentum exchange between the wave and the basic state does not depend on the winding sense of the spiral at the Lindblad resonances, so a leading and a trailing wave train can exchange angular

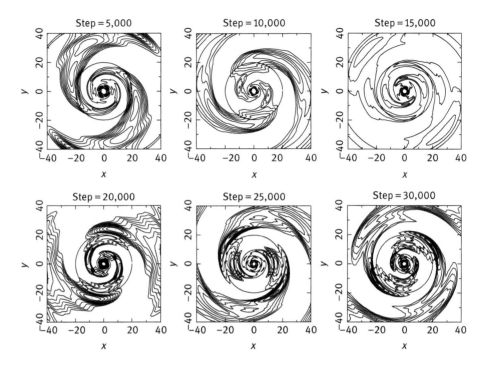

Figure 5.13: Morphological evolution of the $m = 2$ component of the density of the N-body spiral/bar mode. The softening parameter is $a_{\text{soft}} = 0.25$. The circle indicates the reference radius of 20. Each frame is individually scaled to have 10 contours. Only positive contours are plotted (adapted from Zhang 2016).

momentum with the basic state in the same way. Sellwood & Binney (2002) studied the so-called "radial migration" process of stars due to resonant interaction of waves and stars near the corotation region, yet they conclude that the end result of this interaction is such that the entire population of interacting stars do not lose or gain angular momentum, so there is no systematic radial mass flow that leads to bulge building through this so-called "radial migration" process. These resonant interactions thus are of entirely different nature than the collective interactions between the wave mode and the basic state stars that we studied in this work. Resonant interaction is top-down, that is, stars passively respond to an enforced smooth potential, whereas the collective interaction is sideways, that is, individual stars develop correlations in their motion through the mediation of collective instabilities and collisionless shocks.

For unstable global modes, however, their surface density and kinematics are such that they support a characteristic distribution of PDPS between the potential spiral and the density spiral of the mode. This phase shift distribution in turn leads to a secular torque by the modal potential on the density, and as a result a secular radial mass flow pattern as we had mentioned above. The global self-consistency requirement for these modes (since the modes are self-organized and self-sustained) is what brought on the effective singularity condition at the spiral/bar wave crest, which leads

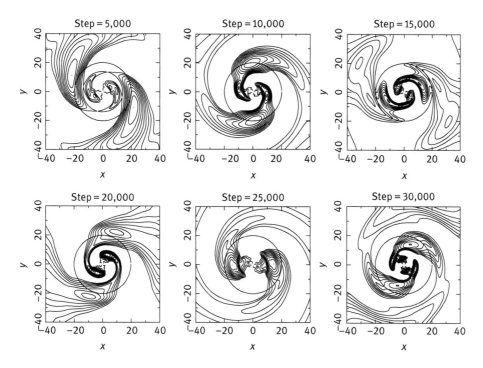

Figure 5.14: Morphological evolution of the $m = 2$ component of the (negative) potential of the N-body spiral/bar mode. The softening parameter is $a_{\text{soft}} = 0.25$. The circle indicates the reference radius of 20. Each frame is individually scaled to have 10 contours. Only positive contours are plotted (adapted from Zhang 2016).

to coordinated secular evolution of the basic state of the galactic disk, as well as the constant revision of the wave modal shape to accommodate the changing basic state.

For transient wave trains in an overstable disk, on the other hand, the global self-consistency condition is not enforced, and it is not necessarily that the spiral arms are the site of effective singularity and collective dissipation. The no-global-self-consistency condition means no QSS, thus the transient patterns continuously evolve (in a haphazard fashion, rather than in a coherent fashion as in the modal scenario). The fact that some researchers succeeded in producing a continued sequence of wave decay and reemergence within this scenario (Baba 2013; D'Onghia et al. 2013) does not mean all observed spirals are such recurrent transients, especially not those grand design two-armed spirals which account for about 50% of disk galaxies (Pettitt, Tasker, & Wadsley 2016), nor those galaxies that possess long-lived resonance features such as nuclear, inner and outer rings.

If our goal is to obtain the coordinated radial mass flow pattern that is responsible for the secular growth of bulge and galaxy morphological evolution along the Hubble sequence, then we cannot avoid choosing an unstable basic state to global modal formation, and employing the collective interaction of these modes with their parent basic state of the disk, such as is done in the current work. To model galaxies

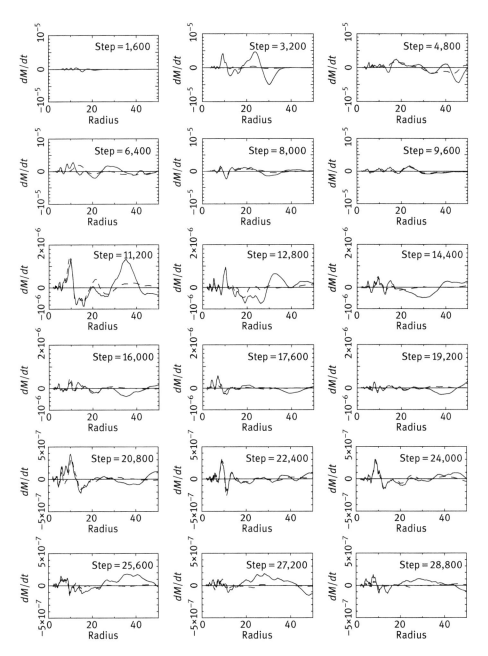

Figure 5.15: Radial mass flow rates for the N-body spiral/bar mode with $a_{\text{soft}} = 0.25$. Solid lines: full Fourier components. Dashed lines: $m = 2$ Fourier components. Note that the maximum positive vertical scale changes from 10^{-5} for the first two rows, to 2×10^{-6} for the second two rows, and to 5×10^{-7} for the third two rows, respectively (adapted from Zhang 2016).

is oftentimes more than to produce a visual appearance of spirals or bars, but rather to bring out the intrinsic common dynamics underlying the formation and evolution of a large class of these galaxies.

To study global collective effect thoroughly is inherently more challenging than to study local effects one at a time. However, without taking a globally self-consistent view, we will not be able to arrive at the forced singularity condition (in some sense, nature itself is *forced* into arriving at the collective dissipation process in these singularities when given conflicting requirements that cannot be resolved by a smooth differentiable solution). When taking a local view, it is like the fable of the blind men and elephant, we might mistake a tail, a trunk, a tusk, as the essence of the elephant. And even if we take stock of all of these pieces, we still cannot pile them together to arrive at a *live* elephant. Anything that is alive is the result of co-evolution between the system and the environment. The co-evolution process is when the correlations are gradually set up. These multitudes of correlations among the constituent parts are what give the system its livelihood and its evolutionary dynamics.

In this context we see that the so-called action-angle approach adopted by many previous practitioners of galactic dynamics (Binney & Tremaine 2008 and the references therein; Sellwood 2014 and the references therein), which follows a venerable tradition in classical mechanics and classical kinetic theory, exactly ignored this correlation among the constituent particles. It produces a system that is harmonic by design (that is what the angle part represents – that is, sinusoidal oscillations; and the action part represents conserved integrals). The validity of the action-angle approach depends on the assumption that most of the orbits supporting the galactic potential are made up of so-called regular orbits that respect the conserved integrals under the smooth average potential; whereas we learned in the current work that most of the orbits in the self-organized density wave potential are severely perturbed (by the collisionless shock) and thus are almost always chaotic (see Figure 3.24), and these chaotic orbits are associated with a grainy potential incorporating correlations. The closed-up harmonic system described by the action-angle approach will not display secular behavior (apart from orbit resonances, which is a local effect), at least not the type involving global collective dissipation processes which produce interparticle correlations that lead to mass inflow inside corotation and outflow outside.

These said, it is not the current author's assertion that the action-angle approach has no merits in studying galactic dynamical problems. There are at least three scenarios that this approach (or the associated phase space distribution function approach governed by collisionless Boltzmann equation) is in fact applicable: (1) when the system involved is in steady state and is not unstable to the growth of global modes. (2) When the system is unstable to the emergence of modes, but its behavior is first analyzed in the linear regime. In that case we are studying the modal growth regime with no secular evolution of the basic state. (3) In non-self-consistent studies of, say, gas response to an applied stellar potential, then the stellar potential itself can be regarded as smooth. One needs to be cautious about interpreting the results of this approach when, say, the self-gravity of the gas becomes important. It such cases the often-large phase shift between the stars and gas derived from the non-self-consistent analysis are not physical, and disappears when the self-gravity of the gas is included in the study (Lubow et al. 1986; Balbus 1988).

5.2 Broader Implications

In this section we discuss the implications of the current work on the self-organized structure formation process in a broader range of physical systems, as well as on the cosmological evolution of galaxies and structures.

5.2.1 Self-Organization in Nonequilibrium Systems and the Formation of Singularity Hierarchy

Understanding the self-organization (or morphogenesis) behavior of nonlinear, far from equilibrium systems is among the most challenging of problems of contemporary research. More than half a century ago, Richard Feynman closed volume II of "The Feynman Lectures on Physics" by declaring: "The next great era of awakening of human intellect may well produce a method of understanding the *qualitative* content of equations. Today we cannot. Today we cannot see that the water flow equations contain such things as the barber pole structure of turbulence that one sees between rotating cylinders. Today we cannot see whether Schrodinger's equation contains frogs, musical composers, or morality – or whether it does not" (Feynman, Leighton, & Sands 1964).

Among studies on morphogenesis, one of the most prophetic is that of Ilya Prigogine. In his theory of "dissipative structures," Prigogine emphasized the entropy-production-enhancing function of self-organized global patterns in far-from-equilibrium systems, as well as the constructive role of dissipation in maintaining these patterns (Prigogine 1980). To paraphrase this theory, most of the self-organized structures in nature come in roughly two types: (1) equilibrium structures, which are formed through equilibrium phase transitions. The examples of equilibrium structure formation include the phase transition of water to form ice when the temperature is lowered to 0°C, as well as the formation of minerals of a specific crystal structure when the environmental temperature and pressure satisfy a range of conditions (i. e., the formation of diamond crystal from carbon in a high-pressure environment); (2) nonequilibrium structures, which are formed in systems far from equilibrium, and which (in addition to sharing some common features with structures formed in equilibrium phase transitions) have the added feature that these self-organized structures are generally in a dynamical equilibrium state, meaning they are sustained through the competition of growth and decay tendencies (as highlighted by the well-known fluctuation-dissipation theorem, proved mostly at close-to-equilibrium regimes but is also valid at far-from-equilibrium situations for self-organized dissipative structures). There is in general also a continuous flux of energy and entropy through the system to maintain the nonequilibrium fluctuation. The formation of self-organized dissipative structures in nonequilibrium systems often serves the important function of greatly accelerating the speed of entropy evolution of the parent systems (Prigogine 1980)[†].

As we have seen in this work, the hierarchy of self-organization processes in a nonequilibrium system often leads to a series of effective singularities in the dynamics, which allow *emergent* new dynamics to form that cannot be *deductively* derived from the differential formulation one starts the analysis with. A *synthetic* approach uniting the various local aspects to achieve global self-consistency will need to be adopted.

Although the mechanism for the emergence of new dynamics through "spontaneous breaking of gauge symmetry" had been routinely proposed in high energy physics, it was mostly used in a model context, rather than derived from first principles. In such applications, perturbation expansion around a symmetry-breaking potential minimum (ground-state) of the model Lagrangian function is performed when studying excitations from such a ground state. This perturbation expansion generates a new effective Lagrangian with its associated new dynamics (for example, it could give mass to electrons through the Higgs mechanism); yet no self-consistent evolution of the global background environment (corresponding to the basic state of the galaxy problem), which is responsible to the form of the original Lagrangian in the first place, can be studied through such local algebraic manipulations. Quoting condensed matter physicist P.W. Anderson: "... the concept of broken symmetry has been borrowed by the elementary particle physicists, but their use of the term is strictly an analogy, whether a deep or a specious one remaining to be understood" (Anderson 1972).

On the other hand, the fact that, as of now, there have been very few examples of self-organized dissipative structures in low-energy physical systems analyzed from first principles (in Prigogine's work, chemical clock was used as one prominent example) is partly due to the intrinsic complexities of such problems, that is, the many degrees of freedom of the components, and the correlations among the components, which invalidate many basic assumptions underlying the usual kinetic-theory approach for treating many-particle systems. One of the most important assumptions

---

[†] The transportation of entropy to its environment is the chief means a dissipative structure can maintain a low or constant entropy state locally despite it being a very efficient engine at the local production of entropy. This aspect also resolves the paradox of how nature can generate complex biological entities such as human beings as a result of nonequilibrium evolution, despite being governed by the second law of thermodynamics universally. The study of nonequilibrium dissipative structures tells us that entropy-increasing evolution does not always mean the rush toward homogeneity everywhere, at least not for open, many degree-of-freedom, far-from-equilibrium systems.

used in the derivation of the Boltzmann kinetic equation is the so-called "molecular chaos" assumption (see, e. g., Kreuzer 1981 for a detailed description of the BBGKY procedure for the derivation of the collisional Boltzmann equation, which we briefly outline in the Appendix of this monograph), or the assumption that particle collisions are uncorrelated. This assumption is crucial to Boltzmann's arriving at his famous H-theorem, or that thermodynamic entropy (which is of the opposite sign as Boltzmann's H-function) never decreases in nonequilibrium processes. The interparticle correlations, however, are the necessary ingredient for obtaining self-organized behavior, and their reintroduction into the study of nonequilibrium dynamics allows local-entropy-decreasing processes to be admitted into the analyses.

Accompanying the emergence of global patterns, in an originally featureless system past its instability threshold, there is also the spontaneous emergence of new meta-laws. The torque-integral/mass-flow rate equation we have confirmed using N-body simulations in this work is an example of emergent meta-law. The original derivation of this law required the assumption of global self-consistency between the various physical processes in the galaxy, which is a logical necessity; as well as the assumption of QSS of the wave mode, which is an empirically well-supported hypothesis.

Given these two very reasonable basic assumptions, the outcome of the secular evolution of the basic state of the galactic disk is inevitable, given that the pattern's matter distribution and kinematics naturally lead to the characteristic PDPS pattern which implies a secular torque action by the wave mode on the basic state, and thus secular energy and angular momentum exchange between the two components. The operation of this exchange process is necessarily dissipative (as there is the irreversible conversion of stellar orbital energy into random motion energy of stars), and this operation is accomplished through the density-wave collisionless shock.

Generalizing what we have learned in this work, we realize that in nature irreversibility may be the rule rather than the exception, even though the differential forms of governing equations we commonly employ are all time-reversible. We now realize that these differential laws are likely to be idealizations, and their origins may be emergent under the general boundary conditions of the evolving universe. In this sense laws and meta-laws (or fundamental laws and emergent laws) have no clear distinction. It all depends on the scale we conduct our observations and analyses in.

We also realize that effective singularity hierarchies (as reflected, e. g., in the spontaneous emergence of density wave modes, and the attainment of new closure relations at the QSS) may be the general mechanism nature employs to separate the different domains of physical and biological sciences. Within a given hierarchy, deductive approach may be used to arrive at new inferences. However, when we are crossing the boundary of a given hierarchy (such as in the study of the emergence and maintenance of global spiral modes, as well as the secular evolution of the basic state of the galactic disk as a result of its interaction with these modes), we need to rely on empirical evidence (such as the QSS hypothesis of the mode), meta-principles (such as the second law of thermodynamics and the theory of dissipative structures), as well as reasonable global closure relations (such as global energy and angular momentum balance considerations) in order to establish a new axiomatic structure for subsequent analyses.

We now also have a tentative answer to the question Feynman raised: Schrodinger's equation most likely does not *directly* predict the existence of frogs, musical composers, and morality. Since the physical universe is organized in a hierarchical fashion with effective singularities separating the different hierarchies, and since emergent laws governing the higher-level phenomena cannot be *deductively* derived from the lower-level "fundamental laws," the claim that complex phenomena of the universe can be logically derived from a simple set of fundamental equations is unfounded. Instead, the differential forms of equations governing given regimes of physics are expected to be idealizations that have limited range of applicability and will break down at instability/singularity front of the next emerging hierarchy. In such cases, guided by empirical evidence and physical intuition, new global-self-consistency conditions and new meta laws (formulated perhaps as a new set of differential equations with their new range of applicability) need to be obtained to guide the analyses.

The universe since the Big Bang is in a process of irreversible evolution. If we take into account the subtleties of quantum mechanics, no process is truly reversible (i. e., no physical system can become truly isolated). Yet we have learned in this work that some processes are more irreversible than others, that is, those that involve self-organized structures that drastically increase the rate of entropy production, generate levels of irreversibility exponentially faster than processes that are passive. Dissipation is an important element of the formation of nonequilibrium self-organized structure because it helps to stabilize the dynamical equilibrium state by damping out random noise fluctuations.

5.2.2 Implications on the Cosmological Evolution of Galaxies

The secular evolution scenario describes mainly a morphological transformation process of individual galaxies. By itself this does not uniquely specify a cosmology. However, through its implications on the galaxy evolutionary pathways, the secular evolution scenario gives strong hints at the essential elements of any cosmology that is to be compatible with it. For example, the secular evolution scenario favors late-type disk galaxies to form relatively early in the cosmic history, and these late-type galaxies subsequently undergo morphological transformation along the (reverse) Hubble sequence mainly due to the mediation of *internal* dynamical processes related to global structures such as spirals, bars, and 3D twisted isophotes. Environmental effects can accelerate the evolution speed of individual galaxies, but these environmental effects operate mainly through the mediation of internal global instability structures, and not through the actual merging of galaxies.

5.2.2.1 Challenges Faced by the Lambda Cold Dark Matter Paradigm

The hierarchical clustering/cold dark matter (CDM) paradigm, including its most recent Lambda-CDM (LCDM) variant (Ostriker & Steinhardt 1995; Liddle 2003 and the references therein; Mo et al. 2010 and the references therein), has been the most popular paradigm for structure formation over the past few decades. Historically, CDM type of theories were invented partly to get around the problem that there was not sufficient time for the seeds of anisotropies observed on the cosmic microwave background, about one part in 10^5, to grow into the nonlinear structures we see today by gravitational means alone, since this would require primordial seeds of one part in $\sim 10^3$ at the time of recombination ($z = 1,000$). Furthermore, the standard Big Bang nucleosynthesis model also requires a significant amount of nonbaryonic dark matter (Dar 1995; Fields, Freese, & Graff 2000).

Both of the above problems in fact do not necessarily require the existence of nonbaryonic CDM. For example, the former problem can be solved in a primordial turbulence scenario (von Weizsacker 1951a; Gamow 1952; Ozernoy 1974a, 1974b; Bershadskii & Sreenivasan 2002, 2003; Bershadskii 2004), which leads to structure formation due not solely to gravity, but also to initial (possibly supersonic) velocities of the primordial matter[†]; while the latter situation can change significantly if the fundamental constants in the early universe vary from the values they have now, if the laws are themselves emergent and evolve with the evolving universe. (see further discussions in Section 5.2.2.4).

During past few decades of intensive search by international teams using a diverse range of instruments, no direct evidence of the existence of the purported nonbaryonic CDM particles has been found. Furthermore, growing evidence points to serious incompatibilities of the predictions of the LCDM paradigm with the accumulating observational results both on the large scale and on the individual galaxy level.

In general, the LCDM paradigm prescribes a medium for structural formation that is too clumpy on small scales, yet too smooth on large scales. For example, the predicted but unobserved cusps for the central density profiles of low surface brightness galaxies (de Blok 2010), as well as the predicted but unobserved overabundance of small satellites of normal galaxies (Klypin et al. 1999), are both due to the overclumpiness of the assumed LCDM medium on small scales. So is the problem of the formation of many small disks, which is related to the problem of rapid angular momentum loss of the small clumps of disk material due to dynamical friction (White & Frenk 1991).

On the other hand, the over-smoothness of the LCDM medium on largest scales underlies its inability to account for the early formation of quasars and first galaxies, as well as the early formation of giant high redshift clusters (Francis et al. 1997;

[†] This should not come as a surprise: even galactic star formation is not due solely to gravity: most of the stars form at spiral arms, now understood to be the site of collisionless shocks produced by the potential-density phase shift condition.

Steidel et al. 1998; Williger et al. 2002; Coe et al. 2013; Oesch et al. 2016). In a purely bottom-up structure formation scenario such as the LCDM, it is very difficult to account for the observed alignment of the spin axis of the bright galaxies in a cluster (Ozernoy 1974a and the references therein; Plionis et al. 2003); the alignment of substructures in a cluster (Plionis & Basilakos 2002); as well as the observed galaxy-cluster-supercluster alignment effect on the largest scale (West 2001). Other known observational evidence that conflicts that of the LCDM predictions include the large-scale velocity flows (Watkins et al. 2009; Kashlinsky et al. 2009), and the profiles of cluster halos (Broadhurst et al. 2005; Umetsu & Broadhurst 2008).

In the LCDM paradigm, large galaxies are formed out of the mergers of smaller ones, and therefore should form last. Recent studies of high z galaxies have shown that exactly the opposite is observed: that is, the larger the mass of a galaxy, the shorter the time scale of its formation (Boissier et al. 2001; Thomas, Maraston, & Bender 2002). This trend is the same as the so-called downsizing trend of galaxy formation, not only in terms of the star-formation history but also in terms of the mass assembly history (Cowie et al. 1996; Kodama et al. 2004; Cimatti et al. 2006; and the references therein). This trend is in direct conflict with the CDM-type bottom-up assembly scenario. While the former trend (downsizing in star formation) can perhaps be accounted for by proposed mechanisms such as AGN or star-formation feedback (see, e. g., de Lucia et al. 2006; Croton et al. 2006 and the references therein), the latter trend of mass-assembly downsizing (Cimatti et al. 2006) is in direct conflict with the mass assembly history predicted by LCDM, which is known to significantly underpredict the extremely red objects (ERO) observed in the early universe (Daddi, Cimatti, & Renzini 2000 and the references therein), with EROs likely to be mostly large, mature galactic systems.

Peebles (2002) compared the current state of cosmology with the state of physics at the turn of the nineteenth/twentieth century, and commented that several known problems of the CDM theories could potentially turn out to be the same type of "Kelvin-level clouds" which a century ago ushered in the quantum revolution. These problems include "the prediction that elliptical galaxies form by mergers at modest redshifts, which seems to be at odds with the observation of massive quasars at $z \sim 6$; the prediction of appreciable debris in the voids defined by L_* galaxies, which seems to be at odds with the observation that dwarf, irregular, and L_* galaxies share quite similar distributions; and the prediction of cusp-like dark matter cores in low surface brightness galaxies, which is at odds with what is observed" (Peebles 2002). Other more recent confrontations of LCDM predictions with observations are reviewed in Walker (2014).

The secular evolution scenario we have presented in this work poses its own serious challenge to the LCDM paradigm. The magnitudes of the mass flow rates we have derived for physical galaxies (i. e., those examples we have calculated in Chapter 4) appear large enough to be able to result in significant Hubble-type transformation during the past Hubble time (i. e., 1 – 10 solar masses a year of radial mass accretion corresponds to $10^{10} - 10^{11}$ solar masses during the past 10^{10} years, enough to build up a massive central bulge in a Milky-Way-type galaxy in the lower mass flow rate cases, and a disky elliptical in the higher mass flow rate cases). The calculated mass flow rates are especially large for galaxies experiencing strong tidal perturbations (i. e., NGC 3627 and NGC 5194), due to the large-amplitude, open-density wave patterns excited during these strong interactions, and due to the effective quadratic dependence of the radial mass flow rates on both the wave amplitude and on the pattern pitch angle.

From the galaxy rotation curves and mass flow rates we have presented in the last chapter, we see that along the Hubble sequence from late- to early galaxy types, more and more of the inner rotation curve is contributed by luminous baryonic matter (a previously known observational fact), and furthermore, this growth of the importance in the luminous matter in the central region of galaxies along the inverse Hubble sequence is likely to be a direct result of secular radial mass accretion process (a result that is new). A natural question then arises: what happens to the nonbaryonic CDM in the central region of a galaxy as this evolution proceeds? The nonbaryonic CDM particles could not simply be displaced by the baryonic matter since a well-known result of the LCDM numerical simulations is the so-called *adiabatic compression of dark matter halos* (e. g., see Sellwood & McGaugh 2005 and the references therein; Binney & Tremaine 2008, pp. 337–338), that is, as the baryons become more centrally concentrated, the dark matter particles should be dragged into the central potential wells thus formed, and also become centrally concentrated. Thus, taking together the facts of the observed decreasing importance of dark matter in the central regions of galaxies of progressively earlier Hubble types; the newly demonstrated importance of secular mass inflow process in transforming galaxy morphology from the late to the early Hubble types; and the expected adiabatic compression process *if* the CDM particles are indeed present, together imply that the main composition of the galactic dark matter could not be nonbaryonic CDM particles.

5.2.2.2 On the Baryonic Composition of Galactic Dark Matter

We propose that at least on the scale of galaxies and clusters, the dark matter is mainly composed of nonluminous Massive Compact Halo Objects (MACHO) of a continuous mass spectrum, a small portion of which at the substellar mass range had previously been revealed by microlensing observations (Alcock et al. 2000; Koopmans & de Bruyn 2000; Shvartzvald et al. 2016).

One may argue that the current microlensing surveys have not uncovered enough baryonic dark matter to account for what is needed to reproduce the galactic rotation curves (Tisserand et al. 2007), and that the extrapolated initial mass function (IMF) from star formation in the galactic plane does not seem to produce enough mass to close the gap either (e. g., see Najita, Tiede, & Carr 2000 and the references therein). We comment that the current microlensing observations may still be incomplete (especially regarding the environments this approach has so far been applied to); and some of the discrepancies from the IMF argument may be due to the fact that the forms of IMF in different environments are different, that is, in the disk plane the star formation is caused by spiral density wave shocks, thus producing a family of stars that weighs heavily toward solar and higher-mass stars. The structures formed in the quiescent environments of the halo and outer galaxy, as well as formed primordially through processes such as violent relaxation and primordial turbulence, on the other hand, may weigh more toward the lower-mass spectrum. Thus we cannot simply use the IMFs derived from star clusters in our own Galaxy to infer a universal IMF in more primordial formation environments, since the Galactic star clusters themselves are heavily reprocessed entities which possess higher degrees of dynamical relaxation/dynamical evolution. It is the process of reprocessing that allows the Galactic clusters to rid much of the primordial material, including presumably the dark baryons of substellar mass spectrum.

This scenario for the segregation of mass spectra in different environments is supported by the following known observational facts (see in addition Freeman & McNamara 2006 for a thorough historical survey of the dark matter research up to the publication of that book): (1) Dark matter is known to be more abundant in objects of primordial composition as well as lower degrees of dynamical relaxation, such as dwarf galaxies and low-surface-brightness galaxies (de Blok 2010 and the references therein), as well as other large but diffused galactic systems (van Dokkum et al. 2016 and the references therein). On the other hand, objects such as globular clusters, which are of similar masses as the smallest dwarf galaxies, but of much higher degrees of dynamical relaxation (as reflected in their much smaller diameters as compared to dwarf galaxies), are known to be almost devoid of dark matter. (2) On the scale of galaxy clusters, dark matter was sometimes found to be about 10 times as massive as luminous matter. On the individual galaxy level, excluding the very early Hubble types, dark matter is often between 10 and 100 times the mass of the luminous matter, *with the fraction of dark matter correlating strongly with the relaxation/evolution stage of a galaxy, rather than simply with its mass* (Freeman & McNamara 2006). These observed ratios of the dark matter versus luminous matter are all significantly higher than the cosmic nonbaryon-to-baryon fraction of 5:1 as inferred from Big Bang nucleosynthesis, as well as from the model-fitted CMB observations. This shows that one cannot avoid taking into account the baryonic composition of a significant fraction of dark matter, even if one adheres to Big Bang nucleosynthesis and the conventional wisdom regarding the matter composition of the universe. This conclusion of the inevitability of the presence of dark baryons has also been reached previously from several other lines of investigations (Persic & Salucci 1992; Fukugita, Hogan, & Peebles 1998; McGaugh et al. 2010).[†]

[†] The recent discoveries of a significant portion of the missing baryons as filaments between galaxies, and as hot gas in clusters, do not by themselves solve the missing baryon problem within individual galaxies. This is so especially for dwarf galaxies, where as much as 99% of the mass are undetected,

The current proposal for a significant fraction of the galactic dark matter to be composed of baryons makes the secular evolution picture more complete and more coherent: since most MACHOs have smaller masses, they can be displaced to the outer region of a galaxy as the galaxy's dynamical relaxation advances as a result of density-wave-induced radial mass flow – this is a kind of generalized equipartition or dynamical friction effect. This is supported by the observational fact that for early-type galaxies, the dark matter is preferentially distributed in the outskirt of a galaxy, whereas for later Hubble types the dark matter is more concentrated to the galactic central region. A small fraction of the dark baryons, that is, the fraction that resides close to the disk, could also be ignited through actual collisions while passing through the density-wave-crest collisionless shock, or be incorporated into the star-forming material, a process that also helps with the decrease of M/L ratio observed along the reverse Hubble-sequence, as the secular evolution progresses.

The new proposal for the nature of galactic dark matter also makes the Hubble-sequence galaxy structural properties fit comfortably together: The early-type galaxies have cuspy central mass distribution because the baryons there are more concentrated through secular evolution. In certain early-type galaxies, such as NGC 4736, it has been found that the contribution of dark matter to the total mass budget is very small (Jalocha et al. 2008). The late-type galaxies, including dwarf galaxies and low-surface-brightness galaxies, on the other hand, are dark-matter dominated but were often found to have core-shaped mass distribution, contrary to the LCDM prediction of central cusps (de Blok 2010). In the current scenario, since in late-type galaxies secular evolution has just been launched, not enough baryonic matter resides in the central region of a galaxy. Despite the dominance of dark matter over luminous matter in the central regions of late-type galaxies, these baryonic dark matter do not lead to cuspy density distributions since they have large random velocities.

The baryonic composition for dark matter fits well many known dark matter observations on large scales, such as the colliding Bullet Cluster (1E 0657-558, Clowe et al. 2006) and the Train-Wreck Cluster (Abell 520, Mahdavi et al. 2007; Jee et al. 2014), which are difficult to reconcile with the standard LCDM paradigm. For example, in the Bullet Cluster, the inferred dark matter distributions cluster with stars in the outer two lobes, and not with the gas in the middle of the collision remnant; whereas in the Train-Wreck Cluster dark matter clusters with both the stars in the outer two lobes

---

and the dark components in dwarfs are not likely to be in gaseous form. Dwarfs are known to consist mainly of the primordial material, i.e., with a closer-to-cosmic abundance for its mass content, compared to massive galaxies, which makes it even more compelling to assume that most of the dark mass in dwarfs are in fact dark baryons.

Furthermore, if stars and various forms of gas can already account for the entire cosmic baryon budget according to the Big Bang Nucleosynthesis (BBN) model, the contribution of significant additional dark baryons from individual galaxies will overrun the BBN baryon budget and cause a serious conflict with it: this makes the implications of our current work all the more prescient. Evidence from the current work suggests that the BBN model in its current form may not be suitable for constraining the baryon fraction of the universe, and the majority of galactic dark matter may in fact consist of baryons.

and with the gas in the central clump. A single nonbaryonic CDM species thus would have difficulty in reproducing both of these observed distributions. In our new picture, since the baryonic dark matter is hypothesized to be of MACHO type with a wide range of mass spectra (with its detailed composition possibly changing with environment), it can display variable degrees of dissipation. In the Bullet Cluster, the mass spectrum of the baryonic dark matter is likely to lean toward substellar; whereas in the Train-Wreck Cluster a more continuous spectrum from the substellar to the superdusty were likely to be present, and the collision dynamics serves as a graded sieve to separate and rearrange these mass components along a continuous path. Rather than due to the varying properties of self-interacting nonbaryonic dark matter as were hypothesized in models of colliding clusters, it is more likely that the varying properties of baryonic dark matter itself are what were responsible for producing the varying effective cross sections of the matter which formed these collisional remnants.

We comment here further that the reason for the preference of MACHOs as candidates for a significant fraction of the galactic dark matter, as opposed to nonluminous gaseous baryons that reside in the galaxy disk (e. g., Pfenniger & Combes 1994) is due, first of all, to the trend of properties seen along the Hubble sequence, that is, the late-type disk galaxies have the highest dark matter contribution to their rotation curves, and yet their (luminous) disks are the lightest. This is hardly consistent with the disk-dark-baryon hypothesis if the dark and luminous baryons come in fixed fraction. Secondly, from the modal theory of Bertin et al. (1989a, 1989b), density wave morphology along the Hubble sequence is linked to the galaxy's basic state properties. The fact that late-type galaxies have open and multiarmed spiral patterns requires that their parent basic state has a dynamically nonresponsive halo, that is, it requires the dark baryons to reside in a spherical configuration, consistent with the MACHO composition of dark baryons.

The observations of the Bullet Cluster are also inconsistent with the dark matter being made solely of a dissipative gas component (either cold or hot gas). Nor are the observations of the Bullet and Train-Wreck Clusters taken together consistent with the dark matter being mainly black holes/neutron stars, or else massive neutrinos. The latter, as a hot-dark-matter candidate, also conflicts with the observational evidence that on the galactic (and oftentimes also cluster) scales the dark matter is found to cluster mostly with the luminous matter (Freeman & McNamara 2006).

5.2.2.3 Formation of Realistic Galaxy Disks in the Cosmological Context

The present work concerns mostly with the long-term evolution of the morphology of individual galaxies once a realistic basic state (i. e., the initial-boundary condition for galaxy evolution) has been specified, and it does not specifically address the formation of galaxies in the cosmological context. However, some of our results do shed light on the possible reasons for the past difficulty of forming galaxies with realistic rotation curves and morphologies in the LCDM cosmological simulations (Navarro 2010; White 2009, 2010).

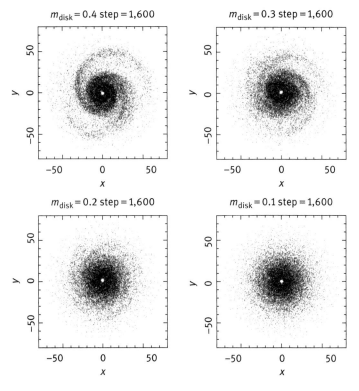

Figure 5.16: N-body morphology of the spiral bar mode with differing disk mass fraction, and with a computation grid similar to that used in Zhang (1998), at time step 1,600. Pattern speed is 0.011 radian per step.

In Figure 5.16, we plot the results of the modal morphology at time step 1,600 (when the spiral pattern is close to being most prominent), from a simulation using a computation grid similar to that used in Zhang (1998), i.e., 55 radial divisions and 64 azimuthal divisions, and a softening length of 1.5. We use a set of basic state specification that varies the active disk to rigid halo mass ratio, with the rigid bulge mass held fixed at 0.1, and with the total galaxy mass held fixed at a normalized value of 1. The four panels in Figure 5.16 have $m_{disk} : m_{halo} : m_{bulge}$ varying from 0.4 : 0.5 : 0.1, to 0.3 : 0.6 : 0.1, to 0.2 : 0.7 : 0.1, and finally to 0.1 : 0.8 : 0.1, respectively. The length scale of the disk, halo, bulge in the normalized units are 10, 5, 1, respectively, as in Section 3.2.1.

From the four panels of Figure 5.16, we see that as the ratio of the mass of the rigid halo component to the active disk component increases, with the mass of the rigid bulge component itself held fixed, the spiral mode (shown near its maximum modal amplitude stage) decreases progressively in prominence. For the last two frames, i.e., for the mass of the disk having 20% and 10% mass ratio, respectively, in terms of the

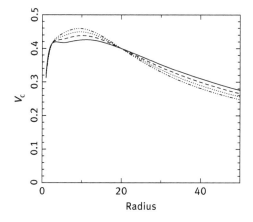

Figure 5.17: Rotation curves for the four runs with differing active disk mass. Solid line: $m_{\text{disk}} = 0.4$; dashed line: $m_{\text{disk}} = 0.3$; dotted line: $m_{\text{disk}} = 0.2$; dash-triple-dotted line: $m_{\text{disk}} = 0.1$. The total galaxy mass is normalized to 1.

total galaxy mass, the spiral pattern is barely discernible. The rotation curves of these four runs are given in Figure 5.17, which are seen to have varied little. Decreasing the softening length does increase the prominence of the spiral patterns in all four frames, but the trend of decreasing spiral prominence with decreasing disk-to-spheroidal ratio is unchanged when varying softening.

Yet, from observations, we know that many light disks (e. g., our own Galaxy's disk) do contain prominent spiral structures. How do we explain this discrepancy? We note here that the simulations presented in this work are done in two-dimensional configurations, with the bulge and halo both held as dynamically inactive. There is no back-response of the halo and bulge to the formation of the global spiral/bar modes in the active disk. In real galaxies, if we take into account of the fact that the bulge and halo components are likely both made of baryons, we can expect the active participation of both these components to the formation of global modal instabilities, i. e., both the bulge and halo of a disk galaxy are expected to be partially rotationally supported, skewed, and participate in the self-organization and collective dissipation of the single global modal set that unites the dynamical behavior of the disk and the spheroidal components.

Therefore, we must interpret the active disk mass presented in the simulations of this work as incorporating also the dynamical effect of those portions of the bulge and halo that are actively participating in the global modal dynamical behavior of the entire galaxy.

The LCDM halos, on the other hand, contain a high proportion of the nonbaryonic dark matter particles that are pressure supported, thus cannot participate in the self-organization and collective dissipation process of the global density wave modes. This explains why past LCDM simulations of disk galaxy formation either produced spiral structures accompanied with high central baryon content and (unrealistically) high central rotation curves, or else produced realistic rotation curves but no spiral structures when the baryonic content is maintained as per the LCDM prescriptions (Navarro 2010; White 2010 and the references therein). The triaxiality of the LCDM halos also leads to the secular tumbling of the embedded baryonic disks within the CDM halo potentials (Yurin & Springel 2015), further endangering the long-term survival of especially the nested density wave modal sets (i. e., bar within spiral, or vice versa), observed to have overwhelming presence in nearby grand design disk galaxies, such as NGC 4321, which we have analyzed in Chapter 4.

Our above view is consistent with the recent observations of Ibata et al. (2013) of a plane of dwarf galaxies corotating with the Andromeda galaxy, a fact that a pressure-supported dark halo would have difficulty in reproducing. Similar observations by Pawkowski, Pflamm-Altenburg, & Kroupa (2012) for the distributions and kinematics of satellites, globular clusters, and streams of the Milky Way Galaxy also pose severe challenges to the standard LCDM galaxy formation scenario.

The difficulty of using the LCDM prescriptions to form disk galaxies with realistic properties thus appears to be intrinsic, and *ad hoc* procedures of using artificially imposed feedback processes to get around the known difficulties of LCDM (see, e. g., Brooks 2014 and Vogelsberger 2014, as well as the references therein) ended up mimicking baryon-dominated galaxy dynamics. In a way, the recent progress on the formation of disk galaxies in the LCDM context only highlighted the close ties of the observed galaxy morphology and kinematics, as well as their inferred evolution history, to the baryonic compositional characteristics of their galaxy mass.

We comment here that our assertion that a Hamiltonian system is capable of exhibiting dissipation-like behavior can apply to the dark-matter component of the mass as well, in principle, just as it applies to the stars that are nominally dissipationless. The real issue here is not whether the mass component is baryonic or not, but rather a matter of initial conditions. Our particle simulations presented in Chapter 3 are able to display self-organization and collective dissipation behavior, because our initial-boundary condition (i. e., the basic state) specification is that of a differentially rotating disk that allows the spontaneous emergence of a global density wave mode. Such an initial condition is in fact the end stage of a dissipative collapse of a primordial galactic clump, which involved the dissipation of the gas component that subsequently formed stars. A spherically distributed particle cloud, even made with stars and not with non-baryonic dark matter, and without mass segregation, will not spontaneously form a differentially rotating disk. Thus, the possibility for the secular morphological evolution of galaxies through density-wave-mode-induced collective dissipation process in fact implicitly benefited from the earlier dissipational process of disk formation (as well as the formation of skewed baryonic halo, thick disk, and primordial bulge, as star formation proceeds within the skewed parent gas distributions in these galactic components).

The non-baryonic dark halo, on the other hand, because of its heavy mass compared with the baryon content, did not benefit much from this earlier dissipative collapse process to form the galactic disk. So the lack of participation of the CDM dark halo in the collective process of the disk is not because of impossibility as a matter of principle, but rather because of the actual characteristics of the halo (i. e., its massiveness and its initial three-dimensional distribution, which is a high-entropy configuration). The dissipation-like behavior possibly exhibited by a Hamiltonian system is necessarily tied to a low-entropy initial configuration (i. e., think of the paradox of the mixing of gas initially confined to the left compartment of a two-compartment box, with the partition subsequently removed. The gas particles can be governed by reversible equations, but the initial condition dictates that the entropy will increase irreversibly after the partition is removed). Any self-gravitating system (including the dark halo) has an arrow of time (Lynden-Bell & Wood 1968), but the speed for entropy-increasing evolution can be very low without its being amenable to manufacturing entropy-production-enhancing dissipative structures.

5.2.2.4 Generalize Mach's Principle and the Origin of Fundamental Laws

As we have seen, the challenges faced by the LCDM paradigm are significant. In other branches of natural sciences and in mathematics, similar challenges had resulted in scientific paradigm shifts in the past. Besides the examples of the quantum revolution and the discovery of relativity theories, we also know that a fundamental paradox in set theory, similar to the so-called barber's paradox, discovered by Bertrand Russell at the beginning of the 20th century, resulted in a paradigm shift in the foundation of mathematical logic; and Godel's incompleteness theorem in the 1930s revolutionized our views on the limit of mathematical proof and the limit of the axiomatic construction approach. The challenges faced by the LCDM paradigm may turn out to be of the same nature, as Peebles had earlier speculated. This crisis, however, is also a potential opportunity, as it may serve as the impetus for us to update our theories regarding fundamental aspects of nature.

Many previous workers in the cosmological community have made attempts at finding alternatives to the CDM-based cosmological theories. Most of these, such as the hot or warm dark matter theories, as well as the self-interacting dark matter theories, are still based on the idea of the nonbaryonic dark matter composition. The results and implications of the current work, on the other hand, prompt us to look beyond the confines of the current cosmological studies to reflect on the state of knowledge of our fundamental physical theories as a whole.

As we have seen in the previous chapters, new meta-laws (i. e., the volume–torque closure relation) governing the evolution of galaxy disks spontaneously emerge when a new hierarchy of organization (i. e., the quasi-steady density wave modes) appears. On the scale of the whole universe, we could likewise speculate whether all of our physical laws might be emergent, together with the emergence of patterns of organization in the natural world.

The success of applying the ideas of spontaneous symmetry breaking in both condensed matter and high-energy physics gives us confidence to generalize this concept to other branches of physical sciences, but with a significant caveat: the past application of spontaneous symmetry breaking in quantum field theories is a purely local procedure – as the quantum field theory itself mostly is, which may account for the need of renormalization. Whereas in the current work, we have seen that spontaneous symmetry breaking and the emergence of new meta-laws are oftentimes a global and distributed process: the torque and mass-flow rate equations are expressed in terms of basic state and density-wave modal parameters, which are determined by a global feedback cycle and distributed dissipation processes. Furthermore, the tight inter-relations between the emergent laws and the matter distribution in the substratum in such a globally self-consistent treatment also lead to the co-evolution of the emergent laws and the substratum with time, an element that the local field theories by nature could not take into account. It is thus obvious that our current model for fundamental interactions of nature, i.e., the standard model of elementary particle physics, had not exhausted the novelties of interactions in nature, because of its intrinsic inability to account for globally emergent processes in a self-consistent manner. Without a globally consistent model of the universe, our patchy theories in various physical disciplines are inevitably left with many empirically-determined structure constants. These models with fitted constants have validity in their respective domains of application, of course. Our cautionary note here is mainly against unwarranted complacency with regard to the current state of our knowledge of the fundamental aspects of physical interactions, to point out the limitations of local treatments in determining the values of fundamental constants, as well as to highlight the prospect of these constants changing with time.

Examining closer in this direction, the fact that all fundamental laws in our universe are Lorentz-covariant (i. e., they satisfy Lorentz transform relations) points to the likelihood of a relational origin of these laws. As is well known, the Lorentz transform relations apply to flat space-time (i. e., they are special-relativistic transform relations). Observational cosmology points to the likelihood of our universe being nearly flat. Yet how did our fundamental laws become aware that the universe is flat on a global scale, if the laws themselves are not emergent (i. e., relationally generated) from the underlying matter and energy distribution? The possible globally emergent nature of fundamental laws is also reflected in the action principle which underlies the derivations of these laws; in the universality of fundamental constants appearing in these laws; as well as in the quantized nature of fundamental exchanges, which hints for its origin at a global resonant cavity of finite size, likely the universe itself. The hypothesis that the origin of natural laws lies in the matter and energy distribution in the whole universe, has been referred to as a "Generalized Mach's Principle" by the current author, in reference to a more restricted version due to Ernst Mach positing the origin of gravity and inertia in the global matter distribution.

Until we obtain an understanding of the origin of natural laws from this global and relational point of view, it is premature to claim that we have in our possession all the fundamental principles, and what is needed from here on is only to carry out "precision cosmology." We need to keep an open mind of the provisional nature of our cosmological models. Some existing models of the universe may demonstrate the exclusion of the baryonic composition of dark matter, but models are merely models, they are not the final arbiter of truth.

5.2.2.5 Secular Evolution as an Emerging Paradigm for the Cosmological Evolution of Galaxies

It is in the spirit of the aforementioned considerations that the emerging secular evolution paradigm could supply an added reality check when developing and testing our cosmological models. In the results presented here, a clear picture emerges of the secular morphological transformation of galaxies driven mostly by collective effects mediated by large-scale density wave modes. This evolution is a well-coordinated quasi-steady evolution, with the formation and transformation of large-scale density wave modes compatible with the evolving basic state configurations. We thus expect that at every stage of the transformation from the late Hubble type to the early Hubble type, there should be a good correspondence between the basic state properties and the characteristics of the density-wave modes, except during brief periods in a galaxy's life when it encounters strong gravitational perturbation from a companion galaxy. New equilibrium states are expected to be restored once the perturbation ceases and the galaxy adjusts its mass distribution and modal pattern to be once again mutually compatible – the reason these new equilibrium states are nearly always possible is because for every basic state configuration there is almost always a set of unstable modes corresponding to it (Bertin et al. 1989a, 1989b).

As a result of this coordinated co-evolution of the basic state of the disk and the density-wave modes, over the major span of a galaxy's lifetime the effective evolution rate depends on the stage of life a galaxy is in. Late- or intermediate-type galaxies, having larger amplitude and open-density-wave modes in their outer disks, would lead to larger mass flow rates and thus faster secular morphological transformation rates; whereas for early-type galaxies the secular evolution activity is shifted to the central region of a galaxy. In general, the secular mass flow rate in the outer disks of early-type galaxies is low, and this is reflected in the modal characteristics as well, i. e., early-type galaxies generally have either tightly wrapped spiral arms, or straight bars, both correspond to small phase shift and thus small secular mass flow rate.

On the other hand, the secular evolution rate also depends on the interaction state of a galaxy, with the galaxy in group or cluster environments generally having larger evolution rates due to the large-amplitude, open, spiral and bar patterns excited during galaxy interactions. We have shown earlier in this text that interaction-enhanced, density-wave-mediated evolution appears to underlie the so-called morphological Butcher–Oemler effect (Butcher & Oemler 1978a, 1978b) in rich clusters, as well as the morphology-density relation originally discovered by Dressler (1980), and subsequently shown to hold over more than 4 orders of magnitude in mean density, spanning environments from field and groups, onward to poor clusters and rich cluster outskirts, all the way to the dense cluster central region. The universality of such a correlation is in fact a powerful illustration of the unifying role played by "nurture-assisted nature" type of processes during galaxy evolution. The dependence of morphology on environmental density shows that nurture plays an unquestionable role in determining the average evolution rate; yet the well-known inverse correlation of environmental density with merger rates shows that the effect of the environment is not chiefly in the form of violent cannibalism, but rather through mild tidal perturbations. Even the moderately violent and episodic "harassment" type of direct damage to galaxy morphology is known to be ineffective at transforming the morphology of large disk galaxies: they are shown to mostly lead to the transformation of small disks to dwarf spheroidals. These leave the most obvious driver for the observed morphological transformation of galaxies in different environments, and the origin of the morphology-density relation, as the tidally enhanced density waves. However, as we see in the current work, even for interacting galaxies like M51 and NGC 3627, the environment exerts its effect through the innate mechanism already present in individual galaxies, i. e., the density waves that are excited during the interaction are mainly the *intrinsic modes*, and the effect of interaction only enhanced the amplitudes of these modes.

The flip side of this correlation of evolution rate with environment is that some disk galaxies in isolated environments could have very small secular evolution rate throughout the span of a Hubble time, which explains the observational fact that many large disk galaxies in the fields were found to have evolved little during the past few Gyr. Still, isolated environment cannot be automatically equated to a slow evolution rate. One of the most impressive examples of mass redistribution we have encountered is the strongly barred galaxy NGC 1530, which lies in a surprisingly pristine environment, i. e., nearly totally isolated. Yet NGC 1530 has by far the largest mass flow rate of all the galaxies we have calculated, of more than 100 solar mass per year, due to its large surface density and the presence of a strong set of bar-spiral modal structure. This shows that the initial conditions that galaxies inherited at birth are likely to have played an important role as well in determining the subsequent morphological evolution rate.

The secular morphological transformation of galaxies along the Hubble sequence also implies that the required external gas accretion rate to sustain the current level of star formation rate in disk galaxies can be much reduced from the amount previously sought: as a galaxy's Hubble type evolves from late to early, more and more of its store of primordial gas will be exhausted, and its star formation rate will naturally decline. But this is precisely the observed trend of star formation rate in galaxies along the Hubble sequence, i. e., early-type disk galaxies do not have nearly as much star formation activity as late-type galaxies. Galaxies thus do not have to sustain their "current" level of star formation rate over cosmic time, since what is current today will be history by the next phase of their morphological evolution.

Finally, we note that the continuous evolution across the S0 boundary into disky ellipticals may erase the distinction between pseudo-bulges and classical bulges, i. e., galaxies such as the Milky Way were previously thought to have classical bulges because of its $r^{1/4}$ central density profile, yet the building up of the Galactic Bulge is most likely through the secular mass accretion process over the past 10 billion years, and the disky bulge can further relax into the $r^{1/4}$-shaped bulge with time. In the end, as has already been pointed out by Franx (1993), there might not be a clear distinction between early-type disks and disky ellipticals, and only a gradual variation of the bulge-to-disk ratio. The recent result from the ATLAS^{3D} project of a significant disk component in all low-mass ellipticals (Cappellari et al. 2013) provides strong support to this continuous evolution trend from late-type disk galaxies all the way to disky ellipticals. Buta et al. (2010)'s result from analyzing the internal structures of early-type galaxies also provides supports for continued secular evolution among early-type disk galaxies, which further highlights the role of *stellar* mass accretion, since the early-type galaxies have progressively less gas.

The secular evolution mechanism presented in this work is expected to be one of the most important mechanisms in producing the Hubble sequence. This is both because it is a physical mechanism intrinsic to the individual galaxy and is operating throughout the lifetime of a galaxy – including those that have been through a merger – irrespective of the environment it is in; and also because of the recognition of spiral structure in disk galaxies as an example of the general class of "dissipative structures" in nonequilibrium systems. The important function of any dissipative structures is to *greatly accelerate* the speed of entropy evolution of the underlying nonequilibrium system, and this special function is also achieved by the galactic spiral structure, i. e., the emergence of a galactic spiral structure greatly accelerates the speed of entropy evolution of a disk galaxy (Zhang 1992) along the Hubble sequence from the late to the early types.

The above being said, we want to emphasize also that many other previously proposed mechanisms of galaxy formation and evolution, i. e., the variations in the initial density perturbation (Eggen, Lynden-Bell, & Sandage 1962), the initial dissipative collapse of a proto-galactic clump (White & Rees 1978), and subsequent angular momentum acquisition via tidal torques (Hoyle 1953; Fall & Efstathiou 1980), as well as mergers of spiral galaxies, satellite and gas accretion, are also expected to be operating with varying strengths in contributing to the formation of the Hubble sequence, in addition to the internal secular evolution process that the current monograph has been focusing on. Furthermore, some of the amorphous proto-galaxies seen in the Hubble Deep Fields (HDFs) may be a class in their own right, i. e., they are neither the merger product of two late-type disks, nor are galaxies that are evolving into disk galaxies. Some of these objects may bypass the phase of formation of a well-defined thin disk, and evolve into spheroidal systems with the assistance of some three-dimensional global instability structures. This is supported by the observational fact that many of the high-z amorphous galaxies in the HDF images possess twisted isophotes in their luminosity distribution (Fasano, Filippi & Bertola 1997). A central thread that connects all of the above galaxy evolution scenarios, however, is the principle of entropy-increasing evolution for an open and nonequilibrium system, and the ability for self-gravitating systems to manufacture "tools" to enhance the rate of entropy production and transport, these tools being either quasi-two-dimensional as in disk galaxies, or quasi-three-dimensional as in the twisted isophotes in disky ellipticals and in amorphous proto-galaxies.

6 Concluding Remarks

We are now coming to the end of this monograph, in which we have presented a diverse range of astrophysical phenomena that can be explained and placed in a coherent context by a deepened understanding of cooperative behaviors of gravitational many-body systems governed by emergent laws of self-organization and collective dissipation.

The current work highlights the fact that in dealing with complex systems, sometimes a purely reductionist, mean-field approach will not suffice. We must supplement the deductive analytical procedures with a synthetic approach, by employing the wealth of empirical information to constrain our analyses, and by constantly re-examining the rules we have inherited from few-body dynamics as well as from continuum treatment. Finally, we can verify our intuitive understanding through numerical simulations – though this endeavor itself needs to be guided by deepened physical understanding at all stages; and through comparison with observational results.

As advocated in Zhang (2016), the current work supports the trend of a paradigm shift of "From Being to Becoming" (using the title of Prigogine's 1980 book), the "Being" type of study is the older paradigm of using action-angles to describe a system that is immutable to change (apart from local resonant interactions), whereas the "Becoming" type of study is one which incorporates the inherent self-organizing power of a many-body system, which drives its long-term evolution through coordinated interactions of its component parts. The coordination which makes possible the correlation among component particles is not done in a top-down fashion, but is controlled by the system's intrinsic instability dynamics, which is akin to the "invisible hand" analogy used by economists, for understanding the long-range order which can emerge out of seemingly random interactions of component systems responding only to short-term and local interests. This paradigm shift is aligned also with the one mentioned by Feynman we quoted in the previous chapter, that of the switch from the study of *quantitative dynamics* to *qualitative dynamics*: the qualitative dynamics *is* the emergent dynamics of a self-organized dissipative structure arrived at through spontaneous symmetry breaking. Methods of our former inheritance, of taking things apart, studying one aspect at a time, without worrying about the correlations among the component parts; or of employing only the linear and deductive dynamics, would not be adequate in the studying of evolutionary problems of many degree-of-freedom systems possessing self-organized dissipative structures.

Nature and its laws are organized into hierarchies. This is the underlying reason that we can divide the physical sciences into different branches. Across the boundaries of these hierarchies, singularity-like behavior halts the validity of laws that apply in one hierarchy, and new laws emerge in the domains beyond. The singularity behavior in the case of global density wave patterns in galaxies is reflected in the collisionless-shock nature of spontaneously formed density wave modes, which makes laws that were valid prior to the emergence of these patterns cease to be so. To derive emerging laws governing the behavior of structures after the emergence of singularities, we need to rely on a combination of empirical evidence and global self-consistency requirement of the system dynamics as a whole. In the case of galaxies, the major piece of empirical evidence is that density wave patterns appear to be quasi-steady modes in the underlying basic state of galactic disks. The global self-consistency requirement then leads to the conclusion that the angular momentum deposition/pickup by the wave mode will be used to feed the secular mass redistribution of the basic state stars and gas. This conclusion is supported by the manifested potential-density phase shift distributions in observed galaxies, as well as the correspondence of phase-shift positive-to-negative zero crossings with galaxy resonance features. In order for the secular evolution process to proceed in a quasi-stationary configuration, we found that we had to accept not only the failure of the continuum (differential) formulation once the collisionless shock forms, but also the emergence of a new closure relation governing the global angular momentum exchange and transport. The underlying physical process is in essence similar to the concept of spontaneous breaking of gauge symmetry in generating new physical laws, though high-energy physicists used this procedure mostly in a local and model context, which precluded the consideration of self-consistent evolution of the background environment, i.e., the universe itself. Symmetry breaking and the resulting singularity hierarchy are nature's ways to provide order and stability amidst a wealth of possibilities and chance factors.

The emerging field of research on the self-organized behaviors in gravitational many-body systems promises many rich returns. To make further progress, we can be aided by the fruitful research already conducted in other physical sciences dealing with many-body cooperative behavior, ranging from fluid turbulence research to condensed matter physics. By the same token, the lessons we learned in galaxy studies can also "give back" to the studies of these other complex systems, in especially the aspect of evolution of the basic state (i.e., the underlying global boundary condition), as part of the process of self-organization and collective dissipation which serves to maintain a nonequilibrium quasi-steady perturbation. We will find that all physical systems behave in surprisingly similar *trends* in terms of their long-term entropy evolution process, partly a result of global organizational principles such as those governing the evolution of dissipative structures first discovered by I. Prigogine and coworkers. These global principles or meta-laws often are insensitive to the detailed microscopic laws governing the component-level behaviors (this brings to mind the analogy with the different representations of an identical mathematical group: in fact, mathematical groups and algebras capture many aspects of broken symmetry in the natural world, with each successive broken symmetry reducing the number of degrees of freedom of the previous organizational level, and with the internal properties of a given organizational level reflected in the structure constants of, say, a Lie-algebra).

We may also find that the practice of enforcing global self-consistency requirement, such as that used by Kolmogorov in arriving at the eddy-cascade scaling relations for fully developed turbulent flows in the inertial range, or that used in the current work in deriving the closure relations at the quasi-steady state of the density-wave modes, is also likely a universally valid approach for dealing with self-organized phenomena. This approach is akin to our common practice of using energy, momentum, angular momentum, and mass conservation relations to study the evolutionary behavior of a particle system, without always needing the Newtonian equations of motion to solve detailed particle trajectories – though in this latter case the two approaches are equivalent and both can provide unique insights, whereas in the case of crossing of the boundaries of a singularity hierarchy the global approach is often the only viable one when the differential/mean-field procedure fails at the effective-singularity shock front. We emphasize also that the continuum approach, even with its deficiency in terms of modeling the microscopic behaviors of particles within a global-instability shock front, is nonetheless needed (especially when coupled with the global balance-equation point of view) when we attempt to obtain a structured understanding of self-organizational phenomena. A pure particle approach (such as N-body simulations) tends to lump all processes together without separating the different organizational hierarchies. Furthermore the continuum approaches, as we have demonstrated in Chapter 2, often possess inherent ability to indicate its own failure at the instability threshold, signalling the necessity of introducing new, global perspectives. Therefore, we should employ all our available intellectual arsenal when attacking the self-organization problems in complex systems.

Once we have established a firm foundation for the explanation of cooperative behaviors in galaxies, we also need to take the conclusions we draw from it seriously, and explore their implications to our understanding of large-scale structure and cosmological processes. In doing so, the insights we gained on galaxy-scale processes will help us make sure-footed progress toward understanding wider problems in the physical universe.

7 Appendix: Relation to Kinetics and Fluid Mechanics

In this Appendix we attempt to place the dynamics of nonequilibrium phase transition, as applied to galaxies containing density wave modes, in the context of the well-known treatments in classical kinetic theory and in fluid mechanics. This process will help us see more clearly where the new breakthrough was made with respect to traditional treatments. We will also discuss the nature of emergent laws in the context of the natural hierarchies of organization in the physical universe.

7.1 Foundation of Kinetic Theory: The Boltzmann Equation

One of the central aims of the classical kinetic theory is the description of the process of a slow approach to equilibrium of a perturbed system, after the initial fast transients have died down and an approximate continuum description in terms of the N-particle distribution function in phase space can be employed. The founding father of the classical kinetic theory is Ludwig Boltzmann, who established the statistical foundations of the classical kinetic theory of a dilute gas (Boltzmann 1872), in the form of the celebrated Boltzmann equation with its demonstrated property of the secular increase of entropy.

Self-gravitating galactic systems obviously are not equivalent to the dilute gas problem investigated by Boltzmann, yet past galactic dynamical studies employing various approximate forms of the Boltzmann equation and its moment-equation descendants had met with successes in many applications. We will now first take a closer look at the derivation and the underlying assumptions of the Boltzmann equation, in order to appreciate the reasons for its successes and limitations in treating galactic problems.

7.1.1 Outline of the Derivation of the Boltzmann Equation Through the BBGKY Hierarchy

Even though the original derivation of the equation bearing his name was made by Boltzmann using heuristic approaches, a modern and more rigorous derivation makes use of the so-called BBGKY hierarchy (Bogolyubov 1946; Born & Green 1946; Kirkwood 1946; and Yvon 1937). Detailed discussions of this derivation can be found in Green (1952), Uhlenbeck & Ford (1963), Harris (1971), or Kreuzer (1981). In what follows we give a brief outline of this modern derivation to highlight the crucial assumptions made in arriving at the well-known form of the so-called "Boltzmann collision term," and why the physics at the far-from-equilibrium regime that is relevant to the self-organized dissipative structures could not have been incorporated in Boltzmann's classical treatment.

A statistical description of an N-particle system can be obtained from its ensemble probability function $\rho(\vec{x}_1, \vec{x}_2, ..., \vec{x}_N, \vec{p}_1, \vec{p}_2, ..., \vec{p}_N, t)$, which represents the probability of finding an ensemble element in a region of the $6N$-dimensional phase space $(\vec{x}_1, \vec{x}_2, ..., \vec{x}_N, \vec{p}_1, \vec{p}_2, ..., \vec{p}_N)$ at time t. The equation of motion for this probability function is the well-known Liouville's equation

$$\frac{\partial \rho}{\partial t} = \sum_{i=1}^{N} \left[(\nabla_{x_i} H_N) \cdot (\nabla_{p_i} \rho) - (\nabla_{p_i} H_N) \cdot (\nabla_{x_i} \rho) \right] \equiv \{H_N, \rho\} \tag{7.1}$$

where H_N is the hamiltonian of the system (defined later in equation (7.4)), and the curly brackets denote the Poisson bracket of the relevant variables. This equation is a direct result of Hamilton's dynamical equations and the continuity equation for the evolution of the phase space density.

To derive the kinetic equation of Boltzmann from Liouville's equation, a systematic procedure through the so-called BBGKY hierarchy is employed. At the lowest level of this hierarchy, an N-particle distribution function $f_N(\vec{x}_1, ..., \vec{x}_N, \vec{p}_1, ..., \vec{p}_N, t)$ is defined, which is proportional to the probability function ρ in Liouville's equation, with the normalization

$$\int f_N(\vec{x}_1, ..., \vec{x}_N, \vec{p}_1, ..., \vec{p}_N, t) d^3x_1, ..., d^3x_N, d^3p_1, ..., d^3p_N \equiv V^N, \tag{7.2}$$

where V is the physical volume of the system, and the integration is over the entire phase space volume. In general, a reduced l-particle distribution is defined as

$$f_l(\vec{x}_1, ..., \vec{x}_l, \vec{p}_1, ..., \vec{p}_l, t) \equiv$$
$$V^{l-N} \int f_N(\vec{x}_1, ..., \vec{x}_N, \vec{p}_1, ..., \vec{p}_N, t) d^3x_{l+1}, ..., d^3x_N, d^3p_{l+1}, ..., d^3p_N. \tag{7.3}$$

Assuming the system of N particles is interacting through two-body mutual potentials of $V_{i,j} = V(\vec{x}_i, \vec{x}_j)$ as well as the external potential $\Phi(\vec{x}_i)$, the Hamiltonian of the system is thus

$$H_N = \sum_{i=1}^{N} \left[\frac{p_i^2}{2m} + \Phi(\vec{x}_i) \right] + \sum_{1 \leq i < j \leq N} V_{i,j}. \tag{7.4}$$

Making use of the exchange symmetry of the distribution functions with respect to particle indices as well as their normalization properties, after partial integration of the original Liouville equation over $(N - l)$ coordinates the derived BBGKY hierarchy of equations become

$$\frac{\partial f_l}{\partial t} = \{H_l, f_l\} + \frac{N-l}{V} \int \sum_{i=1}^{l} (\nabla_{x_i} V_{i,l+1}) \cdot (\nabla_{p_i} f_{l+1}) d^3 x_{l+1} d^3 p_{l+1}, \tag{7.5}$$

where

$$H_l \equiv \sum_{i=1}^{l} \left[\frac{p_i^2}{2m} + \Phi(\vec{x}_i) \right] + \sum_{1 \leq i < j \leq l} V_{i,j}, \tag{7.6}$$

and the curly brackets once again denote the Poisson bracket.

The set of BBGKY hierarchy equations is formally equivalent to the Liouville's equation. To solve an equation for reduced distribution function at a lower l, however, a higher order $(l + 1)$ reduced distribution function is required, which goes on until order N, and this procedure thus couples reduced distribution functions of all orders. Therefore, in order to make use of the economizing power of the hierarchy, ways to truncate the hierarchy need to be sought. This was indeed how the BBGKY hierarchy was previously used to derive the Boltzmann equation, as we will continue to outline below.

The lowest-order ($l = 1$) equation in the hierarchy can be written as

$$\left[\frac{\partial}{\partial t} + \frac{\vec{p}}{m} \cdot \nabla_x + m\vec{F}(x) \cdot \nabla_p \right] f_1(\vec{x}, \vec{p}, t) = \left(\frac{\partial f}{\partial t} \right)_c, \tag{7.7}$$

where $\vec{F}(\vec{x}) = -\nabla \Phi(\vec{x})$ is the external force field, and where the right-hand side is the collision integral

$$\left(\frac{\partial f}{\partial t} \right)_c = \frac{N}{V} \int \nabla_x V(\vec{x}, \vec{x}_1) \cdot \nabla_p f_2(\vec{x}, \vec{p}, \vec{x}_1, \vec{p}_1, t) d^3 x_1 d^3 p_1, \tag{7.8}$$

where in the numerator on the right-hand side of the equation we have used N to replace $N - 1$ for large N.

From here on, it is customary to make the simplifying assumption that

$$f_2(\vec{x}, \vec{p}, \vec{x}_1, \vec{p}_1, t) = f_1(\vec{x}, \vec{p}, t) \cdot f_1(\vec{x}_1, \vec{p}_1, t), \tag{7.9}$$

which is the so-called "molecular chaos" assumption that ensures that the colliding particles have no correlations.

With this assumption, the collisional integral can be carried out to become

$$\left(\frac{\partial f}{\partial t} \right)_c = \nabla_x \Phi_{\text{mean}}(\vec{x}, t) \cdot \nabla_p f_1(\vec{x}, \vec{p}, t) \tag{7.10}$$

where

$$\Phi_{\text{mean}}(\vec{x}, t) \equiv \frac{N}{V} \int V(\vec{x}, \vec{x}_1) f_1(\vec{x}_1, \vec{p}_1, t) d^3 x_1 d^3 p_1 \tag{7.11}$$

is the so-called "mean field potential," which represents the average influence of all the other particles on a test particle. The resulting first equation of the hierarchy becomes the so-called Vlasov equation:

$$\left\{\frac{\partial}{\partial t} + \frac{\vec{p}}{m} \cdot \nabla_x + [m\vec{F}(\vec{x}) - \nabla_x\Phi_{mean}(\vec{x}, t)] \cdot \nabla_p\right\} f_1(\vec{x}, \vec{p}, t) = 0, \qquad (7.12)$$

where $\vec{F}(\vec{x})$ represents the external force field, and $\nabla_x\Phi_{mean}(\vec{x}, t)$ represents the mean field due to the particles of the system.

This equation is sometimes also referred to as the collisionless Boltzmann equation, or, in the context of galaxy dynamics, the fundamental equation of stellar dynamics (Binney & Tremaine 2008). The equation, when solved self-consistently with respect to the single-particle distribution function, is highly nonlinear.

Here we see that the BBGKY procedure, carried out in the Vlasov limit, is akin to a renormalization procedure that enfolds the global/environmental influence into a local effective field of interaction.

The same molecular chaos assumption, however, can lead also to the collisional form of the Boltzmann equation, if we make the *additional* assumptions of short-range binary interaction of the dilute gas ensemble and coarse graining. The short-range binary interaction assumption allows the truncation of the BBGKY hierarchy at the second reduced distribution function f_2 level (since the three-particle contribution to the collision integral is neglected and only pairwise collision/scattering processes will be considered in the derivation of the form of the collision integral), and the coarse-graining assumption assumes a homogeneous distribution function (i.e., retaining only its velocity dependence but erasing the positional dependence of the distribution function when modeling the local scattering process) over the range of interaction, thus further introducing the statistical element into the formulation.

With these approximations, and using the standard scattering geometry, the collisional form of the Boltzmann equation can be derived from the truncated BBGKY hierarchy to be (Harris 1971; Kreuzer 1981; Liboff 2003)

$$\left\{\frac{\partial}{\partial t} + \frac{\vec{p}}{m} \cdot \nabla_x + m\vec{F}(\vec{x}) \cdot \nabla_p\right\} f_1(\vec{x}, \vec{p}, t) =$$

$$\frac{N}{V} \int |\vec{p} - \vec{p}_1| \sigma(\Omega, |\vec{p} - \vec{p}_1|)[f_1(\vec{x}, \vec{p}', t)f_1(\vec{x}, \vec{p}_1', t) - f_1(\vec{x}, \vec{p}, t)f_1(\vec{x}, \vec{p}_1, t)]d^3p_1 d\Omega, \qquad (7.13)$$

where the primes represent the same variables after the collision, Ω is the (momentum-dependent) solid angle of particle collision/scattering, and $\sigma(\Omega, |\vec{p} - \vec{p}_1|)$ is the particle-scattering cross section, which depends on both the scattering angle Ω and on the magnitude of the relative initial momentum $|\vec{p} - \vec{p}_1|$.

A good heuristic derivation of the collisional Boltzmann equation can be found in Shu (1992). In terms of the heuristic derivation, the two terms in the Boltzmann collision integral correspond to the phase space density variation as a resulting of scattering *into* and *out* of the local phase space volume (the primed quantity corresponds to the inverse scattering processes of the unprimed quantity). The heuristic approach, on the other hand, blurs the exact juncture where the irreversibility element was introduced into the original reversible Hamiltonian formulation. This might had been part of the reason why Boltzmann's original heuristic derivation of his eponymous equation failed to convince some of his colleagues of the reality of an arrow of time in systems governed by the reversible dynamical equations. The BBGKY approach, on the other hand, makes it clear that the statistical or probabilistic element is introduced during the *coarse graining* of the phase space volume in the particle-scattering treatment. This procedure sets the collisional version of the Boltzmann equation apart from the mean-field treatment of the Vlasov equation. even though both effectively utilized the gradients of the distribution function as the driving force of evolution in the absence of external or internal forces.

Incidentally, one cannot simply set the collisional cross section to zero in the collisional version of the Boltzmann equation to arrive at the collisionless version of the Boltzmann equation, since the latter also contains the mean-field contribution in the driving force field in the left-hand side of the equation, whereas for the former the interparticle force field (apart from gravity) is reflected in the collision integral itself, and the mean-field contribution is not explicitly included in the left-hand side of the equation. This difference reflects the essential difference in a reversible (fine grained) versus a statistical (coarse grained) treatment of the evolution of the distribution function. In the study of the growth and propagation of spiral density waves/modes, these two points of view in fact lead to similar results for the wave profile (apart from the wave-particle resonance regions), which is not too surprising since, as we have seen previously, the mean-field term of the collisionless Boltzmann equation in fact derives from the treatment of the collisional integral in the BBGKY hierarchy, just as Boltzmann's collisional term. They are different variants for truncating the BBGKY hierarchy at the second order, reflecting essentially the same dynamical agent driving the evolution of the distribution function in the absence of external or internal force field.

Both versions of the Boltzmann equation, as we have seen, are under the heavy-handed assumption of molecular chaos, or lack of particle correlation before particle collisions. This assumption is *ad hoc*, and its introduction in fact serves the purpose of emphasizing one type of evolution (i. e., from inhomogeneous to homogeneous, whether in the form of phase-mixing in the collisionless case, or in the form of uniformizing the distribution function in the collisional case, in the absence of internal or external force field) at the expense of other types of evolution behavior (i. e., from homogeneous to inhomogeneous in localized regions of an open system, such as during the formation of self-organized dissipative structures – though for this latter process to operate, the gradients of the distribution function are expected to interact with other types of forces, such as gravity, to lead to the formation of self-organized structures). Ultimately, we come to realize, even some of our best theoretical constructs reflect merely prejudiced views that are suitable to the description of only a limited range of phenomena. We need to be cautious about generalizing the range of validity of even a well-founded theory.

7.1.2 Growth of Instability and the Arrow of Time

Boltzmann equation with the collision term predicts the increase of entropy S defined by

$$S \equiv -k \int f \ln f d^3x d^3p, \qquad (7.14)$$

as long as there is secular changes happening in the system (the only time entropy remains the same is at a thermodynamical equilibrium state). Boltzmann equation has served as the foundation of the kinetic theory analyses (Lifshitz & Pitaevskii 1981), as well as the starting point for its successive approximations, that is, expansions in orders of the mean free path to system size, with the zeroth-order approximation the Eulerian fluid equations, and the first-order approximation the Navier–Stokes equations (see further discussions in Section 7.2).

The Boltzmann collision term, in its derivation through the BBGKY hierarchy, explicitly excludes correlated interactions through the "molecular chaos" assumption. The fact that correlated interactions are not incorporated in Boltzmann's collision term is also reflected through the fact that using Boltzmann's equation one can prove the *local* irreversible increase of entropy whenever secular changes are occurring (i. e., if the system is not already at local thermodynamic equilibrium, or LTE. See Liboff 2003); whereas for dissipative structures that are in a *dynamical* equilibrium state, local entropy can remain nearly constant when there is secular evolution of the underlying basic state, at the expense of the environmental entropy increase through the flux of entropy export by the dissipative structures themselves.

The so-called *stellar-dynamical equation*, which is simply Vlasov equation in the context of the self-gravitating galactic systems, had been used extensively in the early years of the studies of density wave theories (e. g., Toomre 1964, 1969; Lynden-Bell & Kalnajs 1972; Lin & Shu 1964, 1966). As it turned out, the stellar dynamical equation predicts similar behavior for the linear asymptotic calculation of density waves, compared to that using the simpler moment-equation descendants of the collisional-form of the Boltzmann equation, that is, the Eulerian fluid equations (or the Poisson–Euler set, to be discussed in Section 7.2), though there are some differences in behavior between the two treatments at the wave-particle resonance locations (Shu 1992).

In the application of both the stellar dynamical equation and the Eulerian fluid equations in the linearized density-wave calculation in a galactic configuration, growing wave perturbations as well as unstable global modes can be obtained in the linear regime. This at the first sight seems to contradict both the reversible nature of the Vlasov equation, as well as the irreversible entropy-increasing nature of the collisional Boltzmann equation from which the Eulerian fluid equations are derived. In fact Lynden-Bell and Ostriker (1967) had pondered this very question several decades ago in the form of the so-called "anti-spiral theorem," which concluded that reversibility must be reflected in an equal probability of the growth of the leading and trailing type spiral patterns, and the fact that predominantly trailing patterns are observed in physical galaxies implies that there is irreversible dissipation occurring in galaxies, such as that contributed by the gas component.

We now know that the ultimate source of irreversibility lies in the low-entropy initial condition of the Big Bang, from which subsequent nonequilibrium evolution can occur to increase the entropy of the entire universe, including to manufacture local dissipative structures to accelerate the rate of entropy production. This low-entropy cosmic initial condition places the gravitational matter in a widely separated spatial arrangement, which leaves room for the growth of instability. And the instability-growth itself contains the seed of irreversibility even for a Hamiltonian many-body system placed in an open environment, so an explicitly dissipative gas component is not needed for the predominant trailing forms of density wave patterns to form and be observed.

The fact that we can use the reversible Vlasov equation or the locally entropy-increasing collisional Boltzmann equation and its moment equation descendants to calculate the growth of instabilities, whose existence contradicts the very premises of these equations, appears to be nature's tendency to join the different regimes of physical phenomena through effective singularity hierarchies, as we have discussed in the main body of the text (this behavior also underlies the concept of spontaneous symmetry breaking: i.e. the original equation set can have symmetries not respected by the solutions of the equations set). What these equations predict are states whose complete description requires the introduction of elements – such as particle correlations which are crucial to the spontaneous growth of any instabilities; as well as the irreversible dissipation process which always accompanies any realistic growth of inhomogeneities, even for nominally Hamiltonian systems – that are missing in these equations themselves. But these global fluctuation states are nonetheless compatible with these dissipationless equations as long as we work in regimes of infinitesimal growth (i. e., linear modal calculation), or in situations where quasi-steady-state and global self-consistency conditions had not been enforced (i. e. for nonlinear but steepening shock waves, or else for transient and non-steady fluctuations). Sooner or later though, the deficiency of these approximate formulations announces its arrival by one form or another of singularity conditions (i. e., the blow-up of wave amplitude during nonlinear modal growth, or the infinite steepening of the shock front when solved as an initial value problem), signaling the inevitable need for a complete, self-consistent treatment.

7.2 From Kinetic Theory to Fluid Mechanics

The most typical moment equations derivable from integrating the original collisional Boltzmann equation are the equations for the conservation of mass, momentum, and energy, which together give five linearly independent equations for the 13 variables, $\rho, u_i(3), P, \pi_{i,j}(5)$, and $F_i(3)$, where ρ is the physical space matter density, u_i are the three Cartesian components of the physical space velocity, P is the isotropic pressure, $\pi_{i,j}$ are the five independent components of the viscous stress tensor, and F_i are the three components of the conduction heat flux (de Groot & Mazur 1962; Kreuzer 1981).

Due to the disparity between the number of unknowns compared to the number of independent equations, there is the need for the so-called *closure relations* derived from physical arguments in order to close the equation set and solve them analytically.

As discussed in Shu (1992), the closure equations are usually more easily obtainable at the *extremes* of physical parameter regime. For example, when the mean free path l for collision is much smaller than the macroscopic length scale L, we are in the regime of local thermodynamic equilibrium, characterized by the well-known Maxwell–Boltzmann velocity distribution. Through the so-called Chapman–Enskog procedure, to zeroth order in the expansion in terms of l/L, the eight additional closure relations are $\pi_{i,j} = 0$ and $F_i = 0$, which amounts to neglecting the diffusive effect, and we arrive at the Eulerian fluid equation set:

$$\frac{\partial \rho}{\partial t} + \nabla \cdot (\rho \vec{u}) = 0, \tag{7.15}$$

$$\frac{\partial \vec{u}}{\partial t} + (\vec{u} \cdot \nabla)\vec{u} = -\nabla \mathscr{V} - \frac{1}{\rho}\nabla P, \tag{7.16}$$

$$\rho \left(\frac{\partial \mathscr{E}}{\partial t} + \vec{u} \cdot \nabla \mathscr{E} \right) = -P \nabla \cdot \vec{u}, \tag{7.17}$$

together with the equation of constitution

$$\rho \mathscr{E} = \frac{3}{2} P = \frac{3}{2} n k T, \tag{7.18}$$

where k is the Boltzmann constant, $\mathscr{E}$ is the energy density, P is pressure, n is the number of particles per unit volume, $\mathscr{V}$ is the gravitational potential, and T is the thermodynamic temperature.

The above equations, plus the Poisson integral (which together forms the Poisson–Euler set), have been used extensively in the derivation of linear and global density wave modal properties, notably by Lin and his collaborators (Lin & Lau 1979 and the references therein). The fact that the Eulerian fluid equations (which assume full collisions among the particles constituting the medium) can be used to represent the behavior of the nominally collisionless stellar medium (i. e., the fact that we can use the continuum or fluid approximation for the nominally collisionless assembly of stars) is at least partly due to the fact that in disk galaxies there exists an angular momentum barrier, so orbiting particles execute epicyclic motion around their *mean orbital radius* while circulating the galaxy. The near-collision condition at the density wave crest also helps to maintain the quasi-continuum nature of the orbiting stellar particles (this is reflected partly in the age–velocity–dispersion relation of the solar neighborhood stars, which we have argued in the main text is due to the continued thermalization effect of the spiral-density-wave collisionless shock).

Even though collisions among particles are implicitly assumed in the Eulerian fluid equation set, its lack of dissipation and transport effects is at odds with the true behavior of particles at the density wave crest. This is the underlying reason that by employing the Poisson–Euler equation set we can at most observe the growing global modes at the linear regime, or else the steepening of the azimuthal profile of these modes at the nonlinear regime in an iterative solution procedure when solved as an initial value problem, but cannot derive the dissipative secular evolution effect due to the interaction of the density wave modes and the basic state of the galactic disk.

The stellar dynamical equation, as we had commented previously, is reversible, thus cannot be used to treat the collisionless-shock phenomenon either, since, as we have commented before, the phrase "collisionless shock" is a bit of a misnomer as the processes happening at the spiral shock front is akin to true collisions or scatterings.

Another quantity usually defined in the context of the continuum treatment of a system of particles is the specific entropy s

$$s \equiv c_v \ln(P\rho^{-\gamma}), \tag{7.19}$$

where c_v is the specific heat at constant volume and $\gamma = 5/3$ for monatomic idea gas. It can be shown that for such a gas in the Eulerian regime the specific entropy s satisfies

$$\frac{Ds}{Dt} = 0 \tag{7.20}$$

(Shu 1992), which shows that such a gas goes through only adiabatic changes. This shows again that within the Eulerian regime one cannot obtain the secular evolution effect which requires *nonadiabatic* changes in the underlying matter (i.e., that is essentially what the term "secular" in our usage implies).

To the next order in the perturbation expansion with respect to l/L, we obtain the Navier–Stokes set of equations (Chapman & Cowling 1961; Liboff 2003; Shu 1992), which is otherwise identical to the Eulerian set except that the momentum equation now becomes:

$$\frac{\partial \vec{u}}{\partial t} + (\vec{u} \cdot \nabla)\vec{u} = -\nabla \mathscr{V} - \frac{1}{\rho}\nabla P + \frac{1}{\rho}\nabla \cdot \overline{\overline{\pi}}, \tag{7.21}$$

where the viscous stress tensor $\overline{\overline{\pi}}$ is usually defined as

$$\pi_{ik} \equiv \mu D_{ik} + \beta(\nabla \cdot \vec{u})\delta_{ik}, \tag{7.22}$$

where μ and β are the coefficients of shear and bulk viscosity, respectively ($\beta = 0$ in the Navier–Stokes set of equations, see Liboff 2003), and where

$$D_{ik} \equiv \frac{\partial u_i}{\partial x_k} + \frac{\partial u_k}{\partial x_i} - \frac{2}{3}(\nabla \cdot \vec{u})\delta_{ik} \qquad (7.23)$$

is the rate of deformation tensor. There is also a corresponding modification of the energy balance equation to include the effect of heat conduction and viscous dissipation.

The Navier–Stokes equation set explicitly contains terms that correspond to transport phenomena (i. e., viscosity and heat conduction). However, in the practical usage of this equation set, the coefficients of viscosity and heat conduction are either determined empirically from measurements of the fluid properties, or else represented formally as in the α parameter model of the viscous accretion disk theory (see Section 2.6.5). The collective effects we are interested in, on the other hand, cannot emerge directly in a local/perturbative treatment as in the derivation of the transport coefficients. The past treatment of the secular evolution in the gas component uses essentially this type of transport approximation, with the coefficient of viscosity determined essentially arbitrarily (since microscopic particle mean free path or gas viscosity is orders of magnitude too small to be able to account for the galactic secular evolution effect, some form of macroscopic mean free path had to be assumed in the effective kinematic-viscosity expression). The assumed effective gas viscosity is similar in spirit to the α-parameter approach, that is, a level of effective viscosity is assigned for what is needed to accomplish the rate of evolution, without being able to provide an underlying physical justification. A quantitative calculation for the effective gravitational viscosity of the galactic gas component participating in secular evolution can only be obtained through the globally self-consistent potential-density phase shift closure relation. The past lack of correct treatment of the gas component led to uncertainties as to exactly how great an effect the gas component actually has on the secular evolution of galaxies, and the results from the different research groups often disagree from one another.

The opposite regime to the small l/L regime used to obtain the Eulerian fluid equations and the Navier–Stokes equations, is when the mean free path l is much greater than the macroscopic length scale L of the problem. This is the optically thin regime of the radiative transfer problems.

7.3 Nonequilibrium Phase Transition and Galaxy Evolution

The intermediate regime, which Shu (1992) called "the really awkward regime $l \sim L$ (the state of affairs, e. g., in planetary rings)", is also the regime relevant to the collective dissipation process at the crest of density wave modes, a process we had focused on in this monograph. Recall that in Chapter 2 we had shown that the effective particle mean free path l within the two-fluid gravitational instability at the solar neighborhood is close to 1 kpc, which is of the same order as the spiral arm width, which is also of the same order as the effective system length scale L. It is indeed in this gray area that the emergent new physics comes into play.

As I. Prigogine aptly titled one of his popular books, *From Being to Becoming: Time and Complexity in the Physical Sciences* (Prigogine 1980), the study of nonequilibrium structure formation is the study of the process of "from being to becoming," or the process of generating "order out of chaos" (Prigogine & Stengers 1984). Our scientific tradition in the past millennia had mostly focused on solving problems of "being" or the description of stationary states. Whereas the problems of "from being to becoming" are focused on the gray area between one (nonequilibrium, quasi-) stationary state and another, which is where the frontier of science lies, and where some of the most significant new understandings and syntheses are expected to be made in the upcoming decades. The transition process between quasi-stationary states does not necessarily have to happen in a chaotic or random fashion, as in the galaxy merger scenario, but can be a well-coordinated process driven by internal dynamics, as we have seen in this work in the coordinated modal morphological evolution in response to the changing basic state characteristics as a result of secular evolution.

While nonequilibrium thermodynamics and nonequilibrium statistical mechanics as subjects of modern scholarly research have been in existence for many decades, most of these studies confined themselves to the linear regime (as in the well-known works of Onsager, Fokker-Planck, and Wiener-Khinchine on the subject of transport phenomena. See, for example, the discussions in de Groot & Mazur 1962). The far-from-equilibrium and highly nonlinear regimes as pertain to the studies of dissipative structures, including the studies of convection, turbulence, biological morphogenesis, as well as the structure formation processes in disk galaxies in the universe, are significantly more difficult and challenging. Among these studies, the study of the role of dissipative structures in the cosmic evolution has until now received less attention than similar studies in other branches of physical sciences. Prigogine himself stated: "However, the questions, What is irreversibility on the cosmic scale? Can we introduce an entropy operator in the framework of a dynamical description in which gravitation plays an essential role? are formidable ones. I prefer to confess my ignorance (Prigogine 1980, p. 214)". We now know that a gravitational arrow of time does exist, and the dissipative structure theory can be applied successfully to the self-organization problems in galactic system. We expect that extensions of such concepts and approaches can also be used toward addressing broader structure formation problems in the cosmological context.

7.4 The Proper Choice of Analytical Hierarchies

The hierarchies we choose to construct our physical theories oftentimes reflect or mirror the natural hierarchies in the organization of physical phenomena. This kind of choice is thus a proper one since, as we have noted, natural hierarchies are often the result of nonequilibrium phase transitions that spontaneously break symmetry and generate organization.

However, there are instances where our classical choice of the organization of hierarchies in physical theory does not adapt very well to the problems we are trying to solve. One good example is the hierarchical organization from kinetic theory to fluid mechanics. As we have seen, in applying the Chapman–Enskog procedure to obtain successive approximations to the Boltzmann kinetic equation, as we traverse the hierarchy through the Eulerian equations, the Navier–Stokes equations, and so on, nowhere will we be able to derive the collective dissipation behavior as reflected in the self-organized galactic density wave patterns. This is because the Chapman–Enskog procedure is a *local* procedure and is a treatment for smooth flows that obey differential equations. At the collisionless shock front, it is indeed the differential treatment that becomes invalid, and the requirement for a global treatment is necessitated.

To put it another way, the Chapman–Enskog procedure is valid when we are interested in the successive approximations within one gross layer of the hierarchy of organizations, and its successive approximations correspond to the different orders of the perturbation expansions with respect to the ratio of mean free path to system size. However, when dealing with problems of self-organization in the far-from-equilibrium regime, we effectively experience "quantum leaps" in the characteristics of the phenomena (a phase transition is by nature discontinuous), and the perturbation expansion approach will fail when these discontinuous changes take place. For example, the need for renormalization in quantum electrodynamics occurs exactly where the perturbation procedure fails, and in light of the discussion we have presented here, this is an indication that a new organization hierarchy has formed that requires us to go beyond the existing physical theory.[†]

The renormalization procedure (including the renormalization group approach) allows the integration of correlated global/environmental influence into the treatment of local dynamics directly. In modern-day high-energy physics, many *effective field theories* are constructed under the same guiding principles (Weinberg 1995).

Thus we see, the question of how to choose a proper theoretical hierarchy depends on the type of physical problems we are trying to solve, and on the actual regimes/energy scales that are relevant. In dealing with new physical problems, or in taking a fresh new look at existing ones, we have to decide first and foremost whether we are crossing the boundary of the organization hierarchy of the existing physical theories or not, that is, whether we can treat the problem as a perturbation from a well-known solution and regime, or else we must enter a new regime governed by a new set of emerging dynamics. From our experience so far, we see that the onset of a global instability, which evolves toward a singularity condition in the equation set used, is a sure sign of the beginning of the breakdown of the assumptions of the underlying equation set, so that the detailed microscopic processes underlying the formation of such instabilities cannot be treated by such an equation set, even when going to the full nonlinear orders.

[†] In light of the above discussions, we see that the search for a Wavefunction of the Universe is likely to be a futile attempt as a matter of principle: the existence of the singularity hierarchy in the organization of physical laws and phenomena means that a single-level continuum formulation cannot be the correct description for our Universe.

References

Abraham, R. et al. 1996, *The Astrophysical Journal*, vol. 471, p. 694.
Adams, T.F. 1977, *The Astrophysical Journal Supplement Series*, vol. 33, p. 19.
Alcock, C. et al. 2000, *The Astrophysical Journal*, vol. 542, p. 281.
Ambartzumian, V.A., & Gordeladse, S.G. 1938, *Bull. Abastumani Obs.*, vol. 2, p. 37.
Anderson, P.W., 1972, *Science*, vol. 177, issue 4047, p. 393.
Andredakis, Y.C., Peletier, R.F., & Balcells, M. 1995, *Monthly Notices of the Royal Astronomical Society*, vol. 275, p. 874.
Antonov, V.A. 1962, *Vest. Leningr. Gos. Univ.*, vol. 7, p. 135.
Appleton, P.N., & Hughes, D.H. 1988, in *Dust in the Universe*, ed. M.E. Bailey & D.A. Williams (Cambridge: Cambridge University Press).
Arp, H.C. 1966, *Atlas of Peculiar Galaxies* (Pasadena: CIT Press).
Athanassoula, E. 2003, *Monthly Notices of the Royal Astronomical Society*, vol. 341, p. 1179.
Baba, J., Saitoh, T.R., & Wada, K. 2013, *The Astrophysical Journal*, vol. 763, p. 46.
Bahcall, J.N. 1984, *The Astrophysical Journal*, vol. 287, p. 926.
Balbus, S.A. 1988, *The Astrophysical Journal*, vol. 324, p. 60.
Balbus, S.A., & Cowie, L.L. 1985, *The Astrophysical Journal*, vol. 297, p. 61.
Balogh, A., & Treumann, R.A. 2013, *Physics of Collisionless Shocks* (New York: Springer).
Barbanis, B., & Woltjer, L. 1967, *The Astrophysical Journal*, vol. 150, p. 461.
Barbuy, B., & Grenon, M. 1990, in *Bulges of Galaxies*, eds. B.J. Jarvis & D.M. Jerndrup (Garching: ESO), p. 83.
Barnes, J.E., & Hernquist, L. 1991, *The Astrophysical Journal*, vol. 370, p. L65.
Barnes, J.E., & Hernquest, L. 1992, *Annual Reviews of Astronomy & Astrophysics*, vol. 30, p. 705.
Barr, J.M., Bedregal, A.G., Aragon-Salamanca, A., Merrifield, M.R., & Bamford, S.P. 2007, *Astronomy and Astrophysics*, vol. 470, p. 173.
Bell, E.F., & de Jong, R.S. 2001, *The Astrophysical Journal*, vol. 550, p. 212.
Bender, R., & Saglia, R.P. 1999, in *Galaxy Dynamics*, eds. D. Merritt, J.A. Sellwood & M. Valluri (San Francisco: ASP), p. 113.
Bender, R., Surma, P., Dobereiner, S., Mollenhoff, C., & Madejsky, R. 1989, *Astronomy and Astrophysics*, vol. 217, p. 35.
Bershadskii, A., & Sreenivasan, K.R. 2002, *Physics Letters A*, vol. 299, p. 149.
Bershadskii, A., & Sreenivasan, K.R. 2003, *Physics Letters A*, vol. 319, p. 21.
Bershadskii, A. 2004, *Physica A*, vol. 335, p. 611.
Bertin, G. 1983, *Astronomy and Astrophysics*, vol. 127, p. 145.
Bertin, G. 2014, *Dynamics of Galaxies*, 2nd edn. (Cambridge: Cambridge University Press).
Bertin, G., Lin, C.C., Lowe, S.A., & Thurstans, R.P. 1989a, *The Astrophysical Journal*, vol. 338, p. 78.
Bertin, G., Lin, C.C., Lowe, S.A., & Thurstans, R.P. 1989b, *The Astrophysical Journal*, vol. 338, p. 104.
Bertin, G., & Romeo, A.B. 1988, *Astronomy & Astrophysics*, vol. 195, p. 105.
Binney, J., & Lacey, C. 1988, *Monthly Notices of the Royal Astronomical Society*, vol. 230, p. 597.
Binney, J., & Tremaine, S. 2008, *Galactic Dynamics*, 2nd edn. (Princeton: Princeton University Press).
Block, D.L., Buta, R.J., Knapen, J.H., Elmegreen, D.M., Elmegreen, B.G., & Puerari, I.P. 2004, *The Astronomical Journal*, vol. 128, p. 183.
Bogolyubov, N.N. 1946, *Journal of Physics* vol. 10, p. 256.
Boissier, S., Boselli, A., Prantzos, N., & Gavazzi, G. 2001, *Monthly Notices of the Royal Astronomical Society*, vol. 321, p. 733.
Boltzmann, L. 1872, *Sitzungsberichte der Akademie der Wissenschaften, Wien, II*, vol. 66, p. 275 [English translation in S.G. Brush, *Kinetic Theory*, vol. 2, *Irreversible Processes*, pp. 88–175 (Oxford: Pergamon Press), 1966].

Born, M., & Green, H.S. 1946, *Proceedings of Royal Society of London*, vol. A188, p. 10.
Boselli, A., & Gavazzi, G. 2006, *The Publications of the Astronomical Society of the Pacific*, vol. 118, p. 517.
Bower, R.G., Kodama, T., & Terlevich, A. 1998, *Monthly Notices of the Royal Astronomical Society*, vol. 299, p. 1208.
Bower, R.G., Lucey, J.R., & Ellis, R.S. 1992, *Monthly Notices of the Royal Astronomical Society*, vol. 254, p. 601.
Broadhurst, T.J., Takada, M., Umetsu, K., Kong, X., Arimoto, N., Chiba, M., & Futamase, T. 2005, *The Astrophysical Journal*, vol. 619, p. L143.
Brooks, A. 2014, Talk presented in the Eighth Harvard-Smithsonian Conference on Theoretical Astrophysics, *Debates on the Nature of Dark Matter,* https://www.cfa.harvard.edu/events/2014/sackler/.
Burgers, J.M. 1948, *Advances in Applied Mechanics*, vol. 1, p. 171.
Buta, R.J., Laurikainen, E., Salo, H., & Knapen, J.H. 2010, *The Astrophysical Journal*, vol. 721, 259.
Buta, R.J., Vasylyev, S., Salo, H., & Laurikainen, E. 2005, *The Astronomical Journal*, vol. 130, p. 506.
Buta, R.J., & Zhang, X. 2009, *The Astrophysical Journal Supplement Series*, vol. 182, p. 559.
Butcher, H., & Oemler, A. Jr. 1978a, *The Astrophysical Journal*, vol. 219, p. 18.
Butcher, H., & Oemler, A. Jr. 1978b, *The Astrophysical Journal*, vol. 226, p. 559.
Byrd, G.G., & Howard, S. 1990, in *Dynamics and Interactions of Galaxies*, ed. R. Wielen (Heidelberg: Springer), p. 128.
Byrd, G.G., Valtonen, M.J., Sundelius, B., & Valtaoja, L. 1986, *Astronomy and Astrophysics*, vol. 166, p. 75.
Cappellari, M. et al. 2013, *Monthly Notices of the Royal Astronomical Society*, vol. 432, p. 1862.
Carlberg, R.G. 1987, in *Nearly Normal Galaxies*, ed. S.M. Faber (New York: Springer), p. 129.
Carlberg, R.G., Dawson, P.C., Hsu, T., & Vandenberg, D.A., 1985, *The Astrophysical Journal*, vol. 294, p. 674.
Carlberg, R.G., & Freedman, W.L. 1985, *The Astrophysical Journal*, vol. 298, p. 486.
Carlberg, R.G., & Sellwood, J.A. 1985, *The Astrophysical Journal*, vol. 292, p. 79.
Ceverino, D., Dekel, A., Mandelker, N., Bournaud, F., Burkert A., Genzel, R., & Primack, J. 2012, *Monthly Notices of the Royal Astronomical Society*, vol. 420, p. 3490.
Chandrasekhar, S., & Munch, G. 1952, *The Astrophysical Journal*, vol. 115, p. 103.
Chapman, S., & Cowling, T.G. 1961, *The Mathematical Theory of Nonuniform Gases* (Cambridge: Cambridge University Press).
Chung, A., van Gorkom, J.H., Kenney, J.D.P., Crowl, H., & Vollmer, B. 2009, *The Astronomical Journal*, vol. 138, p. 1741.
Cimatti, A., Daddi, E., & Renzini, A. 2006, *Astronomy and Astrophysics*, vol. 453, p. L29.
Cisternas, M. et al. 2011a, *The Astrophysical Journal*, vol. 726, p. 57.
Cisternas, M., 2011b, *The Astrophysical Journal*, vol. 741, p. L11.
Clowe, D. et al. 2006, *The Astrophysical Journal*, vol. 648, p. L109.
Courteau, S. 1996, in *New Extragalactic Perspectives in the New South Africa*, eds. D.L. Block & J.M. Greenberg (Dordrecht: Kluwer), p. 255.
Coe, D. 2013, *The Astrophysical Journal*, vol. 567, p. 672.
Cohen, J.G. 2002, *The Astrophysical Journal*, vol. 567, p. 672.
Combes, F., & Sanders, R.H. 1981, *Astronomy and Astrophysics*, vol. 96, p. 164.
Conselice, C.J., Kershady, M.A., Dickinson, M., & Papovich, C. 2003, *The Astronomical Journal*, vol. 126, p. 1183.
Conselice, C.J., Wilkinson, A., Duncan, K., & Mortlock, A. 2016, *The Astrophysical Journal*, vol. 830, p. 83.
Contopoulos, G. 1980, *Astronomy and Astrophysics*, vol. 31, p. 198.
Contopoulos, G., & Grosbol, P. 1986, *Astronomy and Astrophysics*, vol. 155, p. 11.
Contopoulos, G., & Grosbol, P. 1988, *Astronomy and Astrophysics*, vol. 197, p. 83.

Contopoulos, G., & Mertzanides, C. 1977, *Astronomy and Astrophysics*, vol. 61, p. 477.
Contopoulos, G., & Voglis, N. 1996, in *Barred Galaxies*, eds. R. Buta, D.A. Crocker & B.G. Elmegreen, ASP Conf. Series (San Francisco: ASP), vol. 91, p. 321.
Cooper et al. 2006, *Monthly Notices of the Royal Astronomical Society*, vol. 370, p. 198.
Cooper et al. 2007, *Monthly Notices of the Royal Astronomical Society*, vol. 376, p. 1445.
Couch, W.J., Barger, A.J., Smail, I., Ellis, R., & Sharples, R.M. 1998, *The Astrophysical Journal*, vol. 497, p. 188.
Couch, W.J., Ellis, R.S., Sharples, R.M., & Smail, I. 1994, *The Astrophysical Journal*, vol. 430, p. 121.
Couch, W.J., Matthew, M.C., & De Propris, R. 2004, in *Clusters of Galaxies: Probe of Cosmological Structure and Galaxy Evolution*, eds. J.S. Mulchaey, A. Dressler & A. Oemler (Cambridge: Cambridge University Press), p. 42.
Cowie, L.L., Songaila, A., Hu, E.M., & Cohen, J.G. 1996, *The Astronomical Journal*, vol. 112, p. 839.
Croton, D.J., Springel, V., White, S.D.M., De Lucia, G., Frenk, C.S., Gao, L., Jenkins, A., Kauffmann, G., Navarro, J.F., & Yoshida, N. 2006, *Monthly Notices of the Royal Astronomical Society*, vol. 365, p. 11.
Daddi, E., Cimatti, A., & Renzini, A. 2000, *Astronomy & Astrophysics*, vol. 362, p. 45.
Dahari, O. 1985, *The Astrophysical Journal Supplement Series*, vol. 57, p. 643.
Dame T.M., Elmegreen B., Cohen, R., & Thaddeus, P. 1986, *The Astrophysical Journal*, vol. 305, p. 892.
Dar, A. 1995, *The Astrophysical Journal*, vol. 449, p. 550.
Darwin, C. 1859, *On the Origin of Species*, 1st ed., reprinted by Wideside Press, 2003.
Das, M., & Jog, C. J. 1995, *The Astrophysical Journal*, vol. 451, p. 167.
Debattista, V.P., & Sellwood, J.A. 2000, *The Astrophysical Journal*, vol. 543, p. 704.
de Blok, W.J.G. 2003, in *Galaxy Evolution: Theory & Observations*, eds V. Avila-Reese, C. Firmani, C.S. Frenk, & C. Allen, *Revista Mexicana de Astronomìa y Astrofìsica* (Serie de Conferencias), vol. 17, p. 17.
de Blok, W.J.G. 2010, *Advances in Astronomy*, No. 789293.
de Groot, S.R., & Mazur, P. 1962 *Non-Equilibrium Thermodynamics* (Amsterdam: North Holland).
De Simone, R.S., Wu, X., & Tremaine, S. 2004, *Monthly Notices of the Royal Astronomical Society*, vol. 350, p. 627.
Dehnen, W., & Binney, J.J. 1998, *Monthly Notices of the Royal Astronomical Society*, vol. 298, p. 387.
De Lucia, G., Springel, V., White, S.D.M., Croton, D., & Kauffmann, G. 2006, *Monthly Notices of the Royal Astronomical Society*, vol. 366, p. 499.
de Vaucouleurs, G. 1959, in *Handbuch der Physik*, vol. 53, ed. S. Flugge, (Berlin: Springer-Verlag), p. 275.
de Vaucouleurs, G. 1975, in *Stars and Stellar Systems*, vol. 9, *Galaxies and the Universe*, eds. A. Sandage, M. Sandage & K. Kristian (Chicago: University of Chicago Press), p. 309.
Djorgovski, S.G., & Davis, M. 1987, *The Astrophysical Journal*, vol. 313, p. 59.
D'Onghia E. 2015, *The Astrophysical Journal*, vol. 808, p. 8.
D'Onghia E., Vogelsberger, M., & Hernquist, L. 2013, *The Astrophysical Journal*, vol. 766, p. 34.
Donner, K.J., & Thomasson, M. 1994, *Astronomy and Astrophysics*, vol. 290, p. 785.
Dressler, A. 1980, *The Astrophysical Journal*, vol. 236, p. 351.
Dressler, A., Lynden-Bell, D., Burstein, D., Davies, R.L., Faber, S.M., Terlevich, R.J., & Wegner, G. 1987, *The Astrophysical Journal*, vol. 313, p. 42.
Dressler, A., Oemler, Jr. A., Butcher, H.R., & Gunn, J.E. 1994, *The Astrophysical Journal*, vol. 430, p. 107.
Dressler, A., Oemler, A. Jr., Couch, W.J., Smail, I., Ellis, R.S., Barger, A., Butcher, H., Poggianti, B.M., Sharples, R.M., 1997, *The Astrophysical Journal*, vol. 490, p. 577.
Dressler, A., Oemler, A. Jr., Poggianti, B.M., Smail, I., Trager, S., Shectman, S.A., & Couch, W.J. 2004, *The Astrophysical Journal*, vol. 617, p. 867.
Dressler, A., Smail, I., Poggianti, B.M., Butcher, H., Couch, W.J., Ellis, R.S., & Oemler, A., Jr. 1999, *The Astrophysical Journal Supplement Series*, vol. 122, p. 51.

Duschl, W.J., Hofmann, K.H., Rigaut, F., & Weigelt G. 1995, *RevMexAASC*, vol. 2, p. 17.
Efstathiou, G., David, M., Frenk, C.S., & White, S.D.M. 1985, *The Astrophysical Journal Supplement Series*, vol. 57, p. 241.
Eggen, O.J., Lynden-Bell, D., Sandage, A.R. 1962, *The Astrophysical Journal*, vol. 136, p. 748.
Elbaz, D., Daddi, E., Le Borgne, D., Dickinson, M., Alexander, D.M., Chary, R.-R., Starck, J.-L., Brandt, W.N., Kitzbichler, M., MacDonald, E., Nonino, M., Popesso, P., Stern, D., & Vanzella, E. 2007, *Astronomy and Astrophysics*, vol. 468, p. 33.
Ellis, R. 1997, *Annual Review of Astronomy & Astrophysics*, p. 35.
Ellis, R.S., Smail, I., Dressler, A., Couch, W.J., Oemler, A., Jr., Butcher, H., & Sharples, R.M. 1997, *The Astrophysical Journal*, vol. 483, p. 582.
Elmegreen, B.G., 1979, *The Astrophysical Journal*, vol. 231, p. 372.
Elmegreen, B.G., 1989, *The Astrophysical Journal*, vol. 344, p. 306.
Elmegreen, B.G., 1995, *Monthly Notices of the Royal Astronomical Society*, vol. 275, p. 944.
Elmegreen, B.G., 1998, arXiv:astro-ph/9811289.
Elmegreen, B.G., & Elmegreen, D.M. 1983, *Monthly Notices of the Royal Astronomical Society*, vol. 203, p. 31.
Elmegreen, B.G., & Elmegreen, D.M. 1985, *The Astrophysical Journal*, vol. 288, p. 438.
Elmegreen, B.G., & Elmegreen, D.M. 1989, in *Evolutionary Phenomena in Galaxies*, eds. J.E. Beckman & B.E.J. Pagel (Cambridge: Cambridge University Press), p. 83.
Elmegreen, B.G., & Scalo, J. 2004, *Annual Review of Astronomy and Astrophysics*, vol. 42, p. 211.
Erwin, P., & Sparke, L.S. 1999, in *Galaxy Dynamics*, ASP Conf. Ser. vol. 182, eds. D. Merritt & J.A. Sellwood (San Francisco: ASP), p. 243.
Evrard, A.E. 2004, in *Clusters of Galaxies: Probes of Cosmological Structure and Galaxy Evolution*, eds. J.S. Mulchaey, A. Dressler & A. Oemler (Cambridge: Cambridge University Press), p. 1.
Faber, S.M., & Jackson, R.E. 1976, *The Astrophysical Journal*, vol. 204, p. 668.
Faber, S.M., et al. 1987, in *Nearly Normal Galaxies*, ed. S.M. Faber (New York: Springer), p. 175.
Faber, S.M. et al. 1997, *The Astronomical Journal*, vol. 114, p. 1771.
Falceta-Gonçalves, D., Bonnell, I., Kowal, G., Lepine, J.R.D., & Braga, C.A.S. 2015, *Monthly Notices of the Royal Astronomical Society*, vol. 446, 973.
Fall, M., & Efstathiou, G., 1980, *Monthly Notices of the Royal Astronomical Society*, vol. 193, p. 189.
Fasano, G., Filippi, M., & Bertola, F. 1997, in *The Second Stromlo Symposium: The Nature of Elliptical Galaxies*, ASP Conference Series, v.116, 557.
Fasano, G., Poggianti, B.M., Couch, W.J., Bettoni, D., Kjaerggard, P., & Moles, M. 2000, *The Astrophysical Journal*, vol. 542, p. 673.
Feinstein, A. 1995, *Revista Mexicana de Astronomía y Astrofísica Serie de Conferencias*, vol. 2, *The Eta Carinae Region: A laboratory of Stellar Evolution*, p. 57.
Feynman, R.P., Leighton, R.B., & Sands, M. 1964, *The Feynman Lectures on Physics* vol. II, 41–12.
Fields, B., Freese, K., & Graff, D. 2000, *The Astrophysical Journal*, vol. 534, p. 265.
Filippenko, A.V., & Sargent, W.L.W. 1985, *The Astrophysical Journal Supplement Series*, vol. 57, p. 503.
Fleck, R.C. Jr., 1980, *The Astrophysical Journal*, vol. 242, p. 1019.
Fleck, R.C. Jr., 1981, *The Astrophysical Journal*, vol. 246, p. 151.
Fleck, R.C. Jr., 1982, *The Astrophysical Journal*, vol. 261, p. 631.
Fleck, R.C. Jr., 1996, *The Astrophysical Journal*, vol. 458, p. 739.
Foyle, K., Rix, H.-W., & Zibetti, S. 2010, *Monthly Notices of the Royal Astronomical Society*, vol. 407, p. 163.
Francis, P.J., Woodgate, B., Williger, G., & Malumuth, E., 1997, *Bulletin of the American Astronomical Society*, vol. 191, p. 312.
Franx, M. 1993, in IAU Symp. 153, *Galactic Bulges*, eds. H. Dejonghe and H.J. Habing (Dordrecht: Kluwer), p. 243.

Franx, M. 2004, in *Clusters of Galaxies: Probes of Cosmological Structure and Galaxy Evolution*, eds. J.S. Mulchaey, A. Dressler, and A. Oemler (Cambridge: Cambridge University Press), p. 196.
Freeman, K. 1970, *The Astrophysical Journal*, vol. 160, p. 811.
Freeman, K., & McNamara, G. 2006, *In Search of Dark Matter* (New York: Springer).
Friedli, D., & Martinet, L. 1993, *Astronomy and Astrophysics*, vol. 277, p. 27.
Frisch, U. 1995, *Turbulence: The Legacy of A.N. Kolmogorov* (Cambridge: Cambridge University Press).
Frogel, J.A., Quillen, A.C., & Pogge, R.W. 1996, in *New extragalactic perspectives in the New South Africa*, eds. D.L. Block and J.M. Greenberg (Dordrecht: Kluwer), *Astrophysics & Space Science*, vol. 209, p. 65.
Fujimoto, M. 1968, in IAU Symposium 29, *Nonstable Phenomena in Galaxies*, ed. V. Ambartsumian (Yerevan: Armenian Acad. Sci), p. 453.
Fujita, Y., & Nagashima, M. 1999, *The Astrophysical Journal*, vol. 516, p. 619.
Fukugita, M., Hogan, C.J., & Peebles, P.J.E. 1998, *The Astrophysical Journal*, vol. 503, p. 518.
Fuller, R.B. 1975, *Synergetics: Explorations in the Geometry of Thinking* (New York: Macmillan).
Gamow, G., 1952, *Physical Review D*, vol. 86, p. 251.
Geach, J.E. et al. 2006, *The Astrophysical Journal*, vol. 649, p. 661.
Gebhardt et al. 2000, *The Astrophysical Journal*, vol. 539, p. L13.
Genzel, R. et al. 2006, *Nature*, vol. 442, p. 786.
Gilmore, G., King, I. R., & van der Kruit, P. C. 1990, *The Milky Way as a Galaxy*, Saas-Fee Advanced Course Lecture Notes, No. 19, (Mill Valley: University Science Books).
Glansdorff, P., & Prigogine, I. 1971, *Thermodynamic Theory of Structure, Stability and Fluctuations* (New York: Wiley-Interscience).
Gliese, W. 1969, *Veroffentl. Astron. Rechen-Inst. Heidelberg*, No. 22.
Gnedin, O.Y., 1999, in *Galaxy Dynamics*, eds. D. Merritt, J.A. Sellwood, & M. Valluri (San Francisco: ASP), p. 495.
Gnedin, O.Y. 2003a, *The Astrophysical Journal*, vol. 582, p. 141.
Gnedin, O.Y. 2003b, *The Astrophysical Journal*, vol. 589, p. 752.
Gnedin, O.Y., Goodman, J., & Frei, Z. 1995, *The Astronomical Journal*, vol. 110, p. 1105.
Goldreich, P., & Nicholson, P.D. 1989, *The Astrophysical Journal*, vol. 342, p. 105.
Goldreich, P., & Lynden-Bell, D. 1965, *Monthly Notices of the Royal Astronomical Society*, vol. 130, p. 125.
Goldreich, P., & Tremaine, S. 1979, *The Astrophysical Journal*, vol. 233, p. 857.
Goto, T. 2005, *Monthly Notices of the Royal Astronomical Society*, vol. 359, p. 1415.
Goto, T., Okamura, S., Yagi, M., Sheth, R.K., Bahcall, N.A., Zabel, S.A., Crouch, M.S., Sekiguchi, M., Annis, J., Bernardi, M., Chong, S.-S., Gomez, P.L., Hansen, S., Kim, R.S.J., Knudson, A., McKay, T.A., & Miller, C.J. 2003a, *Publications of the Astronomical Society of Japan*, vol. 55, p. 739.
Goto, T., Okamura, S., Maki, S., Bernardi, M., Brinkmann, J., Gomez, P.L., Harvanek, M., Kleinman, S.J., Krzesinski, J., Long, D., Loveday, J., Miller, C.J., Neilsen, E.H., Newman, P.R., Nitta, A., Sheth, R.K., Snedden, S.A., & Yamauchi, C. 2003b, *Publications of the Astronomical Society of Japan*, vol. 55, p. 757.
Goto, T., Yamauchi, C., Fujita, Y., Okamura, S., Sekiguchi, M., Smail, I., Bernardi, M., & Gomez, P.L. 2003c, *Monthly Notices of the Royal Astronomical Society*, vol. 346, p. 601.
Goto, T., Yagi, M., Tanaka, M., & Okamura, S. 2004, *Monthly Notices of the Royal Astronomical Society*, vol. 348, p. 515.
Grabelsky, D.A. Cohen, R.S., Bronfman, L., & Thaddeus P. 1988, *The Astrophysical Journal*, vol. 331, p. 181.
Green, H.S. 1952, *The Molecular Theory of Fluids* (Amsterdam: North-Holland).
Gunn, J.E., & Gott, J.R., 1972, *The Astrophysical Journal*, vol. 176, p. 1.

Haan, S., Schinnerer, E., Emsellem, E., Garca-Burillo, S., Combes, F., Mundell, C.G., Rix, H.-W. 2009, *The Astrophysical Journal*, vol. 692, p. 1623.

Haken, H. 2004, *Synergetics* (New York: Springer).

Harris, S. 1971, *An Introduction to the Theory of the Boltzmann Equation* (New York: Holt, Rinehard abd Winston), p. 108.

Haynes, M.P., Giovanelli, R., & Morton, S.R. 1979, *The Astrophysical Journal*, vol. 229, p. 83.

Helfer, T., Thornley, M.D., Regan, M.W., Wong, T., Sheth, K., Vogel, S.N., Blitz, L., & Bock, D.C.-J. 2003, *The Astrophysical Journal Supplement Series*, vol. 145, p. 259.

Hernandez, O., et al. 2005, *The Astrophysical Journal*, vol. 632, p. 253.

Hernquist, L. 1990, in *Dynamics and Interactions of Galaxies*, ed. R. Wielen (Heidelberg:Springer), p. 108.

Heyer, M. 1998, *The Astrophysical Journal*, vol. 502, p. 265.

Hockney, R.W., & Eastwood, J.W. 1988, *Computer Simulation Using Particles*.

Hohl, F. 1973, *The Astrophysical Journal*, vol. 184, p. 353.

Hohl, F., & Hockney, R.W. 1969, *Journal of Computational Physics*, vol. 4, p. 306.

Hollenbach, D.J., & Tielens, A.G.G.M. 1999, *Review of Modern Physics*, vol. 71, no. 1, p. 173.

Hopkins, P.F., Bundy, K., Murray, N., Quataert, E., Lauer, T.R., Ma, C.P., 2009, *Monthly Notices of the Royal Astronomical Society*, vol. 398, p. 898.

Hoyle, F. 1953, *The Astrophysical Journal*, vol. 118, p. 513.

Hubble, E. 1936, *The Realm of the Nebulae* (New Haven: Yale University Press).

Hutchings, J.B., & Campbell, B. 1983, *Nature*, vol. 303, p. 584.

Ibata, R.A., et al. 2013, *Nature*, vol. 493, p. 62.

Icke, V. 1985, *Astronomy and Astrophysics*, vol. 144, p. 115.

Illingworth, G., Kelson, D., van Dokkum, P., & Franx, M. 2000, *Astrophysics & Space Science*, vol. 269, p. 485.

Jablonka, P., Gorgas, J., & Goudfrooij, P. 2002, *Astrophysics & Space Science*, vol. 281, p. 367.

Jalocha, J., Bratek, L., & Kutschera, M. 2008, *The Astrophysical Journal*, vol. 679, 373.

Jeans, J.H. 1928, *Astronomy and Cosmology* (Cambridge: Cambridge University Press).

Jee, M.J., Hoekstra, H., Mahdavi, A., & Babul, A. 2014, *The Astrophysical Journal*, vol. 783, p. 78.

Jog, C.J. 1996, *Monthly Notices of the Royal Astronomical Society*, vol. 278, p. 209.

Jog, C.J., & Solomon, P. M. 1984, *The Astrophysical Journal*, vol. 276, p. 114.

Jokipii, J.R., & Lerche, E.N. 1969, *The Astrophysical Journal*, vol. 157, p. 1137.

Jokipii, J.R., & Parker, E.N. 1969a, *The Astrophysical Journal*, vol. 155, p. 777.

Jokipii, J.R., & Parker, E.N. 1969b, *The Astrophysical Journal*, vol. 155, p. 799.

Julian, W.H., & Toomre, A. 1966, *The Astrophysical Journal*, vol. 146, p. 810.

Kalnajs, A.J. 1965, *Ph.D. thesis*, Harvard University.

Kalnajs, A.J. 1971, *The Astrophysical Journal*, vol. 166, p. 275.

Kalnajs, A.J. 1972, *Astrophysical Letters*, vol. 11, p. 41.

Kalnajs, A.J. 1973, *Proceedings of the Astronomical Society of Australia*, vol. 2, issue 4, p. 174.

Kashlinsky, A., Atrio-Barandela, F., Kocevski, D., & Ebeling, H. 2009, *The Astrophysical Journal*, vol. 691, p. 1479.

Kauffmann, G. 1995, *Monthly Notices of the Royal Astronomical Society*, vol. 274, p. 153.

Keene, J., Blake, G.A., Phillips, T.G., Huggins, P.J., & Beichman, C.A. 1985, *The Astrophysical Journal*, vol. 299, p. 967.

Keene, J., Lis, D.C., Phillips, T.G., & Schilke, P. 1997, in *Molecules in Astrophysics: Probes and Processes*, Proc. IAUS 178.

Kenney, J.D.P., Wilson, C.D., Scoville, N.Z., Devereux, N.A., Young, J.S. 1992, *The Astrophysical Journal*, vol. 395, p. L79.

Kennicutt, R.C. et al. 2003, *Publications of the Astronomical Society of the Pacific*, vol. 115, p. 928.

Kirkwood, J.G. 1946, *Journal of Chemical Physics*, vol. 14, p. 180.

Klypin, A.A., Kravtsov, A.V., Valenzuela, O., & Prada, F. 1999, *The Astrophysical Journal*, vol. 522, p. 82.
Kodama, T. et al. 2004, *Monthly Notices of the Royal Astronomical Society*, vol. 350, p. 1005.
Kodama, T., Arimoto, N., Barger, A.J., & Aragon-Salamanca, A. 1998, *Astronomy and Astrophysics*, vol. 334, p. 99.
Kolmogorov, A.N. 1941a, *Dokl. Akad. Nauk SSSR*, vol. 30, p. 299.
Kolmogorov, A.N. 1941b, *Dokl. Akad. Nauk SSSR*, vol. 31, p. 538.
Kolmogorov, A.N. 1941c, *Dokl. Akad. Nauk SSSR*, vol. 32, p. 16.
Koopmans, L.V.E., & de Bruyn, A.G. 2000, *Astronomy & Astrophysics*, vol. 358, p. 793.
Kormendy, J. 1979, *The Astrophysical Journal*, vol. 227, p. 714.
Kormendy, J. 1982, in *Morphology and Dynamics of Galaxies*, 12th Advanced Course of the SSAA, eds. L. Martinet & M. Mayor (Geneva Observatory: Geneva), p. 113.
Kormendy, J., & Bahcall, J.N. 1974, *The Astronomical Journal*, vol. 79, p. 671.
Kormendy, J., Bender, R., & Cornell, M.E., & 2011, *Nature*, vol. 469, p. 374.
Kormendy, J., & Kennicutt, R. 2004, *Annual Review of Astronomy and Astrophysics*, vol. 42, p. 603.
Kornreich, P., & Scalo, J. 2000, *The Astrophysical Journal*, vol. 531, p. 366.
Kowal, G., & Lazarian, A. 2007, *The Astrophysical Journal*, vol. 666, p. L69.
Krall, N.A., & Trivelpiece, A.W. 1973, *Principles of Plasma Physics* (New York: McGraw-Hill).
Kreuzer, H.J. 1981, *Nonequilibrium Thermodynamics and its Statistical Foundations* (Oxford: Oxford University Press).
Kuijken, K., & Gilmore, G. 1989, *Monthly Notices of the Royal Astronomical Society*, vol. 239, p. 605.
Kulsrud, R.M. 1972, in IAU coll. 10, it Gravitational N-Body Problem, ed. M. Lecar (Dordrecht: Reidel), p. 337.
Lada, E.A., Bally, J., & Stark, A.A. 1991, *The Astrophysical Journal*, vol. 368, p. 432.
Laine, S., Shlosman, I., Knapen, J.H., & Peletier, R.F. 2002, *The Astrophysical Journal*, vol. 567, p. 97.
Larson, R.B. 1979, *Monthly Notices of the Royal Astronomical Society*, vol. 186, p. 479.
Larson, R.B. 1981, *Monthly Notices of the Royal Astronomical Society*, vol. 194, p. 809.
Larson, R.B., Tinsley, B.M., & Caldwell, C.N. 1980, *The Astrophysical Journal*, vol. 237, p. 692.
Lau, Y.Y., & Bertin, G. 1978, *The Astrophysical Journal*, vol. 226, p. 508.
Lee, B., Giavalisco, M., Williams, C.C., Guo, Y., Lotz, J., Van der Wel, A., Ferguson, H.C., Faber, S.M., Koekemoer, A., Grogin, N., Kocevski, D., Conselice, C.J., Wuyts, S., Dekel, A., Kartaltepe, J., & Bell, E.F. 2013, *The Astrophysical Journal*, vol. 774, p. 47.
Leung, C.M., Kutner, M.L., & Mead, K.N. 1982, *The Astrophysical Journal*, vol. 262, p. 583.
Levinson, F.H., & Roberts W.W., Jr. 1981, *The Astrophysical Journal*, vol. 245, p. 465.
Liboff, R.L. 2003, *Kinetic Theory: Classical, Quantum, and Relativistic Descriptions* (New York: Springer), 3rd ed.
Liddle, A. 2003, *An Introduction to Modern Cosmology* (London: Wiley).
Lilly, S., Abraham, R., Brinchmann, J., Colless, M., Crampton, D., Ellis, R., Glazebrook, K., Hammer, F., Le Fevre, O., Mallen-Ornelas, G., Shade, D., & Tresse, L., 1998, in *The Hubble Deep Field*, eds. M. Livio, S.M. Fall & P. Madau (Cambridge: Cambridge University Press), p. 107.
Lifshitz, E.M., & Pitaevskii L.P. 1981, *Physical Kinetics* (Oxford: Butterworth - Heinenann).
Lin, C.C. 1967, in *Relativity Theory and Astrophysics Vol. 2: Galactic Structure*, ed. J. Ehlers (Providence: AMS), p. 66.
Lin, C.C., & Lau, Y.Y. 1979, *Studies in Applied Mathematics*, vol. 60, p. 97.
Lin, C.C., & Shu, F.H. 1964, *The Astrophysical Journal*, vol. 140, p. 646.
Lin, C.C., & Shu, F.H. 1966, *Proceedings of the National Academy of Sciences*, vol. 55, p. 229.
Lin, C.C., Yuan, C. & Shu, F.H. 1969, *the Astrophysical Journal*, vol. 155, p. 721; vol. 156, p. 797.
Lin, D.N.C., & Pringle, J.E. 1987, *The Astrophysical Journal*, vol. 320, p. L87.
Lin, D.N.C., Pringle, J.E., & Rees, M.J. 1988, *The Astrophysical Journal*, vol. 328, p. 103.
Lindblad, B. 1963, *Stockholms Obs. Ann.*, vol. 22, No.5.

Liszt, H.S., & Burton, W.B. 1981, *The Astrophysical Journal*, vol. 243, p. 778.
López-Sanjuan, C., Balcells, M., Pérez-Gónzalez, P.G., Barro, G., García-Dabó, C.E., Gallego, J., & Zamorano, J. 2009, *Astronomy & Astrophysics*, vol. 501, p. 505.
Lubow, S.H. 1988, in *Applied Mathematics, Fluid Mechanics, Astrophysics: A Symposium to Honor C. C. Lin*, eds. D.J. Benney, F.H. Shu & C. Yuan (Teaneck: World Scientific), p. 358.
Lubow, S.H., Balbus, S.A., & Cowie L.L. 1986, *The Astrophysical Journal*, vol. 309, p. 496.
Lucentini, J. 2002, *Sky & Telescope*, September, p. 36.
Lynden-Bell, D., & Kalnajs, A.J. 1972, *Monthly Notices of the Royal Astronomical Society*, vol. 157, p. 1.
Lynden-Bell, D., & Ostriker, J.P. 1967, *Monthly Notices of the Royal Astronomical Society*, vol. 136, p. 293.
Lynden-Bell, D., & Wood, R., 1968, *Monthly Notices of the Royal Astronomical Society*, vol. 138, p. 495.
Magnani, L., LaRosa, T.N., & Shore, S. 1993, *The Astrophysical Journal*, vol. 402, p. 226.
Mahdavi, A., Hoekstra, H., Babul, A., & Balam, D.D. 2007, *The Astrophysical Journal*, vol. 668, p. 806.
Mark, J.W.-K. 1974, *The Astrophysical Journal*, vol. 193, p. 539.
Mark, J.W.-K. 1976, *The Astrophysical Journal*, vol. 205, p. 363.
Martínez-García, E.E., Gonzalez-Lopezlira, R.A., & Bruzuel, A.G. 2009, *The Astrophysical Journal*, vol. 694, p. 512.
Martinez-Garcia, E.E., Gonzalez-Lopezlira, R.A., & Bruzuel, A.G. 2011, *The Astrophysical Journal*, vol. 734, p. 122.
McGaugh, S.S., & de Blok, W.J.G. 1998, *The Astrophysical Journal*, vol. 499, p. 41.
McGaugh, S.S., Schombert, J.M., de Blok, W.J.G., & Zagursky, M.J. 2010, *The Astrophysical Journal*, vol. 708, p. L14.
Merritt, D., & Ferrarese, L. 2001, *The Astrophysical Journal*, vol. 547, p. 140.
Mihos, J.C., & Hernquist, L., 1994a, *The Astrophysical Journal*, vol. 427, p. 112.
Mihos, J.C., & Hernquist, L., 1994b, *The Astrophysical Journal*, vol. 437, L47.
Miller, R.H. 1967, *Journal of Computational Physics*, vol. 2, p. 1.
Miller, R.H. 1971, *Astrophysics and Space Science*, vol. 14, p. 73.
Miller, R.H. 1976, *Journal of Computational Physics*, vol. 21, p. 400.
Miller, R., & Prendergast, K.H. 1968, *The Astrophysical Journal*, vol. 151, p. 699.
Mo, H., van den Bosch, F., & White, S.D.M. 2010, *Galaxy Formation and Evolution* (Cambridge: Cambridge University Press).
Monaghan, J.J. 1992, *Annual Review of Astronomy & Astrophysics*, vol. 30, p. 543.
Moore, B. 2004, in *Clusters of Galaxies: Probes of Cosmological Structure and Galaxy Evolution*, eds. J.S. Mulchaey, A. Dressler, and A. Oemler (Cambridge: Cambridge University Press), p. 295.
Moore, B., Katz, N., Lake, G., Dressler, A., Oemler, A., jr. 1996, *Nature*. vol. 379, p. 613.
Moore, B., Lake, G., & Katz, N. 1998, *The Astrophysical Journal*, vol. 495, p. 139.
Moore, B., Lake, G., Quinn, T., & Stadel, J. 1999, *Monthly Notices of the Royal Astronomical Society*, vol. 304, p. 465.
Moran, S.M., Ellis, R.S., Treu, T., Smail, I., Dressler, A., Coil, A.L., & Smith, G.P., 2005, *The Astrophysical Journal*, vol. 634, p. 977.
Moran, S.M., Miller, N., Treu, T., Ellis, R.S., & Smith, G.P. 2007a, *The Astrophysical Journal*, vol. 659, p. 1138.
Moran, S.M., Ellis, R.S., Treu, T., Smith, G.P., Rich, R.M., & Smail, I. 2007b, *The Astrophysical Journal*, vol. 671, p. 1503.
Morris, S.L., Hutchings, J.B., Carlberg, R.G., Yee, H.K.C., Ellingson, E., Balogh, M.L., Abraham, R.G., Smecker-Hane, T.A. 1998, *The Astrophysical Journal*, vol. 507, p. 84.
Murphy, T.W., Jr., Armus, L., Matthews, K., Soifer, B.T., Mazzarella, J.M., Shupe, D.L., Staruss, M.A., & Neugebauer, G. 1996, *The Astronomical Journal*, vol. 111, p. 1025.
Myers, P.C. 1983, *The Astrophysical Journal*, vol. 270, p. 105.
Najita, J.R., Tiede, G.P., & Carr, J.S. 2000, *The Astrophysical Journal*, vol. 541, p. 977.

Navarro, J. 2010, Talk presented in the Sixth Harvard-Smithsonian Conference on Theoretical Astrophysics, *Dynamics from the Galactic Center to the Milky Way Halo*, https://www.cfa.harvard.edu/events/2010/dyn.
Nelan, J.E. et al. 2005, *The Astrophysical Journal*, vol. 632, p. 137.
Nicolis, G., & Prigogine, I. 1977, *Self-Organization in Nonequilibrium Systems* (New York: Wiley-Intersciences).
Noguchi, M. 1987, *Monthly Notices of the Royal Astronomical Society*, vol. 228, p. 635.
Oesch, P.A. et al. 2016, *The Astrophysical Journal*, 819, 129.
Oey, M.S., Parker, J.S., Mikles, V.J., & Zhang, X. 2003, *The Astronomical Journal*, vol. 126, p. 2317.
Ostriker, J.P. 1980, *Comments Ap.*, vol. 8, p. 177.
Ostriker, J.P., & Steinhardt, P.J. 1995, *Nature*, vol. 377, p. 600.
Ozernoy, L.M., 1974a, in *The Formation and Dynamics of Galaxies*, ed. J.R. Shakeshaft (IAU) (Dordrecht: Reidel), p. 85.
Ozernoy, L.M., 1974b, in *Confrontation of Cosmological Theories with Observational Data*, ed. M.S. Longair (IAU) (Dordrecht: Reidel), p. 227.
Papovich, C., Dickinson, M., & Ferguson, H.C. 2001, *The Astrophysical Journal*, vol. 559, p. 620.
Pawlowski, M.S., Pflamm-Altenburg, J., & Kroupa, P. 2012, *Monthly Notices of the Royal Astronomical Society*, vol. 423, p. 1109.
Peebles, P.J.E., 2002, in *Toward an International Virtual Observatory*, Garching, June 2002, arXiv:astro-ph/0209403.
Persic, M., & Salucci, P. 1992, *Monthly Notices of the Royal Astronomical Society*, vol. 258, p. 14.
Persic, M., Salucci, P., & Stel, F. 1996, *Monthly Notices of the Royal Astronomical Society*, vol. 281, p. 27.
Pettitt, A.R., Tasker, E.J., & Wadsley, J.W. 2016, *Monthly Notices of the Royal Astronomical Society*, vol. 458, p. 3990.
Pharasyn, A., Simien, F., & Heraudeau, Ph. 1997, in *Dark and Visible Matter in Galaxies*, ASP Conf. Series 117, eds. M. Persic & P. Salucci (San Francisco: ASP), p. 180.
Peletier, R.F., & Balcells, M. 1996, *The Astronomical Journal*, vol. 111, issue 6, p. 2238.
Pfenniger, D. 1986, *Astronomy & Astrophysics*, vol. 165, p. 74.
Pfenniger, D. 1998, in *Abundance Profiles: Diagnostic Tools for Galaxy History*, eds. D. Friedli, M. Edmunds, C. Robert, and L. Drissen (San Francisco: ASP), p. 237.
Pfenniger, D., & Combes, F. 1994, *Astronomy & Astrophysics*, vol. 285, p. 94.
Pfenniger, D., & Friedli, D. 1991, *Astronomy & Astrophysics*, vol. 252, p. 75.
Phillips, T.G., & Huggins, P.J. 1981. *The Astrophysical Journal*, vol. 251, p. 533.
Plionis, M., Benoist, C., Maurigordata, S., Ferrari, C., & Basilakos, S., 2003, *The Astrophysical Journal*, vol. 594, p. 144.
Plionis, M., & Basilakos, S. 2002, *Monthly Notices of the Royal Astronomical Society*, vol. 329, p. L47.
Plume, R. 1995, *Ph.D. dissertation*, University Texas at Austin.
Poggianti, B.M., Smail, I., Dressler, A., Couch, W.J., Barger, A.J., Butcher, H., Ellis, R., & Oemler, A., Jr. 1999, *The Astrophysical Journal*, vol. 518, p. 576.
Poincaré, H. 1890, *Acta Mathematica*, vol. 13, p. 1.
Poincaré, H. 1893, *Les Méthodes Nouvelles de la Mécanique Celeste* II, (reprinted by Dover, 1957).
Prigogine, I. 1969, *Introduction to Thermodynamics of Irreversible Processes*, 3rd edn. (New York: Interscience).
Prigogine, I. 1980, *From Being to Becoming: Time and Complexity in the Physical Processes* (New York: Freeman & Co).
Prigogine, I., & Stengers, I. 1984, *Order Out of Chaos* (New York: Bantam).
Pringle, J.E. 1981, *Annual Review of Astronomy and Astrophysics*, vol. 19, p. 137.
Quilis, V., Moore, B., & Bower, R. 2000, *Science*, vol. 288, p. 1617.
Quillen, A.C., Frogel, J.A., & González, R.A. 1994, *The Astrophysical Journal*, vol. 437, p. 162.

Rafikov, R.R. 2001, *Monthly Notices of the Royal Astronomical Society*, vol. 323, p. 445.
Rautiainen, P., & Salo, H. 1999, *Astronomy & Astrophysics*, vol. 348, p. 737.
Regan, M.W., Teuben, P.J., & Vogel, S.N. 1996, *The Astronomical Journal*, vol. 112, p. 2549.
Regan, M.W., Vogel, S.N., & Teuben, P.J. 1997, *The Astrophysical Journal*, vol. 482, p. L143.
Roberts, W.W. 1969, *The Astrophysical Journal*, vol. 158, p. 123.
Roberts, M.S. 1975, in *Dynamics of Stellar Systems: Proceedings from IAU Symposium No. 69*, eds. A. Hayli (Dordrecht: Reidel), p. 331.
Roberts, W.W., & Shu, F.H. 1972, *Astrophysical Letters*, vol. 12, p. 49.
Rohlfs, K. 1977, *Lectures on Density Wave Theory* (New York: Springer).
Romeo, A.B. 1990, Ph.D. thesis, SISSA, Trieste, Italy.
Romeo, A.B. 1992, *Monthly Notices of the Royal Astronomical Society*, vol. 256, p. 307.
Romeo, A.B. 1994a, *Astronomy & Astrophysics*, vol. 286, p. 799.
Romeo, A.B. 1994b, *Astrophysics and Space Science*, vol. 216, p. 357.
Romeo, A.B. 1997, *Astronomy & Astrophysics*, vol. 324, p. 523.
Romeo, A.B. 1998, *Astronomy & Astrophysics*, vol. 335, p. 922.
Romeo, A.B., & Falstad, N. 2013, *Monthly Notices of the Royal Astronomical Society*, vol. 433, p. 1389.
Romeo, A.B., & Wiegert, J. 2011, *Monthly Notices of the Royal Astronomical Society*, vol. 416, p. 1191.
Rots, A.H. 1978, *The Astronomical Journal*, vol. 83, p. 219.
Rybicki, G.B. 1971, *Astrophysics and Space Science*, vol. 14, p. 15.
Sanders, D.B., & Mirabel, F.H. 1996, *Annual Review of Astronomy & Astrophysics*, vol. 34, p. 749.
Satyapal, S., Boker, T., Mcalpine, W., Gliozzi, M., Abel, N.P., & Heckman, T. 2009, *The Astrophysical Journal*, vol. 704, p. 439.
Schroder, K., Staemmler, V., Smith, M.D., Flower, D.R., & Jacquet, R. 1991, *Journal of Physics B*, vol. 24, p. 2487.
Schwarz, M.P. 1984, *Monthly Notices of the Royal Astronomical Society*, vol. 209, p. 93.
Schwarzschild, K. 1907. *Gottingen Nachr.*, p. 614.
Scoville, N., & Young, J.S., & Lucy, L.B. 1983, *The Astrophysical Journal*, vol. 270, p. 443.
Seiden, P.E., Schulman, L. S., & Elmegreen, B. G. 1984, *The Astrophysical Journal*, vol. 282, p. 95.
Sellwood, J.A. 1987, *Annual Review of Astronomy and Astrophysics*, vol. 25, p. 151.
Sellwood, J.A. 2011, *Monthly Notices of the Royal Astronomical Society*, vol. 410, p. 1637.
Sellwood, J.A. 2013, *The Astrophysical Journal*, vol. 769, p. L24.
Sellwood, J.A. 2014, *Review of Modern Physics*, vol. 86, p. 1.
Sellwood, J.A., & Binney, J.J. 2002, *Monthly Notices of the Royal Astronomical Society*, vol. 336, p. 785.
Sellwood, J.A., & Carlberg, R.G. 1984, *The Astrophysical Journal*, vol. 282, p. 61.
Sellwood, J.A., & Evans, N.W. 2001, *The Astrophysical Journal*, vol. 546, p. 176.
Sellwood, J.A., & McGaugh, S.S. 2005, *The Astrophysical Journal*, vol. 634, p. 70.
Sersic, J.L. 1968, *Atlas de galaxias australes* (Observatorio Astronomica, Cordoba).
Shapley, A.E., Steidel, C.C., Adelberger, K.L., Dickinson, M., Giavalisco, M., & Pettini, M. 2001, *The Astrophysical Journal*, vol. 562, p. 95.
Shaw, M.A., Combes, F., Axon, D.J., & Wright, G.S. 1993, *Astronomy & Astrophysics*, vol. 273, p. 31.
Shen, J., Rich, R.M., Kormendy, J., Howard, C.D., De Propris, R., & Kunder, A. 2010, *The Astrophysical Journal*, vol. 720, p. L72.
Shlosman, U., Frank, J., & Begelman, M.C. 1989, *Nature*, vol. 338, p. 45.
Shvartzvald, Y. et al. 2016, *The Astrophysical Journal*, vol. 831, p. 183.
Shu, F.S. 1982, *The Physical Universe: An Introduction to Astronomy* (Mill Valley: University Science Books).
Shu, F. 1992, *The Physics of Astrophysics*, vol. 2 (Mill Valley: University Science Books).
Shu, F.H., Milione, V., & Roberts, W.W., Jr. 1973, *The Astrophysical Journal*, vol. 183, p. 819.
Shu, F.H., Yuan, C., & Lissauer, J.J. 1985, *The Astrophysical Journal*, vol. 291, p. 356.

Simkin, S.M., Su, H.J., & Schwarz, M.P. 1980, *The Astrophysical Journal*, vol. 237, p. 404.
Snow, C. 1952, *Hypergeometric and Legendre Functions with Applications to Integral Equations of Potential Theory* (NBS/AMS 19) (Washington DC: National Bureau of Standards).
Solomon, P., Sanders, D.B., & Scoville, N.Z. 1979, *The Astrophysical Journal*, vol. 232, p. 89.
Sparke, L.S., & Sellwood, J.A. 1987, *Monthly Notices of the Royal Astronomical Society*, vol. 225, p. 653.
Spitzer, L. 1978, *Physical Processes in the Interstellar Medium* (New York: Wiley).
Spitzer, L. 1987, *Dynamical Evolution of Globular Clusters* (Princeton: Princeton University Press).
Spitzer, L., & Chevalier, R.A. 1973, *The Astrophysical Journal*, vol. 183, p. 565.
Spitzer, L., & Schwarzschild, M. 1951, *The Astrophysical Journal*, vol. 114, p. 385.
Spitzer, L., & Schwarzschild, M. 1953, *The Astrophysical Journal*, vol. 118, p. 106.
Stanley, H.E. 1971, *Introduction to Phase Transitions and Critical Phenomena* (Oxford: Oxford University Press).
Stark, A.A. 1979, *Ph.D. thesis*, Princeton University.
Stark, A.A., Chamberlin, R.A., Cheng, J., Ingalls, J.G., & Wright, G. 1997, *Review of Scientific Instruments*, vol. 68, issue 5, p. 2200.
Steidel, C.C., Adelberger, K.L., Dickinson, M., Giavalisco, M., Pettini, M., & Kellogg, M., 1998, *The Astrophysical Journal*, vol. 492, p. 428.
Steidel, C.C., Giavalisco, M., Dickinson, M., & Adelberger, K.L., 1996, *The Astronomical Journal*, vol. 112, p. 353.
Stockton, A. 1982, *The Astrophysical Journal*, vol. 257, p. 33.
Stone, J.M., Ostriker, E.C., & Gammie, C.F. 1998, *The Astrophysical Journal Letters*, vol. 508, p. 99.
Stutzki, J., Stacey, G.J., Genzel, R., Harris, A. I., Jaffe, D.T., & Lugten, J.B. 1988, *The Astrophysical Journal*, vol. 332, p. 379.
Sulentic, J.W., Keel, W.C., & Telesco, C.M., eds. 1990, IAU Colloquium 124, *Paired and Interacting Galaxies* (NASA Publ. 3098).
Tagger, M., Sygnet, J.F., Athanassoula, E., & Pellat, R. 1987, *The Astrophysical Journal*, vol. 318, p. L43.
Tanaka, M., Goto, T., Okamurka, S., Shimasaku, K., & Brinkman, J. 2004, in *Outskirts of Galaxy Clusters: Intense Life in the Suburbs*, Proc. IAUC 195, ed. A. Diaferio, p. 444.
Thomas, D., Maraston, C., & Bender, R. 2002, in *Astronomy with Large Telescopes from Ground and Space, Reviews in Modern Astronomy*, vol. 15, p. 219, eds R.E. Schielicke (New York: Wiley).
Thomasson, M. 1989, *Research Report*, No. 162, Department of Radio and Space Science with Onsala Space Observatory, Chalmers University of Technology, Goteborg.
Thomasson, M., Donner, K.J., & Elmegreen, B.G. 1991, *Astronomy & Astrophysics*, vol. 250, p. 316.
Thomasson, M., & Donner, K.J. 1993, *Astronomy & Astrophysics*, vol. 272, p. 153.
Thompson, L.A. 1981, *The Astrophysical Journal*, vol. 244, p. L43.
Tielens, A.G.G.M, & Hollenbach, D.J. 1985, *The Astrophysical Journal*, vol. 291, p. 722.
Tisserand, P. et al. 2007, *Astronomy & Astrophysics*, vol. 469, p. 387.
Toomre, A. 1964, *The Astrophysical Journal*, vol. 139, p. 1217.
Toomre, A. 1969, *The Astrophysical Journal*, vol. 158, p. 899.
Toomore, A. 1977, *Annual Review of Astronomy and Astrophysics*, vol. 15, p. 437.
Toomre, A. 1981, in *Structure and Dynamics of Normal Galaxies*, eds. S. M. Fall & D. Lynden-Bell (Cambridge: Cambridge University Press), p. 111.
Toomre, A., & Toomre, J. 1972, *The Astrophysical Journal*, vol. 178, 623.
Tremaine, S. 1981, in *The Structure and Evolution of Normal Galaxies*, ed. S.M. Fall and D. Lynden-Bell (Cambridge: Cambridge University Press), p. 67.
Tremaine, S., & Weinberg, S. 1984, *The Astrophysical Journal*, vol. 282, p. 5.
Trentham, N.A., 1997, *Ph.D. Dissertation*, University of Hawaii.
Treu, T. 2004, in *Clusters of Galaxies: Probes of Cosmological Structure and Galaxy Evolution*, eds. J.S. Mulchaey, A. Dressler & A. Oemler (Cambridge: Cambridge Univ. Press), p. 177.

Tully, R. B. 1988, *Nearby Galaxies Catalog* (Cambridge: Cambridge University Press).
Tully, R.B., & Fisher, J.R. 1977, *Astronomy & Astrophysics*, vol. 54, p. 661.
Uhlenbeck, G.E., & Ford, G.W. 1963, *Statistical Mechanics*, American Mathematical Society.
Umetsu, K., & Broadhurst, T. 2008, *The Astrophysical Journal*, vol. 684, p. 177.
van der Kruit, P.C., & Freeman, K.C. 1986, *The Astrophysical Journal*, vol. 303, p. 556.
van der Kruit, P.C., & Searle, L., 1981, *Astronomy & Astrophysics*, vol. 95, p. 105.
van der Kruit, P.C., & Searle, L., 1982, *Astronomy & Astrophysics*, vol. 110, p. 61.
Vandervoort, P.O. 1971, *The Astrophysical Journal*, vol. 166, p. 37.
van Dokkum, P.G. et al. 1998, *The Astrophysical Journal*, vol. 504, p. L17.
van Dokkum, P.G. et al. 2011, *The Astrophysical Journal*, vol. 743, p. L15.
van Dokkum, P., Abraham, R., Brodie, J., Conroy, C., Danieli, S., Merritt, A., Mowla, L., Romanowsky, A., & Zhang, J. 2016, *The Astrophysical Journal Letters*, vol. 828, p. 6.
van Dokkum, P., & Franx, M. 2001, *The Astrophysical Journal*, vol. 553, p. 90.
Viscuso, P.J., & Chernoff, D.F. 1988, *The Astrophysical Journal*, vol. 327, p. 364.
Vogelsberger, M., Genel, S., Springel, V., Torrey, P., Sijacki, D., Xu, D., Snyger, G., Bird, S., Nelson, D., & Hernquist, L. 2014, *Nature*, vol. 509, p. 177.
Voglis, N., Stavropoulos, I., & Kalapotharakos, C. 2006, *Monthly Notices of the Royal Astronomical Society*, on-line (astro-ph/0606561).
von Hoerner, S. 1951, *Zeitschrift fur Astrophysik*, vol. 30, p. 17.
von Weizsacker, C.F. 1951a, *The Astrophysical Journal*, vol. 114, p. 165.
von Weizsacker, C.F. 1951b, in *Problems of Cosmical Aerodynamics* p. 158.
von Weizsacker, C.F. 1951c, in *Problems of Cosmical Aerodynamics* p. 200.
Walker, M. 2014, Talk presented in the Eighth Harvard-Smithsonian Conference on Theoretical Astrophysics, *Debates on the Nature of Dark Matter,* https://www.cfa.harvard.edu/events/2014/sackler/.
Walter, F., Brinks, E., de Blok, W.J.G., Bigiel, F., Kennicutt, R.C.Jr., Thornley, M.D., & Leroy, A. 2008, *The Astronomical Journal*, vol. 136, p. 2563.
Watkins, R., Feldman, H.A., & Hudson, M.J. 2009, *Monthly Notices of the Royal Astronomical Society*, vol. 392, p. 743.
Wegener, A. 1929, *Die Entstehung der Kontinente und Ozeane*, first Dover edition of the English translation of the fourth revised edition entitled *The Origin of Continents and Oceans* was published in 1966.
Weinberg, M.D. 1993, *The Astrophysical Journal*, vol. 410, p. 543.
Weinberg, S. 1995, *The Quantum Theory of Fields*, vol. I (Cambridge: Cambridge University Press).
West, M.J., 2001, *Bulletin of the American Astronomical Society*, vol. 33, p. 887.
White, S.D.M., 2009, in *The Galaxy Disk in Cosmological Context*, Proceedings of IAU Symposium 254, eds. J. Anderson, J. Bland-Hawthorn & B. Norstrom, (Cambridge: Cambridge Univ. Press), p. 19.
White, S.D.M. 2010, Summary talk in the Sixth Harvard-Smithsonian Conference on Theoretical Astrophysics, *Dynamics from the Galactic Center to the Milky Way Halo*, https://www.cfa.harvard.edu/events/2010/dyn.
White, S.D.M., & Frenk, C. 1991, *The Astrophysical Journal*, vol. 379, p. 52.
White, S.D.M., & Rees, M.J., 1978, *Monthly Notices of the Royal Astronomical Society*, vol. 183, p. 341.
White, R.L. 1988, *The Astrophysical Journal*, vol. 330, p. 26.
Wielen, R. 1975, in Colloq. CNRS 241, *La Dynamique des Galaxies Spirales*, ed. L. Weliachew (Paris: CNRS), p. 357.
Wielen, R. 1977, *Astronomy & Astrophysics*, vol. 60, p. 263.
Wielen, R. ed. 1990, *Dynamics and Interactions of Galaxies* (Heidelberg: Springer).
Wielen, R., & Fuchs, B. 1990, in *Dynamics and Interactions of Galaxies*, ed. R. Wielen (Berlin: Springer).

Wiggins, S. 2003, *Introduction to Applied Nonlinear Dynamical Systems and Chaos*, 2nd ed. (New York: Springer).
Williams, R.E. 1997, presented in IAU 23rd Genl Assemb., Kyoto.
Williger, G.M., Campusano, L.E., Clowes, R.G., & Graham, M.J. 2002, *The Astrophysical Journal*, vol. 578, p. 708.
Wirth, A., & Shaw, R. 1983, *The Astronomical Journal*, vol. 88, p. 171.
Wolfe, A.M., 2001, in *Galaxy Disks and Disk Galaxies*, eds. J.G. Funes & E.M. Corsini (San Francisco: ASP), p. 619.
Woodward, P.R., 1973, *Ph.D. thesis*, University of California, Berkeley.
Woodward, P.R., 1975, *The Astrophysical Journal*, vol. 195, p. 61.
Wyse, F.G., Gilmore G., & Franx M. 1997, *Annual Review of Astronomy and Astrophysics*, vol. 35, p. 637.
Young, J.S. 1990, in *Second Wyoming Conference on the ISM in Galaxies*, eds. H. Thornson & M. Shull (Dordrecht: Kluwer), p. 67.
Young, J.S., & Scoville, N.Z. 1982a, *The Astrophysical Journal*, vol. 258, p. 467.
Young, J.S., & Scoville, N.Z. 1982b, *The Astrophysical Journal*, vol. 260, p. L11.
Young, J.S., Taccono, L.J., & Scoville, N.Z. 1983, *The Astrophysical Journal*, vol. 269, p. 136.
Yurin, D., & Springel, V. 2015, *Monthly Notices of the Royal Astronomical Society*, vol. 452, p. 2367.
Yvon, J. 1937, *Act. Sci. Ind.*, No. 524 and 543 (Paris: Hermann).
Zeldovich, Y.B. 1970, *A&A*, vol. 5, p. 84.
Zepf. S.E., & Koo, D.C. 1989, *The Astrophysical Journal*, vol. 337, p. 34.
Zhang, X. 1992, *Ph.D. dissertation*, University of California at Berkeley.
Zhang, X. 1996, *The Astrophysical Journal*, vol. 457, p. 125.
Zhang, X, 1997, in Proceedings of the IAU Colloquium 163, *Accretion Phenomena and Related Outflows*, eds. D. Wickramasinghe, G. Birknell & L. Ferrario, ASP Conf. Series, vol. 121, p. 840.
Zhang, X. 1998, *The Astrophysical Journal*, vol. 499, p. 98.
Zhang, X. 1999, *The Astrophysical Journal*, vol. 518, p. 613.
Zhang, X. 2002, *Astrophysics & Space Science* (Kluwer/Springer), vol. 281, p. 281.
Zhang, X. 2003, *Journal of the Korean Astronomical Society*, vol. 36, p. 223.
Zhang, X. 2004, in *Penetrating Bars through Masks of Cosmic Dust*, eds. D.L. Block, I. Puerari, K.C. Freeman, R. Gross, & E.K. Block (Dordrecht: Springer), *Astrophysics & Space Science Library*, vol. 319, p. 317.
Zhang, X. 2008, *The Publications of the Astronomical Society of the Pacific*, vol. 120, issue 864, p. 121 (DOI: 10.1086/527571).
Zhang, X. 2016, *Astronomy and Computing*, vol. 17, p. 86 (www.sciencedirect.com/science/article/pii/S2213133716300828).
Zhang, X., & Buta, R.J. 2007, *The Astronomical Journal*, vol. 133, p. 2584.
Zhang, X., & Buta, R.J. 2012, *Astronomical and Astrophysical Transactions*, vol. 27, issue 2, p. 339.
Zhang, X., & Buta, R. 2015, *New Astronomy*, vol. 34, p. 65 (www.sciencedirect.com/science/article/pii/S1384107614000888).
Zhang, X., Lee, Y., Bolatto, A., & Stark, A.A. 2001, *The Astrophysical Journal*, vol. 553, p. 274.
Zhang, X., Wright, M., & Alexander, P. 1993, *The Astrophysical Journal*, vol. 418, p. 100.
Zwaan, M.A., van der Hulst, J.M., de Bolk, W.J.G., & McGaugh, S.S. 1995, *Monthly Notices of the Royal Astronomical Society*, vol. 273, p. L35.
Zwicky, F. 1956, *Ergebnisse der exakten Naturwissenschaften*, vol. 29, p. 344.

Index

accretion
– and black-hole-mass/bulge-mass correlation 243–245
– and bulge building 5, 170–172, 282
– and color-magnitude relation 233–234
– and disk surface density evolution 64–65
– and evolution of galaxy scaling relations, see galaxy, scaling relations
– and post-interaction secular evolution 222–229
– and the formation of the Hubble sequence 272
– disk analogy 65–66, 296
– disks 64
– due to density waves 64, 65
– flux
  – and 1/r surface density profile 65
– in black-hole/AGN disks 5
– in gas V, 1, 2, 4, 5, 61, 187, 283
– in interacting galaxies 234–243
– in stars 2, 5
– of external gas 282
– rate 60, 67, 68, 95, 146
  – determined from galaxy images 172–179, 181–187
  – equation 61
  – in N-body simulations 95–96, 109–136
action-angle approach
– limitations in treating collective effects 265–266, 284
active galactic nuclei (AGN) 5, 235, 236
– fueling through successive resonances 243
– in bulge-less galaxies 244
adiabatic
– compression of dark matter halos 272
– condition for gas in Eulerian fluid equations 295
– invariance of stellar orbit 4, 10, 21, 67
age
– and velocity dispersion relation of the solar neighborhood stars 3, 60, 147, 188–190, 195, 202, 205, 250
angular momentum
– barrier 5, 15, 294

– conservation 17, 44, 45, 64, 286
– exchange between the basic state and the density wave 10, 11, 15–17, 21, 24–27, 34–57, 59, 62, 64, 86, 115, 120, 125, 146, 150, 152, 157, 163, 169, 187, 188, 195, 216, 248, 250, 251, 261, 268, 284, 285
– flux 8, 34, 38, 42, 46, 48, 93, 165, 169, 181
– transport by density waves 7–9, 11, 15–17, 25, 34–56, 62, 163, 242, 285
anti-spiral theorem 293
arrow of time
– in many degrees-of-freedom Hamiltonian systems VI, 95, 290, 293
AST/RO (Antarctic Submillimeter Telescope/Remote Observatory)
– observations of the Carina Molecular Cloud Complex 197–201
asymptotic stability 95, 144, 256
ATLAS^{3D} project
– results and implications 4, 283
attractor
– of dynamical evolution 107, 144, 227, 234, 244, 256

Bénard problem 14
bar
– and SWING amplifier 8
– and twin-peaks 100
– as density waves/modes VI, 9–11, 17, 21, 26, 28, 29, 34, 50, 94, 107, 108, 147, 150–152, 167, 168, 170, 216, 222, 229, 232, 235, 236, 247, 269, 281, 282
– as dissipative structures 15
– driven spiral 156, 161, 164
– gas accretion in 1, 4
– in N-body simulations 1, 115–116, 118–145, 251–260
– in NGC 1530 153–157
– in NGC 3627 234–243
– in NGC 4321 158–161
– super-fast 108, 120, 162–164
– within-bar 236

bars
- in galaxies 10, 14, 15, 26, 108, 119, 120, 133, 151, 153, 162–164, 167, 229, 232, 236, 243, 265, 281

baryon
- and Big Bang nucleosynthesis 274
- and secular mass accretion 212, 213
- and star formation 233
- as the composition of galactic dark matter 203, 213, 272, 273, 275
- contribution to rotation curves 175, 203, 272
- dark, in the form of disk gas 275
- fraction in LCDM simulations 274
- missing 274

basic state of the galactic disk 275
- and flat rotation curve 66
- and interaction with density waves 2, 8–11, 15–17, 21, 24–27, 34–57, 59, 62, 63, 65, 69, 86, 93, 96, 103, 115, 120, 125, 127, 146, 147, 150, 152, 157, 163, 165, 168, 169, 173, 179, 187–189, 195, 203, 216, 248, 250, 251, 285, 295
- and the secular evolution of mass distribution 2, 6, 63, 69, 93, 94, 96, 98, 144, 147, 164, 189, 190, 247, 251, 256, 258–260, 268, 281, 295
- choice in N-body simulations 67, 71–74, 76, 79, 86, 95, 100, 111, 115, 134, 168
- choice to allow unstable modes 17, 18, 111, 260–266, 281, 284
- definition 1
- in the theory of Lynden-Bell and Kalnajs (1972) 67

BBGKY hierarchy VII, 268, 287–292

bell-shaped
- torque coupling and angular momentum flux curve 17, 34–36, 38, 44, 45, 50, 53, 63, 165, 169, 179

Big Bang
- nucleosynthesis, see nucleosynthesis, Big Bang

black hole
- accretion disk 235, 244
- growth, through nested resonances 245
- mass, correlation with bulge mass 5, 244, 245

Boltzmann
- H-theorem 268

Boltzmann equation
- collisional VII, 68, 268, 287–295, 297
- collisionless VIII, 1, 265, 290, 291

broadening
- of resonances 38, 67, 68

bulge
- and disk connection 203
- and fundamental-plane relation 212, 213
- and Hubble classification 144, 147, 168
- and inner Lindblad resonance 8
- and Q-barrier 9
- and radial migration 262
- and the calculation of potential-density phase shift 152
- building through secular evolution 1, 3–5, 146, 147, 169–172, 177, 181, 208, 259, 263, 271, 282
- building, through gas accretion
  - limitations 4–5
- classical 4, 282
- early formation 149
- in N-body simulations 72, 73, 86, 110–115, 127
- in cluster galaxies 214–232
- mass and black-hole mass correlation 243–245
- possible skewness 149
- pseudo 1, 3, 181, 203, 282

bulge-to-disk ratio (B/D) 168, 171, 214–216, 232, 283

Bullet cluster (1E 0657-558) 274

Butcher-Oemler effect 3, 214–234, 281

Carina molecular complex
- observations, see AST/RO, observations of Carina molecular cloud complex

Chapman-Enskog procedure 294, 295, 297, 298

chemical clock 267

CL 0024+16 220, 224–228, 231

closure relations VIII, 17, 39, 46, 53–55, 59, 70, 102, 109, 251, 268, 269, 284, 285, 294, 296

cloud-in-cell (CIC) method 75, 138

cluster
- galactic 3, 98, 173, 214–235, 270–275, 281, 282
- globular 144, 213, 273
- open 144
- stellar 144, 199, 213, 273

cluster galaxy evolution 217–234
collective effects VII, VIII, X, 1, 2, 5, 10–13, 18, 24, 27, 28, 33, 52, 55–57, 65, 67–71, 74, 77, 94, 97, 98, 102–104, 106–109, 111, 113, 131, 133, 137, 138, 140, 142–144, 147, 148, 162, 164, 168, 170, 172, 183, 195, 210, 262, 263, 265, 266, 281, 284, 296, 297
collisionless
– Boltzmann equation, see Boltzmann equation, collisionless
– disk 54
– shock VIII, IX, 11, 17, 29, 33, 35, 36, 46, 52, 68, 71, 74, 77, 85, 99, 102, 103, 108, 109, 111, 188, 196, 202, 205, 246, 248, 250, 251, 256, 262, 263, 265, 268, 270, 273, 274, 284, 285, 295, 298
– simulation 94
– system 110, 111, 294
color-magnitude relation 215, 217, 230, 233–234
Columbia CO Survey 197, 199
continuum and differential approach in classical mechanics 12, 14, 17, 47, 52, 54, 56, 68, 70, 284–287, 295
corotation resonance (CR) 8–11, 17, 26, 29, 32, 34–36, 38–40, 42, 44–46, 56, 59, 62, 63, 66, 73, 79, 81, 83, 86, 91, 92, 95, 98, 106–109, 111, 115, 120, 123, 125, 127–133, 136, 139, 141, 146, 149, 150, 153, 154, 156, 158, 159, 161–163, 170, 177, 186, 192, 196, 205, 240, 246, 247, 250, 257, 262, 265
– determination by the Potential-Density Phase Shift (PDPS) approach 149–169, 240
cosmic microwave background (CMB) 270, 273
cosmology
– hot dark matter 279
– LCDM 1, 231, 270–272, 274, 275, 277–279
– relation to secular evolution 269
– self-interacting dark matter 279
– warm dark matter 279
COSMOS
– observations of the coevolution of bulges and black holes 5, 244
Cr 228 star cluster 200

damped L_α system 208
damping mechanism

– for the growing density waves/modes 8, 9, 34, 39, 40, 90, 99, 101, 102, 150, 247, 248, 251, 259
dark
– baryons 273–275
– energy 231
– halo 152, 177, 203, 272, 278
 – pressure support 277, 278
– matter 1, 177, 203, 206, 213, 220, 231, 270–275, 277–280
density wave
– and angular momentum transport and deposition, see angular momentum, exchange between the basic state and the density wave; and angular momentum, transport by density wave
– and bell-shaped torque coupling and angular momentum flux, see bell-shaped, torque coupling and angular momentum flux curve
– and closure relations at the quasi-steady state, see closure relations
– and galaxy scaling relations, see galaxy, scaling relations
– and interaction with the basic state, see angular momentum, exchange betwee the basic state and the density wave
– and potential-density phase shift, see potential-density phase shift
– and secular evolution, see secular morphological evolution of galaxies
– and singularity behavior 17, 33, 52, 68, 70, 108, 262, 263, 265, 268, 269, 284, 285, 293, 298
– and the accelerated entropy evolution, see dissipative structures, and the accelerated entropy evolution
– and the conservation of the Jacobi integral, see Jacobi integral, conservation of
– and the over-reflection mechanisms at corotation, see over-reflection mechanisms, for density waves at corotation
– collisionless shock, see collisionless, shock
– corotation radii, see corotation resonance (CR)
– damping mechanism, see damping mechanism, for the growing density waves/modes
– group velocity 7, 42, 44–46

density wave (*continued*)
- in Eulerian fluid formulation VIII, 1, 26, 29–33, 47, 48, 52, 55, 68, 79, 292, 294, 295
- Lin-Shu hypothesis 6
- Lynden-Bell & Kalnajs formulation 7, 8, 16, 34–36, 38, 39, 42, 45–49, 56, 67–69, 181, 292
- modes VII, VIII, 1, 5, 6, 9–13, 16–18, 25, 26, 28, 29, 32–35, 38, 42, 50, 55, 56, 62, 66–70, 73, 79, 85, 89, 94, 99, 101, 103, 106, 107, 111, 115, 117, 119, 120, 126, 127, 135, 137, 141, 143, 144, 146, 147, 149–153, 157, 159, 162–169, 177, 179, 181, 190, 204, 222, 232, 236, 243, 247, 248, 251, 259, 260, 262, 265, 268, 271, 281, 282, 284, 286, 292, 295, 296
  - stabilization VIII, 2, 10, 39
- pattern speed, *see* pattern speed, of density waves
  - determined by Tremaine-Weinberg (TW) method, *see* Tremaine-Weinberg (TW) method, for density-wave pattern speed determination
- resonant interaction, *see* resonance, wave-particle
- SWING mechanism, *see* SWING mechanism, for density wave amplification
- transient 7, 10, 17, 28, 67, 115, 119, 123, 132, 133, 159, 162, 167, 168, 204, 205, 247, 256, 258, 261, 263, 293
- WKBJ treatment 6, 7, 15, 18–21, 33, 57, 58, 78, 79, 83, 101

differential rotation in galaxies 6–8, 209
diffusion
- of stellar orbit 5, 6, 15
- of stellar random velocity 60, 188–192, 196, 204

disk-halo conspiracy 203
dissipative structures
- and the accelerated entropy evolution VII, 11, 14–16, 35, 46, 62, 107, 148, 149, 169, 173, 177, 181, 266, 267, 269, 283, 285, 292, 293, 297

distribution function
- in phase space 265, 287–290
- of stars 48

down-sizing trend of galaxy assembly 148, 234, 271
dust lanes
- at leading edge of spiral arms 81, 85, 99, 157, 158, 236

dwarf galaxies 207, 215, 216, 232, 271, 273, 274, 282
dynamical friction 6, 133, 228, 274

elliptical galaxies
- boxy 213
- dichotomy of properties 213
- disky 4, 5, 174, 177, 179, 183, 212, 213
- scaling relations 147, 209–213
- stellar populations in 171

emergent laws and phenomena VI–VIII, 2, 13, 15, 17, 55, 71, 267–269, 284, 287, 297

entropy
- accelerated evolution of, *see* dissipative structures, and the accelerated entropy evolution
- coarse-grained VI, 107, 290
- direction of evolution VI, 5, 11, 16, 46, 62, 107, 148, 149, 169, 173, 177, 181, 227, 267, 268, 283, 285, 287, 290, 292, 293
- fine-grained VI
- operator 297
- production and transport VII, 11, 15, 35
- specific 295

epicycle
- approximation and parameters 19, 21, 30, 56, 59, 81, 98, 121, 158, 191

equilibrium
- dynamical IX, 6, 34, 36, 39, 75, 99, 109, 165, 167–169, 183, 186, 227, 266, 269, 281, 292
- thermodynamic 292, 294

Eulerian fluid equations VIII, 1, 26, 29, 32, 33, 47, 48, 52, 55, 68, 79, 167, 292, 294, 295, 297

exponential
- decay profile of wave action 40, 43, 46
- disk 64, 73, 146, 193
  - modified 72, 73, 86, 115, 134
  - Sersic 172
- divergence of the orbit from true orbit 94, 143
- growth of wave amplitude 38, 44, 86, 90
- length scale 193

- sensitivity of stellar orbit to perturbations 107
- spacing in the computational grid 74, 139, 142

Faber-Jackson relation, *see* galaxy, scaling relations, Faber-Jackson
fluctuation-dissipation theorem 35, 266, 269
Fundamental-Plane relation, *see* galaxy, scaling relations, Fundamental-Plane

Galaxy
- The (Milky Way) 3, 4, 7, 102, 147, 171, 174, 177, 189–191, 193, 196, 197, 199, 202–204, 241, 243, 271, 273, 278, 282

galaxy
- classification scheme of Hubble V, 10, 144, 147, 168, 281
- color-magnitude relation, *see* color-magnitude relation
- formation
  - in cosmological context 275–279
- harassment 215–217, 219, 221, 222, 226, 232, 282
- merger 1, 3–5, 147, 149, 203, 215, 217–221, 225, 226, 230, 231, 244, 282, 283, 297
  - and the LCDM paradigm 213
  - dry (gas free) 218, 230
  - rate 3, 5, 147, 282
- ram pressure stripping 215, 222, 226, 228, 229, 238
- rotation curves, *see* rotation curves, in galaxies
- scaling relations 147, 209–213, 217, 227, 228
  - Faber-Jackson 147, 209–211, 226
  - Fundamental-Plane 147, 210, 212, 213
  - Tully-Fisher 147, 209–212, 226, 227
- secular evolution, *see* secular morphological evolution of galaxies

gas (interstellar medium, or ISM) V, VII, 1, 2, 4, 5, 8, 14, 21–23, 60, 61, 70, 78, 81, 83, 89, 98–103, 135, 147, 168, 172–174, 182, 183, 186, 187, 194–199, 201, 202, 215, 217–220, 223, 228, 235, 236, 239, 241–244, 246, 248, 250, 265, 282, 283, 287, 293, 296
- role in damping of density waves 8, 99, 101, 102

gas accretion, *see* accretion, in gas
Generalized Mach's Principle
- and the origin of fundamental physical laws 279–280

giant molecular cloud (GMC) complex 23, 191, 194, 198

global self-consistency requirement
- in studying self-organized structures VIII, 12, 18, 24, 29, 33, 55, 65, 70, 79, 123, 126, 162, 164, 166, 171, 248, 251, 258–260, 262, 263, 267–269, 284, 285, 293, 296

group
- of galaxies 3, 85, 98, 173, 216, 223, 227, 230, 232, 234–237, 281

halo
- and the calculation of potential-density phase shift 152
- dark, *see* dark, halo
- gas 215
- in galaxies 273, 275
- in N-body simulations 72, 73, 86, 90, 108, 110, 111, 114, 149
- of clusters 270
- participation in secular evolution 259, 275–278
- possible lost during galaxy interaction in NGC 3627 175
- possible skewness 149, 177, 205

heating
- of disk stars 3, 56, 68, 86, 89, 94, 95, 97, 98, 113, 116, 131, 142, 171, 187–189, 191, 192, 204, 205, 216, 231, 246, 250, 256, 295

HI (neutral hydrogen) 23, 174, 177, 183, 193, 234, 236–239, 241, 242

hierarchical organization
- of nature and its laws 18, 56, 69, 127, 267, 269, 284–287, 297, 298

HII (ionize hydrogen) 161, 198, 199

Hubble
- classification of galaxies, *see* galaxy, classification scheme, of Hubble
- Deep Fields 147, 209, 283
- sequence 2, 4, 5, 9, 11, 17, 62, 111, 147–149, 168, 169, 172, 174, 181, 208, 209, 211, 233, 244, 263, 269, 272, 274, 275, 282, 283
  - early assembly 148
- Space Telescope (HST) observation 3, 147, 148, 209, 283
- time 3, 14, 34, 88, 98, 110, 113, 114, 146, 170, 171, 179, 183, 190, 191, 202, 215, 229, 271, 282
- type evolution, *see* secular morphological evolution of galaxies

Hubble (*continued*)
– types of galaxies V, 1, 3–5, 8, 11, 14, 92, 98, 99, 113, 114, 145, 147, 148, 153, 158, 163, 164, 166–169, 172, 174, 175, 177, 179, 181, 183, 186, 199, 204–208, 210, 211, 214–217, 219–221, 224, 226–228, 230, 233, 234, 243, 244, 248, 269, 271, 281–283
– Ultra Deep Fields 147

IC 2581 (NGC 3293) 199
initial mass function (IMF) 272, 273
instability
– buckling 171
– global VI, VII, 22, 46, 52, 54, 55, 67, 68, 77, 94, 107, 108, 129, 137, 144, 150, 169, 205, 208, 236, 242, 256, 268, 269, 283, 298
– local 5, 10, 11, 17, 18, 21–23, 35, 46, 52, 73, 77, 83, 85, 91, 102, 107, 111, 115, 116, 119, 129, 133, 188, 191, 196, 204, 205, 231, 235, 251, 296
– two-fluid 5, 22, 23, 85, 102, 196, 296
Interstellar Medium (ISM), *see* gas (interstellar medium, or ISM)
interstellar medium (ISM)
– as candidate of dark matter 275
– in Bullet and Train-Wreck Clusters 274
intracluster medium (ICM) 222, 231
isothermal
– primordial clump 203
– sphere 203

Jacobi integral
– conservation of, *see* adiabatic, invariance of stellar orbit
Jeans theory of galaxy evolution V

KAM theorem 108
Kolmogorov scaling laws for fully developed turbulence 12, 55, 86, 196, 251, 285

Lagrangian formulation of dynamics 29, 45, 55, 68
Lambda Cold Dark Matter (LCDM) cosmology, *see* cosmology, LCDM
leap-frog integrator
– in *N*-body simulations 75
lenses
– in galaxies 14

lensing
– micro, and dark halo composition 272
Leo Triplet (NGC 3627, NGC 3628, NGC 3623 234, 236–237
Lindblad
– Bertil
  – and density wave hypothesis 6, 99
Lindblad resonances 7, 8, 38, 68, 81, 88, 92, 100, 118, 135, 139, 156, 157, 235, 239, 240, 242, 261
Liouville's equation 288–290
Liouville's theorem VI
low surface brightness (LSB) galaxies 148, 207, 216, 271, 273
– and cusp-core controversy 274

Mach's Principle
– Generalized 279, 280
MACHO (Massive Compact Halo Object)
– as dark matter candidate 272, 273, 275
mass-to-light ratio (M/L) 64, 151, 152, 154, 173, 174, 211–213, 233, 274
Maxwell-Boltzmann velocity distribution 294
mean free path 66, 85, 91, 102
mean-field approach in classical mechanics IX, 17, 52, 53, 103, 284, 286, 291
merger
– between galaxies, *see* galaxy, merger
Messier Catalogue
– M100 (NGC 4321) 158–161, 164, 168, 174, 175, 177, 179–187, 259
– M51 (NGC 5194) 14, 161, 174, 175, 177, 179–187, 247, 272, 282
– M66 (NGC 3627) 100, 174, 175, 177, 179–187, 234–243, 247, 272, 282
metalicity
– and luminosity relation 233
– of Bulge stars 171
modal theory
– of density waves, 275, *see* density wave, modes
molecular chaos assumption
– in the derivation of Boltzmann equation VII, 268, 289–292
morphology-density relation 216, 230, 232, 281, 282
MORPHS project 225, 229–231
MS 0451 223, 224, 226, 231

N-body simulations 1, 13, 17, 38, 39, 45, 46, 50, 52, 53, 59, 67, 68, 70–145, 149, 156, 163, 167, 170, 171, 177, 181, 187, 189, 192, 251, 259, 260, 264
- and bar formation, see bar, in N-body simulations
- and galactic bulge, see bulge, in N-body simulations
- and galactic halo, see halo, in N-body simulations
- and mass accretion, see accretion, rate, in N-body simulations
- and particle-mesh code, see particle-mesh code, in N-body simulations
- and signature of collisionless shock 77–85
- and the choice of basic state, see basic state of the galactic disk, choice in N-body simulations
- and the role of active disk mass 276–277
- faithfulness 93–95
- grid noise 137–145
Navier-Stokes equations VII, VIII, 52, 292, 294–297
New General Catalogue (NGC)
- NGC 4622 152
- NGC 0628 168, 174, 175, 177, 179–187
- NGC 1073 164
- NGC 1530 153–157, 162, 164, 173, 174, 177, 243, 248, 282
- NGC 253 224
- NGC 3293 see IC 2581
- NGC 3351 164, 174, 175, 177, 179–187, 259
- NGC 3372 199
- NGC 3623 236, 237
- NGC 3627, see Messier Catalogue, M66 (NGC 3627)
- NGC 3628 175, 234, 236, 237
- NGC 4321, see Messier Catalogue, M100 (NGC 4321)
- NGC 4665 162
- NGC 4736 174, 175, 177, 179–187, 259, 274
- NGC 5194, see Messier Catalogue, M51 (NGC 5194)
- NGC 5247 157–158
- NGC 5643 164
Newton's
- third law 55

Newtonian
- equations of gravitational interaction 75, 111, 123, 286
- interaction
  - modified due to softened gravity 111, 139
nonequilibrium
- dynamics VII, 35, 69, 95, 268, 297
- phase transition and dissipative structures VI–VIII, 10, 11, 13, 35, 69, 70, 102, 103, 144, 169, 266, 267, 269, 283, 287, 293, 297
nonlinear
- acoustic waves 33, 78
- approximation/regime 10, 15, 19, 26, 33–35, 39, 40, 45, 46, 50, 56, 62, 65, 78, 79, 83, 84, 88, 90, 98, 101, 103, 104, 116, 129, 150, 151, 163, 167, 181, 248, 250, 251, 258, 266, 270, 293, 295, 297, 298
- system VI, VII, 11, 144
- WKBJ waves 33, 78, 79
nucleosynthesis
- Big Bang 270, 273, 274

open
- system 267, 283, 292
orbit
- adiabatic condition, see adiabatic, invariance of stellar orbit
- chaos 27, 104, 106
- crowding 21, 161
  - in gas streamlines 239, 241
- exponential divergence from true orbit, see exponential, divergence of the orbit from true orbit
- heating, see heating, of disk stars
- parabolic
  - of encounter between NGC 3627 and NGC 3628 237
- passive VI, 10, 103, 104, 106, 108, 109, 120, 158, 162, 241
- periodic 26, 29, 33, 77, 103, 108, 109, 241
- resonant interaction with density wave VII, 8, 30, 40, 46, 67, 68, 73, 108, 115, 121, 139, 164, 165, 247, 261, 262, 284
orbital
- angular momentum 39, 55, 59, 60
- energy 56, 59, 196, 209, 210, 248, 250, 268
- velocity 56, 210

OSUBGS (Ohio State University Bright Galaxy Survey) 153, 164, 166, 177
over-reflection mechanisms
– for density waves at corotation 8, 9, 40, 42, 44, 103, 246

particle-mesh approach
– for N-body simulations 70, 111, 139, 143
pattern speed
– of density waves 19, 26, 56, 62, 83, 86, 88, 95, 104, 117–119, 121–123, 146, 150, 152, 153, 158, 161, 163–166, 179, 187, 240, 248, 258
phase space 46, 191, 204, 230, 265, 287, 288, 291
phase transition
– nonequilibrium, see nonequilibrium, phase transition and dissipative structures
photon-dominated region (PDR) 197
pitch anlge
– finite
  – of spiral pattern 16–19, 21, 23, 29, 33–35, 39, 45, 57, 78, 79, 83, 102, 120, 125, 147, 166, 167, 169, 172, 174, 216, 222, 275, 281
plasma
– collective effects 10
– collisionless
  – shock VIII, 11, 17, 77
Poincaré 8
– recurrence theorem VI
Poisson
– bracket 288, 289
– equation
  – differential form 17, 47, 51, 52, 54
  – WKBJ form 101
– integral 10, 16, 17, 24, 26–29, 32, 34, 40, 55, 68, 79, 103, 151, 164–166, 294, 295
– noise 68, 94, 97, 98, 189, 192
potential-density phase shift (PDPS) VIII, 8, 10, 11, 16, 17, 21, 24–35, 38, 40, 51, 57, 62, 65, 71, 77–79, 83, 84, 99–103, 115, 125–127, 146, 147, 149, 169, 170, 177, 181, 183, 186, 187, 196, 247, 262, 265, 268, 270, 284, 296
– as a method for CR determination 149–169
Prigogine, I. VII, 11, 144, 266, 267, 284, 285, 297
primordial turbulence, see turbulence, primordial

Q
– barrier 8, 9, 73, 74, 135
– parameter, Toomre's
  – of local stability 19, 21, 73, 82, 83, 86, 89, 111, 115
Quasi-stationary spiral structure (QSSS) hypothesis 246, 247
quasi-steady state (QSS)
– of the wave/mode VI, IX, 2, 8, 10–13, 15–18, 24, 26, 28, 34–46, 50, 51, 53–57, 59, 61, 62, 65, 67, 68, 70, 77, 83, 86, 99, 101–103, 119, 120, 122, 125, 144, 150, 152, 153, 159, 163, 165–167, 169, 172, 181, 194, 195, 202, 246–248, 250, 251, 257–259, 263, 268, 269, 281, 286, 293

ram pressure stripping, see galaxy, ram pressure stripping
Rayleigh-Bénard convection 6
redshift (z) 1, 3, 6, 151, 174, 214, 215, 217–220, 223–227, 229–231, 233–235, 270, 271
relational nature of fundamental physical laws 279, 280
relaxation
– behavior in N-body simulations 111
– collective 107
– collisional 107, 114
– dynamical 273
– in clusters 228
– in N-body simulations 76, 77, 94, 110–113, 137, 138, 140, 148
– in plasma systems 10
– in self-gravitating systems 148
– of dynamical systems 10
– time 76, 77, 137, 138
– violent 273
resonance
– corotation (CR), see corotation
  – resonance(CR)
– Lindblad, see Lindblad resonances
  – inner (ILR) 7, 8, 68, 81, 88, 100, 118, 135, 139, 235, 239, 240, 242
  – inner-inner (IILR) 240, 242
  – outer (OLR) 8, 81, 118
  – outer-inner (OILR) 240
– orbit 171, 265
– wave-particle VI, 8, 11, 32, 108, 291, 292

Reynolds stress 47, 53, 54
rings
– in galaxies 14
rotation curves
– in galaxies 65, 111, 174–176, 203, 205, 206, 272
rotation curves in galaxies 274

satellite
– accretion 149, 283
– dwarf spheroidal 215
– Hipparcos 190, 205
– IRAS 197, 199
– small 270
Schwarzschild
– distribution of stellar peculiar velocity 188, 191
secular heating
– of disk stars 59–60, 67, 104, 147, 170, 187–193, 250
– of the ISM 193–201
secular morphological evolution of galaxies V–VIII, X, 1–6, 8, 10–12, 16–18, 21, 24, 25, 28, 33, 35, 51, 52, 56, 57, 62, 63, 67, 69, 70, 77, 91, 93, 94, 98, 103, 104, 107, 109, 111, 115, 125, 134, 144–148, 164, 168–170, 172, 173, 177, 179, 181, 183, 187, 203, 208–210, 212–215, 217, 219, 223, 225–229, 232–234, 243, 244, 246, 247, 251, 258–263, 265, 268, 269, 271, 272, 275, 281–284, 292, 295, 296
– in N-body simulations 1, 70–145
self-organization VII, VIII, 1, 17, 52, 69, 109, 111, 133, 144, 266, 284, 298
Sersic
– index 4
– profile 172
singularity
– at the collisionless shock front 17, 52, 68, 108, 262, 263
– formal
  – in the solutions of the governing equations V, 17, 33, 52, 68, 70, 108, 265, 293, 298
– hierarchy 268, 269, 284, 285, 293
size-line-width relation
– of Galactic molecular clouds (Larson Law) 60, 102, 147, 193–203

smooth-particle-hydrodynamics (SPH) 17, 102
softening
– of gravity in N-body simulations 67, 71, 76, 77, 92, 110–145, 251, 259
solar neighborhood stars
– age-velocity-dispersion relation, *see* age, and velocity dispersion relation of the solar neighborhood stars
soliton/neutral wave
– as density wave solution 7, 8, 14, 99
spiral
– arm
  – as the site of temporary local gravitational instability 10, 11, 17, 19, 21–23, 33, 35, 39, 52, 77, 81, 83, 85, 91, 102, 108, 111
– as density waves/modes, *see* spirals, in galaxies
– bar-driven, *see* bar, driven spiral
– structure
  – generating mechanisms 8
spirals
– in galaxies VI, 4, 8, 10, 14, 15, 21, 24–26, 28, 29, 33–35, 38, 39, 44, 85, 86, 94, 99, 101, 104, 106, 119, 120, 133, 146, 147, 151, 163, 164, 212, 213, 215, 220, 222, 226, 227, 263, 265, 269
Spitzer Space Telescope observations 158, 217, 220, 224, 226
– SINGS (The Spitzer Infrared Nearby Galaxies Survey) Legacy image 158, 161
star formation 272
– in galaxies
  – gradient 227
  – quenching 215, 218, 220, 223, 229, 230
  – rate 1, 218, 220, 224, 230, 282
  – triggered by spiral density waves 171, 216, 223, 270
– time scale 223
stellar accretion, *see* accretion, in stars
stellar dynamical equation, *see* Boltzmann equation, collisionless
stellar population
– aging 5
– evolution 3
– in early type galaxies 171
– in elliptical galaxies 213
– in Galactic Bulge 3

stellar population (*continued*)
– in thin and thick disks of Milky Way 204–205
– old disk 192
– supporting density waves 154
stress
– gravitational 48
– Reynolds, *see* Reynolds stress
– viscous 294
supersonic velocity
– in spiral arms 77, 78, 83, 102, 103, 198
– in turbulence 102, 148, 198
surface of section 104, 105
SWING mechanism
– for density wave amplification 8, 28, 162
symmetry
– breaking of VII, 17, 267, 280, 284, 285, 293, 297
synergetics VII, 2

tidal interaction
– among neighboring galaxies 15, 94, 144, 162, 172, 186, 215, 216, 222, 223, 226–228, 232, 235, 242, 243, 272, 282, 283
time scale
– astronomically significant 65
– black hole accretion 244
– cluster dynamical 231
– for cluster formation and evolution 215
– for cluster galaxy evolution
  – due to harassment 221–222
  – due to merger 220–221
  – due to post-interaction secular evolution 222–223
  – due to ram pressure stripping 222
  – due to various mechanisms 217, 218, 229
– for eddy turnover in the ISM 194
– for energy and angular momentum redistribution
  – due to transport processes 5
– for group and cluster galaxy secular evolution 98
– for noise-induced relaxation 137
– for quasi-stationary spiral modes 34
– for radial mass accretion in galaxies 146
– for secular evolution of the basic state 10
– for stripping/starvation mechanism of cluster galaxy evolution 215
– for transient wave forcing 67
– for turbulence cascade 194, 198
– local dynamical 24, 67, 85, 93, 167, 169, 171, 205
  – for AGN accretion disks 5
– of galaxy formation 271
– two-body relaxation 94
torque
– advective 7, 15, 36, 38, 42, 47, 48, 50, 53, 54, 61, 62, 66, 179, 181
– coupling 17, 34–38, 40, 42, 44–51, 53–56, 61–63, 66, 156, 165, 169, 179, 181, 182
– density wave 10, 24–26, 55, 62, 77, 115, 126, 133, 145, 241, 247, 250, 251, 259, 262, 268
– gravitational 7, 36–38, 42, 46, 50, 53, 54, 61, 62, 66, 179, 181, 182, 234
– tital 243
Tr 14 star cluster 200
Tr 16 star cluster 200
Train-Wreck Cluster (Abell 520) 274
Tremaine-Weinberg (TW) method
– for density-wave pattern speed determination 153
Tully-Fisher relation, *see* galaxy, scaling relations, Tully-Fisher
turbulence
– analogy for stellar energy cascade 46, 55, 86
– and barber pole structure 266
– in interstellar medium 102, 147, 194, 195, 197–199
  – candidate energy-injection mechanisms 194
– magnetohydrodynamic (MHD) 194
– primordial 148, 231, 270, 273
– studies of Kolmogorov, *see* Kolmogorov scaling laws for fully developed turbulence
twin-peak
– and bar, *see* bar, and twin-peaks
twisted
– 3D mass distribution 151
– bars 151, 152
– isophotes 147, 177, 209, 269, 283

Universal Rotation Curve (URC) of galaxies 205–209

velocity dispersion
- and age of stars in the solar neighborhood, *see* age, and velocity dispersion relation of the solar neighborhood stars

virial equilibrium 14, 15, 98

virialization
- of clusters 3, 215, 217, 219, 220, 222–224, 226–228, 231, 232

Vlasov equation, *see* Boltzmann equation, collisionless

WASER (Wave Amplification by Stimulated Emission of Radiation) mechanism
- for density wave amplification 8

wave action 40, 42–46

WKBJ (Wentzel, Kramers, Brillouin, & Jeffries) approximation in density wave solution, *see* density wave, WKBJ treatment

WMAP (Wilkinson Microwave Anisotropy Probe) observations 231

X-ray observations of galaxies 199, 213, 231